THE COASTLINE OF SCOTLAND

THE COASTLINE OF
SCOTLAND

J. A. STEERS

Emeritus Professor of Geography and
Emeritus Fellow of St Catharine's College
University of Cambridge

CAMBRIDGE
AT THE UNIVERSITY PRESS
1973

Published by the Syndics of the Cambridge University Press
Bentley House, 200 Euston Road, London NW1 2DB
American Branch: 32 East 57th Street, New York, N.Y. 10022

© Cambridge University Press 1973

Library of Congress Catalogue Card Number: 72-86419

ISBN: 0 521 08696 5

Printed in Great Britain
at the University Printing House, Cambridge
(Brooke Crutchley, University Printer)

Contents

[v]

Illustrations

FIGURES

FIGURE ACKNOWLEDGEMENTS

Thanks are due to the following for permission to reproduce illustrations –

The Editor, *Aberdeen University Review* for Fig. 52, from the *Aberdeen University Review*, **36**, No. 109 (Autumn 1953):

John Bartholomew and Son Limited for the pullout geological map of Scotland, and for Fig. 15 from *The Bartholomew Road Atlas*:

Blackwell Scientific Publications Ltd for Fig. 18, from H. G. Vevers, The land vegetation of Ailsa Craig, *Journal of Ecology*, **24** (1936), p. 426:

The Council of the Botanical Society of Edinburgh and Professor C. H. Gimingham, for Fig. 55, from *Transactions of the Proceedings of the Botanical Society of Edinburgh*, **36** (1953), Part III, p. 137:

The Principal, Brathay Hall, for Fig. 67, from the Brathay Exploration Group's *Report* booklet, p. 45:

The Buteshire Natural History Society for Fig. 16, from the *Transactions of the Buteshire Natural History Society*, 12 (1945), p. 84:

Figs. 8, 39, 65, 66, and 68 are based upon the Ordnance Survey Map with the sanction of the Controller of Her Majesty's Stationery Office, Crown Copyright reserved:

Edinburgh Geological Society for Fig. 54, from S. Simpson and G. K. Townshend, *Trans. Edinb. Geol. Soc.*, 14 (1950), p. 398; Fig. 56, fom J. C. Howden, *Trans. Edinb. Geol. Soc.*, 1 (1868); Fig. 59, from S. R. Kirk, *Trans. Edinb. Geol. Soc.*, 11 (1924), Plate XLVIII; Fig. 60 from G. A. Cumming, *Trans. Edinb. Geol. Soc.*, 13 (1936), p. 351; Fig. 63, from T. C. Day, *Trans. Edinb. Geol. Soc.*, 10 (1914), Plate XXXV:

Generalstabens Litografiska Anstalt for Fig. 3, from *Geografiska Annaler*, 19 (1937):

The Editors, *Geological Magazine*, for Fig. 6 from C. A. M. King and P. T. Wheeler, The raided beaches of the north coast of Sutherland, Scotland, *Geol. Mag.*, 100 (1963), p. 305:

Mr A. T. Grove for Fig. 58:

Dr Norman Holgate for Figs. 13, 14, 45 (a) and 45 (b) from the *Scottish Journal of Geology*, 5 (1969), pp. 97–139:

The Institute of British Geographers, for Fig. 5, from *Transactions of the Institute of British Geographers*, No. 39 (October 1966):

The Institute of Geological Sciences for Figs. 7, 19 (a), 19 (b), 31 (a), 31 (b) and 60, based on Crown Copyright Geological Survey maps; Figs. 25 and 57 are part of Crown Copyright Geological Survey maps; Figs. 11, 12, 17, 20, 22, 23, 24, 26, 27, 28, 29, 30, 32, 33, 34 (a), 34 (b), 37, 41, 42 (a), 42 (b), 49 and 64 are Crown Copyright Geological Survey diagrams. All this material is published by permission of the Controller, Her Majesty's Stationery Office:

Dr S. Y. Landsberg for Fig. 51, from J. H. Burnett (ed.), *The Vegetation of Scotland* (Oliver and Boyd, 1955):

Professor S. B. McCann for Figs. 4 (a), 4 (b), 4 (c), 9, 10, 35 (a), 35 (b) and 36:

Thomas Nelson and Sons Limited for Fig. 53, from R. Miller and R. Watson, *Geographical Essays in Memory of A. G. Ogilvie* (1959):

Oliver and Boyd and Professor T. N. George for Fig. 1, from G. Y. Craig (ed.), *The Geology of Scotland* (1965), p. 19:

Dr W. Ritchie for Fig. 40, from *The Scottish Geographical Magazine*, 85 (1967), p. 169:

Royal Geographical Society for Fig. 47, from *Geographical Journal*, 90 (1937):

The Royal Scottish Geographical Society for Figs. 2, 21, 40, 46 (a), 46 (b), 50 (a), and 50 (b) from *The Scottish Geographical Magazine*, 58 (1942), 78 (1962), 83 (1967), 30 (1914), 30 (1914), 72 (1956), 72 (1956), respectively:

The Royal Society of Edinburgh for Figs. 43, 44, 48 and 62, from *Transactions of the Royal Society of Edinburgh*, 53 (1923), p. 382, 53 (1923), p. 384, 53 (1923), p. 399, 58 (1935), p. 756, respectively:

The Scottish Academic Press for Figs. 13, 14, 45 (a) and 45 (b), from the *Scottish Journal of Geology*, 5 (1969), pp. 97–139:

Dr J. B. Sissons, for Fig. 61, from *The Evolution of Scotland's Scenery* (Oliver and Boyd, 1967), p. 185:

The Society of Antiquaries of Scotland for Fig. 38, from the *Proceedings of the Society of Antiquaries*, 71 (1936–7), Fig. 1, p. 349:

Professor K. Walton for Figs. 50 (a) and 50 (b), from *The Scottish Geographical Magazine*, 72 (1956), pp. 87 and 89.

PLATE ACKNOWLEDGEMENT

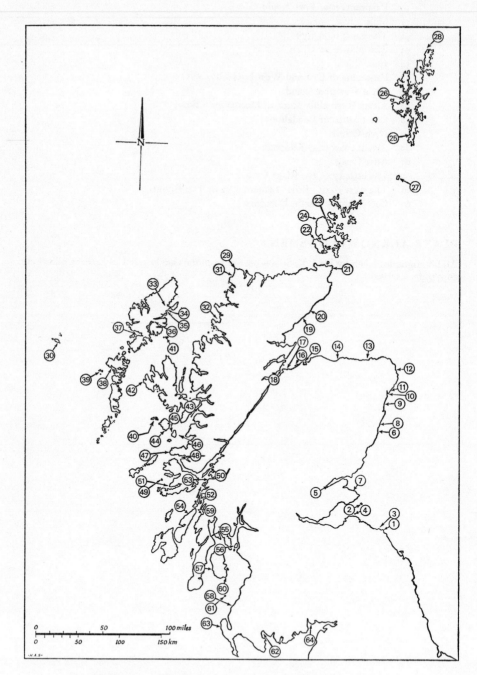

Outline map of Scotland, showing locations illustrated by plates

Preface

Anyone who tries to write about the coast of Scotland and the adjacent islands may well be attempting an almost impossible task. I am fully aware of many of the shortcomings of this volume. The coastline of England and Wales presents fewer difficulties, and it also offers far more obvious opportunity for discussion of the relations between coastal evolution, ecology, archaeology and history. These are by no means lacking in Scotland, but they are largely absent from great lengths of the north and west coasts. On the other hand, the coast of Scotland is magnificent and much of its scenery is unrivalled. To do full justice to it needs a combination of skills almost impossible, in these days of specialized scholarship, to a single writer.

There is, in fact, more than one coast of Scotland. The Inner and Outer Hebrides in a sense triplicate the west coast. Orkney and Shetland may be regarded as separate entities, but Orkney is structurally a part of Caithness. In another sense the coast is a double one. Much of the coast both of the mainland and Inner Hebrides is fringed by raised platforms on which many settlements are built and along which many miles of main road are laid. The present coast, especially in these places, is unspectacular in the sense that the waves have scarcely made an impression on the outer edge of the platform. But to appreciate the coast, old and new must be considered together. The many different sections and variations will be discussed in the body of the book.

To know all the coast well is a lifetime's study. I have had two opportunities of seeing almost all of it, including the islands. Shortly after the 1939–45 war I was asked by St Andrew's House to report on the coast from the point of view of future use and conservation. In the course of several visits I saw all but a few small islands; unfortunately, circumstances have not yet allowed me to visit Colonsay. There is no great need to follow all the intricacies of the west coast; from one or more vantage points one can see often all of a sea-loch, especially with the help of field glasses. On the other hand, it is desirable to walk as much of the coast as possible, and this I have done. Since 1960 I have had the opportunity of another almost total peregrination; The Nature Conservancy asked me to report on those parts of the coast and islands which seemed to be suitable for conservation by that body. Apart from these visits, several private tours have enabled me to visit many places at other times, often under guidance of geographers from Aberdeen, Dundee or St Andrews. I have also worked, in 1937, on the coast between Inverness and Lossiemouth. There are still parts of Orkney and Shetland I should like to see in much more detail.

When I began to consider this book it soon became clear that there was little written on the coast as a whole. But in recent years great strides have been taken. The geographers and ecologists at, especially, Aberdeen, St Andrews, Edinburgh and Glasgow have all made, and continue to make, substantial contributions to our understanding of coastal evolution. They have by no means all worked on the same problems, and so we are now in a position, from a study of their work and that of others, past and present, to appreciate far more fully the problems the coast of Scotland offers. They are many, and all interested in the matter will be particularly grateful to the Geological Survey of Scotland. Some of their first memoirs, now about a century old, still contain some of the best descriptions of cliffs, and many of the more modern ones give a very clear insight into coastal evolution. It is perhaps invidious in this context to mention particular ones, but that on Caithness and that on northern Skye are noteworthy, not least in the light they throw on the evolution of cliffs – a subject all too little examined.

In my *Coastline of England and Wales* I wrote rather long chapters on structure and physiography, the general effects of wave action and the movement of beach material, dunes and salt marshes, and also a reasonably full discussion of vertical movements. It has seemed to me unnecessary to repeat, in any detail, these subjects in this book. The same principles apply in both places and I have contented myself – and I hope anyone who reads this book – with a summary account of raised beaches and vertical movements. These are all important in Scotland, but it seemed to be better to discuss their general nature and refer, for example, to the meticulous work of J. B. Sissons and others, rather than to discuss it in great detail. I am, however, fully aware that much more work is required on the raised beaches and associated problems of the Scottish coast before a completely definitive answer of their origin is available.

Thus it is that the introductory chapter is short. I hope that the sections will suffice and they should be read with the help of geological and physical maps. It may be argued that greater attention should have been given to the structure of Scotland and the nature of the major faults and thrusts. These are discussed in much detail in geological works, and I have been content to take some of their findings. I doubt if a detailed examination is relevant to this volume. On the other hand the recent work on northern Skye, in particular, has thrown a new light on the origin of the Minch, and further research in that area would certainly advance our knowledge of cliff formation in the volcanic areas very considerably.

I have tried to bring together as much information as I could find on the coast of Scotland. I do not doubt that I may have overlooked certain publications – I hope not important ones. The coast offers enormous scope for research work and it is my hope that this book will stimulate it, either by suggestion or by provocation! I am fully aware of much of the work that is in progress; there is scope for still more and in every sense the subject is worthy of it.

The final chapter on conservation has been added at the suggestion of the Cambridge University Press. I was glad to do this; the coast of Scotland is under increasing pressure and unless steps are taken soon, many beaches will be spoiled. Fortunately, the Scottish National Trust, the Nature Conservancy, and the Countryside Commission for Scotland, as well as certain local bodies, are increasingly aware of this matter. In writing this chapter I received generous help from Dr Morton Boyd, Director, Scotland, the Nature Conservancy; Mr J. C. Stormonth Darling, Director, the National Trust for Scotland; Mr J. Foster, Director, the Countryside Commission for Scotland; and Mr B. Gilchrist, Secretary, the Scottish Wildlife Trust.

In conclusion I should like to thank Professor Kenneth Walton of Aberdeen for much help in more ways than one, and also his colleague, Dr William Ritchie, for contributing the section on the Abredeen–Ythan dunes. Mr B. W. Sparks commented most helpfully on Chapter I and Miss J. B. Mitchell on Chapter VII. Dr Alwyn Scarth and Mr L. McLean were responsible for finding the maps of Buddon Ness referred to in Chapter VII, and I am most grateful to them; and Dr Scarth and Dr W. Berry and my son motored me round many miles of coast on the mainland and islands. The photographs are all chosen from the great collection made by Dr J. K. St Joseph for the Committee of Aerial Photography. The Cambridge University Press has not only devoted much care to the volume, but also made some very helpful suggestions. To my wife, I am, as always, much indebted for great help with the proofs, in accompanying me along several parts of the mainland coast and in Orkney and Shetland, and encouragement in many ways.

J. A. STEERS

Cambridge, 1972

AUTHOR'S NOTE

The spelling of place names follows that of the One Inch Ordnance Survey sheets (7th series), except in some figures and quotations taken from other sources. Because there is some inconsistency in the use of diacritical marks in the spelling of Gaelic elements in certain place names on the sheets of the One Inch map, the Ordnance Survey's *Place names on maps of Scotland and Wales* has been used as a guide for this purpose.

Certain botanical names have also been retained as used by authors to whom reference is made.

Chapter I

The coast of Scotland - general conspectus

THE GENERAL NATURE OF THE COAST OF SCOTLAND

It is easy to make some generalizations about the coast of Scotland; it is far more difficult to describe it and explain its formation. An attempt will be made to do this in the following pages, but before concerning ourselves with detail it may help to make a brief survey of its main features.

The first and most striking of these is the difference between the western and eastern coasts. Not only is the west higher, but it is also much indented by deep and narrow sea-lochs, and it is also fringed for part of its length by two lines of islands, the Inner and Outer Hebrides. The eastern coast, on the other hand, is nearly everywhere lower, often faced by fine cliffs, and deeply cut by five main inlets, the Forth, the Tay, the Inner Moray Firth, Cromarty Firth and Dornoch Firth. None of these, except parts of Dornoch Firth, is directly comparable to a western sea-loch. There are two other long stretches of coast which have their own peculiarities. The north coast of Sutherland in its western part is reminiscent of the west coast, but east of the Kyle of Tongue the inlets are on a smaller scale and, as we shall see later, closely related to the structure of the rocks. In the south-west between the Clyde and the Solway, the coastal areas are often low or fringed with cliffs of moderate height and, on the south-facing coast, are penetrated by several widely open bays quite unlike, in both origin and appearance, those characteristic of the other parts of the mainland. These variations are of fundamental significance and their origins will be discussed in later chapters. The many islands on the west coast must be studied individually; some of them, like Skye and Mull, are so close to the mainland that the assumption that the separating straits of sea-water were at one time rivers is easily made. In other words, the inner islands may be regarded as but fragments of the mainland. But whilst this is a reasonable view, it must not be supposed that the separation of the islands was a simple process. How far it is proper to regard the Outer Hebrides in the same way is a more difficult problem.

We have said that the west coast is higher than the east. This is undoubtedly true in general, but the matter needs some analysis. In the north-west from Cape Wrath to Loch Broom the coastal fringe is rugged and intensely glaciated, but within two or three miles, and locally as, for example, around Laxford, for a greater distance, it does not exceed about 400 feet (122 m), and is often less. Farther inland there are high hills and mountains. To the south, north of Gairloch and in the

Applecross peninsula the hinterland of the coast is much higher and plateau-like. The western part of the peninsula between Lochs Carron and Alsh seldom exceeds about 500 feet (152 m), but soon gives place to higher ground near the road to Strome Ferry.

South of Loch Alsh hills and mountains approach closely to the coast. Lochs Hourn and Nevis are two of the finest sea-lochs and are in the midst of true mountain scenery. There is a narrow tract of low land around Morar and Arisaig, and much of the coast of Moidart and Ardnamurchan is by no means high, yet the general impression of the coast is mountainous. This is because in all wide views the high mountains are conspicuous and the long sea-lochs penetrate so far inland that a great deal of high country falls directly to salt water. The presence of Skye and Mull intensify this impression; nevertheless a study of a map shows that their coastal fringe is often flat and low.

South of Oban the coast is rugged and much indented. Locally high hills approach the sea and the distant views of Jura and Islay all help to give the impression of a mountainous coast; but the height of the mountains is beginning to fall. In Knapdale and Kintyre there is a good deal of country exceeding 1,000 feet (300 m) and reaching 1,500 feet (456 m) or more, and the south-west corner near the Mull of Kintyre is a precipitous coast with steep slopes up to 1,400 feet (426 m). Arran is similar; several hills exceed 2,000 feet (610 m), but the coastal rim usually is low and generally flat.

The Ayrshire coast, and that of Galloway and the Solway, is quite different. High ground is clearly visible from the coast, but the coast itself cannot properly be regarded as high.

We must be careful to distinguish between the general appearance of the coast and the particular features. It is only in parts of Inverness and in Argyll that the west coast is truly mountainous; nearly all of it is rugged and broken and backed by higher ground inland. But the fact that, as in the north-west, a plateau-like area at about 400–500 feet (122–152 m) separates cliffs from mountains is of great significance. It is an erosion feature and one of many we find in Scotland.

Before we attempt any correlation between the coastal and inland scenery, attention must be called to some extensive flat areas along the coast. At Gullane and Aberlady on the Firth of Forth there are some fine sandy bays backed by dunes; they are matched by somewhat similar features in and near Largo Bay in Fifeshire. But the first big area of this type is Tents Muir and the smaller peninsula forming the golf-links of St Andrews. On the opposite side of the Tay Barry Links seems to resemble Tents Muir closely, but although there are similarities, it is probable that the outline of Barry Links (Buddon Ness) is controlled by past rather than present conditions (see p. 255). Another extensive flat occurs between Montrose and St Cyrus. In Lunan Bay, just to the south of Montrose, and along the coastal strip, a platform has been produced by wave erosion on which beaches and sands of more than one episode have gathered. Between Aberdeen and Newburgh there are fine dunes, the seaward parts of which are certainly modern. But immediately we cross the Ythan to the Forvie sands we find many traces of former sea-levels, so that we are bound to ask ourselves if we can assume a simple explanation for those south of the Ythan.

In Rattray Bay, Strathbeg Bay and in Fraserburgh Bay there are magnificent beaches and flats, but also many traces at higher levels than the modern beach. This is also true of the far wider expanses of sands and dunes between the Spey and the Lossie, in Burghead Bay, in the Culbin sands and west of Nairn. Farther north, in Caithness, there are extensive beaches and dunes in Sinclair's Bay, Dunnet Bay, Thurso Bay and in some smaller inlets. There are many small but beautiful bay-head beaches on the north coast of Scotland. On the west coast south of Cape Wrath, wide beaches of sand are less common although good examples can be examined at Sandwood Loch, to the west of Kinlochbervie and at Scourie, Bay of Stoer, Achmelvich, Outer Loch Torridon, and in many other places both on the mainland and inner islands. Farther south there are wide sands in Islay, at Machrihanish on the West of Kintyre, and especially around Irvine and Prestwick in Ayrshire. In Wigtownshire the Sands of Luce are well known, and extensive flats are found in Wigtown Bay and around the mouths of the Nith and Annan.

Between all these sandy areas, on any part of the south and east coast, there are cliffs. They are usually of moderate height, but locally may reach 400 or more feet (122 m). They resemble the cliffs which surround many miles of the coast of England; they have been cut largely in sedimentary rocks. On the west coast, north of Skye, the cliffs in the Torridon Sandstone are generally similar, but the presence of large areas of intensely glaciated gneiss and the abundance in Skye and Mull and a few other islands of great sheets of basalt give a distinctly different appearance to the west coast. In many places where sea-loch and outer coast, or narrow channels like the Sound of Sleat and the Sound of Mull, grade into more exposed coasts it is difficult to know if we are dealing with a simple sea-cliff (if such a feature exists!) or a steep slope principally formed by ice advancing along the line of a sea-loch. In the Outer Hebrides there is almost a reversal of the west and east coast pattern of the mainland. With local exceptions many miles of the west sides of The Long Island are low and sandy, whereas the east coast is steep and rocky, but precisely how the steep slope was formed is another matter.

In the foregoing paragraphs several references have been made, in flat areas, to beaches at a higher level than that of the modern beach. Along lines of cliff we also frequently see evidence of change of level, and along many miles of coast, especially in the south-west and parts of the east coast, the cliffs are no longer washed by the sea; in many places they are separated from it by a main road or by fields and settlements. Although (see Chapter III) the level of the foot of the cliffs, relative to present sea-level, varies it is nevertheless clear that in those parts of the coast where such a feature is apparently absent we must nevertheless assume that it is obscured or, because of exposure, has been eroded away. We cannot assume that its absence, even on a long stretch of shore like that south of Cape Wrath, implies that it was not formed in that area. We shall be concerned with the reasons for the variation in level and nature of former beaches in Chapter III. At this point we must emphasize their ubiquity and note that very little of the coast of Scotland is directly the work of the waves on the modern shore. The more we examine the coast of the mainland and of many of the islands the more we shall find that we are dealing with a 'second-hand' coast; a coast the cliffs and other features of which were formed some time, 5,000 years or more, before the present. This phenomenon far from lessening the interest of the Scottish coast, greatly adds to it, and it will be apparent in all parts of this book how essential it is to try to appreciate that vertical movements, some relatively recent and others much older, have played their part in forming the present outline.

THE ROCK SEQUENCE

In order to appreciate the major features of the coast of Scotland and of the islands it is essential to understand, at least in outline, the nature of the rocks and the structure of the country. In age the rocks range from the most ancient, the Lewisian Gneiss, to recent deposits. Moreover, there are great areas covered with igneous rocks. A detailed stratigraphical table is not necessary, but a summary of the rock sequence is required.

Quaternary rocks

These include all the deposits of the Pleistocene Ice, raised beach deposits, peats, river alluvium, etc.

Tertiary rocks

In Scotland these rocks are almost entirely of igneous origin. There are great central intrusions like the Cuillins and Rhum, and vast lava flows. Associated with both of these there are innumerable dykes, often radiating from the central complexes. All these rocks, apart from the dykes, are for the most part in the Inner Hebrides; only locally, as in Morvern, do they appear on the mainland. This great assemblage stands in complete contrast to the Tertiaries of England, where they form the sands and clays of the London and Hampshire Basins and the crags of East Anglia.

Mesozoic rocks

Cretaceous, Jurassic and Triassic sediments are all found, but again there is a great contrast with England and Wales where the Mesozoic beds are fully developed and sweep in great arcs across the country or form extensive low-lying basins in the Permian and Trias.

In Scotland they are found mainly on the west coast and inner islands under the lavas which indeed have protected them. Greensands and chalk of Cretaceous age are present, but do not exceed more than seventy feet (21 m) in thickness.

The Jurassic is represented by all stages from the Lias to the Kimmeridge. Both Jurassic and Cretaceous outcrops also occur on the coast of the Moray Firth near Helmsdale and Brora and in the Tarbat peninsula.

The New Red Sandstone outcrops are scattered. There are small patches at Gruinard Bay, Applecross and in the waist of Skye. There is a larger mass near Elgin, and the main developments are in southern Arran, Stranraer, and between Mauchline and Annan where they rest in basins in the Southern Uplands.

Upper Palaeozoic rocks

There are some minor patches of 'Permian' breccias and sandstones on the Moray Firth. But the great development of rocks of this age include the widespread Carboniferous rocks, with interbedded lavas, in the Midland Valley, and the great spread of Old Red Sandstone on the northern side of the Midland Valley and reaching the coast, usually in fine cliffs, from the Tay to Stonehaven. These Carboniferous beds include productive coals, and also a widespread sandstone series, called the Calciferous Sandstones. The Old Red also covers large areas around the Moray Firth between Buckie and Golspie, and also forms almost the whole of Caithness and the Orkney Islands, and reappears in more limited distribution in Shetland. Outcrops well inland from its present inner margin imply that it once spread over even wider stretches of country. Interbedded with the Old Red sediments are many volcanic rocks. Sometimes these form distinct ranges as the Ochils and Sidlaws; elsewhere they may be contained in the sediments. In both Caithness and the central valley the total thickness of these rocks is of the order of 20,000 feet (6,100 m). There are also hundreds of necks and remains of ancient volcanoes, ranging in size from a few yards in diameter to half-a-mile or more. They are abundant in the Tay and Forth areas, and often have a marked effect on coastal scenery.

Lower Palaeozoic rocks

These include Silurian, Ordovician and Cambrian sediments and, locally in the Ordovician, many lavas. The Silurian and Ordovician are magnificently developed in the Southern Uplands, where they are intensely folded. They are also both found in the Midland Valley and along the Highland Border. They form a thick series, nearly 20,000 feet (6,100 m), in the Silurian of the Southern Uplands, and 14,000 feet (4,280 m) in the Bala stage alone of the Ordovician is a minimum figure. The granite intrusions, e.g. Loch Doon, Criffel, are of the same age as those farther north in the Dalradian and Moinian rocks.

Rocks deposited in the Cambrian and Ordovician periods also occur in the extreme north-west. They are thinly developed, but the different beds, mainly quartzites and limestones, make conspicuous features in the scenery. Their outcrop is long and narrow, and stretches from Loch Eriboll to Loch Kishorn.

Pre-Cambrian rocks

It is simplest to divide these into two areas. In the nort-west there is a great spread of sediments, the Torridon Sandstones, grits, and pebble beds, which are largely unaltered by metamorphism. They are locally covered by the Cambrian beds, and rest with a marked unconformity on the Lewisian Gneiss. The gneisses are highly altered igneous and sedimentary rocks; they cover considerable areas of the mainland, and form almost the whole of The Long Island.

These two groups are separated from the rocks to the east by a major zone of thrusting. East of this belt the Torridonian is completely absent. The great mass of the rocks are referred to as the Moine Schists and granulites. They are some 20,000 feet (6,100 m) thick in places, and rest uncomformably on Lewisian Gneiss. It is now generally assumed that the Moinian rocks are metamorphosed Torridonian. They are, especially in the east, largely covered by Old Red Sandstone, and, beyond the line of the Great Glen, they dip under another extensive Pre-Cambrian group – the Dalradian rocks. These also are metamorphosed, but they were of sedimentary origin and a succession has been established. Some horizons, for example the Loch Tay limestone, can be followed for long distances. We shall (pp. 226 ff.) analyse their nature on the coast of Banff where they form beautiful cliffs.

The Moinian and Dalradian groups have been much folded, both regionally and locally. The minor folding is often apparent in hand-specimens, and in the field earlier folds can often be seen to have been refolded, sometimes more than once. In the Moinian and Dalradian regions there are abundant traces of igneous activity, often in the form of great masses of granite; and also of lavas and smaller intrusions. Most of the granites, referred to as the Newer Granites, were associated in origin with the Caledonian orogeny which gave rise to the great belt of thrusting already mentioned. A glance at the geological map at end of book will show the distribution of these rocks. We are not concerned with their petrology, but we must emphasize the distinction between them and the alkaline intrusions in the Northern Highlands, which are largely syenitic in character, and seem to have been approximately coincident in age with the later phases of the Caledonian folding. The best known and most conspicuous of these makes the beautiful mountain of Ben Loyal.

AN OUTLINE OF THE STRUCTURE OF SCOTLAND
IN RELATION TO THE COAST

Before discussing in general terms the relation of rock type to coastal scenery, it is logical to say something of the main features of the tectonic structure of Scotland. In detail this is a most complicated subject, but from the present point of view little would be gained by a long discussion of the nature of folding and thrusting. Where these closely affect the coastal scenery they will be noted in the body of the book.

Scotland is usually divided into the Northern Highlands, the Central Highlands, the Midland Valley, and the Southern Uplands. The Northern and Central Highlands are clearly separated by the line of the Great Glen from Loch Linnhe to the Inner Moray Firth. The Highland Boundary Fault running from the Clyde to Stonehaven is not only a marked geological break, but also a physiographical one between the Dalradian metamorphics and the sediments and igneous rocks of the Midland Valley. The Southern Uplands Fault is also a well-marked line which for most of its length separates the highly folded Silurian and Ordovician rocks on the south from the Midland Valley. The fault splits into two near its south-western end.

In the north-west the line of thrusting which can be clearly traced from Loch Eriboll to the Sound of Sleat makes a distinct break within the Northern Highlands. The Torridon Sandstone is always on its western side; so, too, is almost all the Lewisian Gneiss. The movement along this line of thrusting has been to the west-north-west. The physiographical break is almost as emphatic as the structural; the intensely glaciated gneiss and the great sheets and detached fragments of the Torridon Sandstone forming such conspicuous features as the plateaux around Applecross, and the great peaks of Liathach, Beinn Eighe, An Teallach, Canisp and Quinag, some with remnants of the Cambrian beds on their summits, form the wildest and most magnificent scenery in Britain.

The movements along some lines of dislocation were formerly assumed to be mainly vertical. This is probably true of the Highland Boundary and Southern Upland Faults. Kennedy (1946) suggested that the major movement along the line of the Great Glen was horizontal. He based this view on the similarities of the granite masses of Strontian and Foyers. These are now sixty-five miles (105 km) apart. If (see pp. 81–6) they were once a single mass, then a great tear-movement has taken place and the Great Glen, as well as much of the Moray Firth and of Loch Linnhe and the Firth of Lorne, owe much of their present-day features to this wrench. Smaller, but by no means insignificant, later movements, possibly in a contrary direction to the initial movement, have taken place along this line of dislocation and are discussed in Chapter IV.

The origin of the Minch is by no means simple. That it is associated with the vast outpourings of lava which form so much of Skye, Mull, Staffa, and parts of Morvern is certain, but (see p. 124) the old view of the collapse of a vast plateau has been considerably modified in recent years, and the more restrained view seems to fit in far better with the coastal features of the Inner Hebrides. That a collapse, no matter on what scale, of a lava plateau area took place involves in part, if not wholly, a tectonic origin for the Minch. This may perhaps be associated with the submarine cliff which follows for long distances the eastern coast of The Long Island. There is, however, another feature pointing to tectonic movement more akin to that in the Great Glen. George (1965) has shown that the nature of the Lewisian Gneiss on the mainland between Loch Laxford and Little Loch Broom is characterized by certain structures which are usually referred to as Scourian since that township is in the district in which they occur. These structures clearly differentiate this part of the gneiss from that to north and south. In the Outer Hebrides (Fig. 1) structures of precisely the same nature as the Scourian on the mainland can be traced from the northern shore of the Sound of Harris to Loch Eynort in South Uist. Here, too, the gneiss to the north and south of this area resembles that to the north and south of the Scourian on the mainland. If (Fig. 1) these areas are plotted on a map, their distribution suggests that there has been a displacement along the line of the Minch, or, in other words, a tear fault like that in the Great Glen. This further implies that, relative to the mainland, the whole of The Long Island has moved. There may have been similar movements on some other major faults, but if so they seem to have had little effect on the coast except perhaps in Loch Fyne and in Upper Loch Etive.

The Great Glen Fault is probably continued to the north-east parallel to the coast of the Black Isle, the Tarbat peninsula and, in a more problematical way, with the trend of the Sutherland–Caithness coast and the eastern side of the Orkney Islands. The later faulting (p. 204) near Brora and Golspie is parallel to it. To the south-west the main fault runs through south-eastern Mull and then to the west of Colonsay, but a branch also appears to run east of Colonsay and through the low ground separating the Rhinns of Islay from the main body of the island. The Moine thrust probably continues along Kyle Rhea, the Sleat peninsula and then skirts the western extremities of Ardnamurchan and Islay. The Highland Boundary Fault traverses the low ground of Bute, is obscured by the igneous complex of Arran, and then possibly skirts the south-eastern end of the Kintyre peninsula.

There is no doubt that major faulting and thrusting does play an important role in giving certain general trends to parts of the Scottish coast. But the faulting, apart from the collapse of the lava plateaux in the Minch, is ancient and the minor, and some major, features of the present coast are, as it were, superimposed on these faults.

The contrast between the western and eastern coasts of the mainland have been described, but at this point attention must be called to the extensive deposits of Old Red Sandstone which are found along many miles of the east coast. These deposits locally extend several miles inland, and mountains such as Ben Griam More and Ben Griam Beg in Sutherland clearly imply a former much more extensive spread of these rocks. It is interesting to speculate if the west–east slope of the

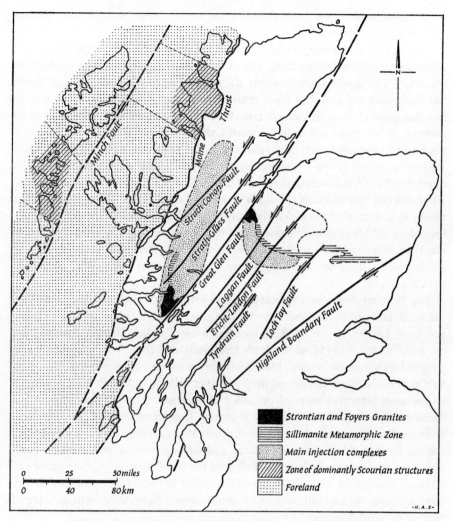

Fig. 1. The main wrench faults in Scotland (after T. N. George)

country existed at that period. However, the presence of the Old Red beds means that, apart from that part of the coast between Buckie and Stonehaven, the Moinian and Dalradian rocks, except very locally on either side of the entrance to Cromarty Firth, are absent for great distances on the east coast.

However, even with this brief survey of the coast in relation to structure we can make certain meaningful generalizations about the nature of the various parts of the coast which should serve as an introduction to the fuller analysis in later chapters. From Fig. 1 it will be seen that the general trend of the major lines of dislocation runs north-north-east to south-south-west, or north-east to south-west.

Some major parts of the coast follow those trends; other parts are in complete discordance with them.

The Lewisian–Torridonian belt from Loch Eriboll to the Sound of Sleat is physiographically a distinct unit. This is far from implying that it is uniform in character! The gneiss when it meets the coast is usually hummocky and forms rounded bluffs and there are often many hummocky islands offshore. Sometimes it makes good cliffs, as at Cape Wrath. In The Long Island cliffs are rather more common in the gneiss especially between Gallan Head and Husinish Point, and at the Butt of Lewis and parts of the east coast. But everywhere the highly glaciated appearance of the gneiss is the feature that imposes itself on the landscape. The overlying Torridon Sandstones may form beautiful cliffs if the bedding is horizontal or inclined landwards. If the dip is seawards the land often slopes gently to the beach; it is sometimes boulder clay covered and the beaches are consequently backed by cobble ridges. The gneiss and the sandstone stand in vivid contrast to one another and the two rock types with their local variations make the coast of north-west Scotland and The Long Island quite unlike that of any other part of the country.

The Moinian rocks cover a great area, and reach the coast in two long stretches, the north coast of Sutherland and the west coast between, roughly, Loch Hourn and Loch Linnhe. In the northern area, from near Whiten Head to Strathy Bay, the trend of the rocks is nearly north and south, and they make a crenulate and irregular coast of great beauty. The various elements respond to differential erosion so that with the minor intrusions, deep-cut river valleys, and sandy bays this is one of the most attractive parts of the coast of Scotland. The western coast in the Moines shows much less relation to structure. It contains some of the finest sea-lochs, and the deepest loch in Scotland, Loch Morar, is only now separated from the sea by a slight sill. Around Morar and Arisaig the low and intricate coast contains some fine sandy beaches. Offshore the Small Isles, with their variety of Tertiary igneous rocks, and the igneous complex at the extremity of Ardnamurchan are quite distinct although, with south-eastern Skye, they form an integral part of the coastal scenery.

The Dalradian rocks are well seen on the coast of Banff and along most of the west coast from Loch Linnhe to the end of the Kintyre peninsula. From the point of view of coastal scenery these two areas stand in marked contrast to one another. In Banff the rocks are highly folded and stand at high angles. They form a line of cliffs, averaging perhaps 100 feet high (30 m). Differential erosion has etched out the less resistant parts of the rocks, but although this still proceeds, it is a factor of the past rather than the present since there are many fragments of raised beaches all along the cliffs.

Beyond Macduff and as far as Aberdeen the coast is in the Dalradian area, but the rocks on or near the coast are much younger and sometimes igneous. It is an interesting part of the coast, but not one that easily, if at all, fits in with any simple structural scheme. Near Gardenstown and Pennan the Old Red reappears to give some striking features; farther east near and beyond Rosehearty the coast is low and rocky. The north-eastern corner of Aberdeenshire consists of fine sandy bays held between low headlands of metamorphic rocks, and around Peterhead there is a mass of granite which southwards gives place to metamorphics which also under-lie the Forvie sands. From the Ythan to Aberdeen is a long line of dunes. We may certainly regard all the Buchan peninsula (*sensu lato*) as basically metamorphic; we can hardly regard the coast as having a close relation to those rocks except locally.

In the west and south-west the *general* relationship of structure and coastal scenery is often most striking. The long sea-lochs, the general trend of Jura and east Islay, Luing and Seil, Lochs Sween, Caolisport and West Loch Tarbert and many other minor features all conform to this pattern. On the other hand, the trend of the rocks cuts right across the Kyles of Bute, Lochs Riddon and Striven, Loch Long, Gare Loch and Loch Lomond. These exceptions should make one wary of trying to find on maps over-simple relations between even major features and 'apparent' structure.

The Old Red Sandstone coast of Caithness and the Moray Firth closely resembles that between Stonehaven and the Tay, but is here considered separately since it has a Moinian foundation. The Old Red forms some of the finest cliffs in Britain; much depends on the nature of the bedding, and on the type of rock which ranges from fine grained sandstones to coarse conglomerates. The rock is often well jointed and spectacular features, geos, are cut along lines of weakness. In Caithness the surface is a rolling plateau, so that in Dunnet Bay and Sinclair's Bay the rock is brought to sea-level. Near Brora and Golspie Mesozoic rocks are faulted down and form a lower coast. Old Red cliffs reappear in Tarbat, but most of the coast of the Moray Firth, except for the Sutors of Cromarty and the narrow fringe of Moinian cliffs in the Black Isle, is low and sandy and formed of extensive sand flats which are, in fact, raised beaches (see p. 216). There is a small outcrop of Permian and Trias between Burghead and Lossiemouth.

The Tertiary volcanic area includes most of Skye, the Small Isles of Inverness-shire, the western end of Ardnamurchan, most of Mull and the Treshnish Isles. From the tectonic as well as the physiographical point of view it is a distinct unit and includes the great lava plateaux and also the central complexes of the Cuillins, Rhum, Ardnamurchan and central Mull. In Skye and Mull in particular, and also in Eigg, the lavas bury sedimentary Mesozoic rocks and the cliff scenery is usually magnificent. There is no other comparable area in Britain and many problems of cliff evolution, with particular reference to faulting, and to marine and glacial erosion, need investigation. It is also a region in which there are superb examples of coastal landslips.

The Highland Boundary Fault and the Southern Uplands Fault are nearly paral-lel to one another. However, because of the north-eastern trend of the coast of Angus and Kincardine, the eastern coast of Midland Valley is longer than the west, and if allowance is made for the Firth of Forth, it is much longer. The rocks in the valley are almost wholly Old Red Sandstone and Carboniferous. These include not only sediments, but also abundant igneous outcrops some of which make conspicuous hill ranges. From Stonehaven to the Tay Old Red sediments contain many interbedded lavas and sills. The cliffs are of the order of 100 feet high (30 m), and (see p. 2) occasionally they give place to extensive sand flats. In Fife and on the southern side of the Firth of Forth there are numerous plugs and vents of former volcanoes. Sometimes they have been sawn through by the waves and show up in the extensive rock platforms which surround most of the Firth. Inland high hills of volcanic origin are conspicuous near Elie and St Monance, and in the south the Bass Rock, North Berwick Law, and Traprain Law are well-known landmarks. The rocks in which all these vents occur are of Carboniferous age. On the western side of the valley the Clyde coast is somewhat similar, but much of it is rather more sheltered. Around Irvine and Ayr there are extensive sands, and the remainder of the coast is nearly always fronted by a raised platform, and the cliffs behind, seldom of any great height, are cut in sediments and included igneous rocks. Arran (p. 98) must be regarded as a separate unit.

The Southern Uplands are mostly formed of Ordovician and Silurian sediments closely folded. On the east coast their outcrop is narrow, but nevertheless the coast from Burnmouth to a mile or two west of Fast Castle, and including the igneous mass of St Abb's Head, is one of the most spectacular in Scotland. The cliffs are high, much cut by ravines, and show both small and larger scale folding to perfection. On the west and south-west Ordovician and Silurian rocks make the coast from Girvan to Balcary Point. From Girvan to Ballantrae there are included igneous rocks. However, the most characteristic feature of almost the whole length of this part of the coast is the fine raised beach platform which is often followed by a road. The cliffs behind are seldom reached by the waves, and are usually covered in vegetation. They are seldom more than 100 feet high (30 m), and because they are only locally wave washed they have rather a dull appearance. Beyond Balcary Point the Criffel Granite is prominent, and the shore is fringed with sand flats and salt marshes.

Frequent mention has been made in this chapter of the raised beaches, and in Chapter III they are analysed more fully. It is relevant to say in this place that they are, in one sense, the newest 'treads' in a long series of steps which rise to the high mountain tops. How long ago the lowest cut platform was formed it is difficult to say; we shall see later that there is good reason for thinking that many low platforms around the coast of Scotland are composite, of more than one age. But there is one very striking feature about them: there is little or no evidence that a new platform is being produced today. The lower platforms occur at various heights, but reach about 120 feet (36 m) as a maximum in Islay and some neighbouring islands. These are the platforms or benches which are spoken of as raised beaches; the platforms may or may not be covered with beach sands and gravels.

At higher levels there are other surfaces which have been produced by erosion, but there is still considerable room for argument whether the erosion was marine or subaerial. We are not concerned with their origin, but we must take account of them, more particularly of the lower ones, because of their effect on coastal scenery.

A number of authors have written on the subject during the last one hundred years or more. The present interest in the matter springs from Hollingworth's (1938) study of the levels in western Britain. In Scotland A. Geikie, B. N. Peach and J. Horne, H. Fleet, A. G. Ogilvie, J. M. Soons, T. N. George, W. G. Jardine and especially Godard in his comprehensive work (1965), have all made valuable contributions. Sissons has briefly but adequately summarized the matter in his book (1967). Unfortunately there is no complete agreement either of their number or their heights, and whatever their origin may have been they, or some of them, have been subsequently tilted so that correlation by height alone is difficult. The note at the end of this section indicates the number and range of surfaces recognized by various authors, and also gives the references to their publications.

Whatever their nature or number it is obvious that the highest, and therefore presumably the oldest, surfaces are the most fragmentary; the lower and especially the Pliocene surface is in all ways more fully developed. This is the surface which Godard puts at 300–600 feet (91–183 m). Moreover, this surface is the one that is best developed near the coast, and the appearance of many parts of the coast largely depends upon it. It is perhaps best seen in Caithness. The cliffs there are cut in a plateau surface which varies in height from sea-level in the large bays to 200 feet (61 m) or more in many lines of cliff. Godard traces this surface all round the coast from the Beauly Firth to Skye. It is broken where river valleys reach the sea but nevertheless is remarkably constant, but the reader must remember that the range of height allowed by Godard is 300 to 600 feet (91–183 m), and that in viewing the coast it is difficult to separate the slopes below 300 feet (91 m) from the surface itself. Nevertheless, around much of Scotland, north and south, the immediate hinterland of the coast is not mountainous, but gently undulating, so that cliff scenery is more often than not characterized by fairly flat tops. If a view of the coast were taken from an aeroplane flying around it this impression would be accentuated. The general idea that the west coast of Scotland is fringed by high mountains is only locally true. There are often many miles of broken and irregular country between the coast and the high interior; this is particularly noticeable between Cape Wrath and Loch Ewe. On the east coast of Sutherland, apart from the downfaulted stretch near Brora and Golspie, high country extends to the coast around Helmsdale and Berriedale. The plateau nature of the country adjacent to the coast is conspicuous from Buckie to Banff and Macduff, and also in Aberdeenshire where, however, the dunes between Fraserburgh and Peterhead, and south of Newburgh deflect attention from the country on their landward side. South of Aberdeen and almost as far as the Border the cliffs are flat-topped, but in Fifeshire and around the Firth of Forth they are lower and irregular and only occasionally give the impression of being cut in a plateau surface. In Galloway and the south-west the cliffs are often flat-topped. Their height varies, and only in a few places, for example near Ballantrae and near Creetown, does ground above 500 feet (152 m) close in with the coast.

All these erosion surfaces are much wider and more extensive than raised beaches, and the highest ones are at thousands rather than hundreds of feet above sea-level. Yet if, as many believe, they are cut by marine action, the difference between them and raised benches (beaches) is only one of degree. This point is illustrated in Argyllshire where there are several plateau features adjacent to the coast around Arisaig. It is the lowest of the platforms that has the most marked effect on the coast in governing cliff height and appearance. Where long views are possible along the coast, or across the Moray Firth and Eddrachillis Bay and other places, it is easy to see how coast, the land adjacent to the coast, and the higher interior country make distinct elements in the scenery.

Erosion surfaces in Scotland have been recognized at many levels by various investigators. The following table summarizes their views:

1938. S. E. Hollingworth (The recognition and correlation of higher level erosion surfaces in Britain, *Quart. Journ. Geol. Soc.*, **94**, 55). South-west Scotland, 2,600–700 (793–820 m), 1,070 (326 m), 730–800 (223–44 m), 500 (152 m) and 400 (122 m) feet.

1959. W. G. Jardine (Post-glacial sea-levels in south-west Scotland, *Scot. Geogr. Mag.*, **80**, 5). 2,600–800 (793–854 m), 1,900–2,000 (580–610 m), 1,700–800 (518–49 m), 1,350–400 (412–27 m), 1,000–100 (200–35 m), 750–850 (229–59 m), 600–700 (183–213 m), 450–500 (137–52 m) and 200 (61 m) feet.

1955. T. N. George (Drainage in the Southern Uplands: Clyde, Nith, Annan, *Trans. Geol. Soc. Glasgow*, **22**, 1). Nith to Clyde. 2,650 (808 m), 2,300 (701 m), 1,670 (509 m), 1,070 (326 m) and 600 (183 m) feet.

1966. T. N. George (Geographic evolution in Hebridean Scotland, *Scot. Journ. Geol.*, **2**, 1). Harris–Jura 3,200 (975 m), 2,400 (732 m) and 1,600 (488 m) feet.

Up to 1930. B. N. Peach and J. Horne in many different publications recognized a High Plateau at 2,000–3,000 feet (610–915 m) with residuals in, for example, Ben Nevis and the Cairngorms rising above it, and an Intermediate Plateau with an upper limit of about 1,000 feet (305 m).

1938. H. Fleet (Erosion surfaces in the Grampian Highlands of Scotland, *Rapp. Comm. Cartog. Surfaces d'Appl. tert. Union géogr. Internat.* **91**). 2,400–3,100 feet (732 m–945 m) (Grampian Main Surfaces), 1,500–2,100 feet (457–641 m) (Grampian Lower Surface), 750–1000 feet (229–305 m) (Grampian Valley Benches). The Cairngorms and a few other high points rise above the Main Surface.

1958. J. M. Soons (Landscape evolution in the Ochil Hills, *Scot. Geogr. Mag.* **74**, 86). Ochils, 1,500–900 (457–579 m) and 750–1,000 (227–305 m) feet.

1930. A. G. Ogilvie (*Great Britain, Essays in Regional Geography*, Cambridge) 1,500–2,000 (457–610 m), 500–1,000 (168–305 m) and 100–500 (31–168 m) feet.

1965. A. Godard (*Recherches de Géomorphologie en Écosse du Nord-Ouest*, Strasbourg). 2,300–3,100 (702–945 m), 2,000–300 (610–701 m), 1,300–2,000 (396–610 m), 600–1,000 (183–305 m) and 300–600 (91–183 m) feet.

The surfaces have been mapped and analysed in various ways. The range of height in any one surface is often great. There is some general agreement about the major surfaces, but there is great need to try to find ways to appreciate the true breaks between the surfaces themselves and also to separate those formed by subaerial and those formed by marine erosion. Still more we need to understand with much more precision than is possible at present exactly how they were formed.

THE COASTAL VEGETATION

In *The Coastline of England and Wales* I described the formation of salt marshes, shingle spits and sand dunes, at some length and also discussed the physiographic effects of the growth of vegetation on them. There is no need to repeat this discussion here. On the other hand there are certain points about the coastal, maritime, vegetation of Scotland which must be made and which are locally very relevant to the nature of the coast.

On the present-day sand beaches the vegetation is similar to that in England and Wales. *Honkenya peploides* is a perennial, but foreshore plants seldom produce a closed community. Most have extensive root systems and in Scotland the commonest forms are *Agropyron junceiforme*, *Ammophila arenaria*, *Elymus arenarius*, *Festuca rubra*, *Atriplex* spp., *Cakile maritima*, *Rumex crispus* and *Salsola kali*. As in England and Wales, these plants tend to spread down the beach in spring and early summer, and then die away in late summer and autumn.

If dunes occur at the back of a sandy beach, their vegetation is similar to that farther south. *Agropyron* and *Ammophila* are particularly important, and *Elymus* is abundant in many dunes. Later the vegetation changes somewhat and *Festuca rubra* is often abundant, and plants such as *Senecio jacobaea*, *Sonchus asper*, *Trifolium repens*, and various mosses spread rapidly. With increasing age the dunes become grey and the vegetation, especially if some way from the beach and sheltered from spray, is more and more characteristic of the adjacent land.

Behind the dunes there is often a wide area of sandy pasture or machair. If the dunes are formed of a high proportion of silica sand, the vegetation is generally similar to that described in the previous paragraphs. If a high proportion of lime is present, as in the true machair of the Hebrides and parts of the west coast, other plants may come in. Since machair is so well developed in The Long Island, an appendix on its nature is given on pp. 189–91.

In any dune or dune pasture area, whether the sand is siliceous or calcareous, there may be a number of hollows, lows, or slacks, between ridges, some of which may have been much flattened because their crests have been blown away. In these slacks the water table is always close to the surface, and in the wetter parts of the year they are filled with shallow lakes. These are common in the machair, and in, for example, the dunes in Luce Bay and many other places. In the damper areas *Salix repens* is often abundant, and on the somewhat better drained sands dune heaths, with *Calluna*, *Erica*, and *Empetrum*, are common in places where the calcareous content of the sands is low. Dune scrub may follow, and in many parts of the east coast of Scotland there are extensive growths of the sea buckthorn (*Hippophaë rhamnoides*).

On shingle ridges which usually, but not always, occur at the head of a beach and below the dunes, the flora is again not unlike that farther south. The proportion of sand in shingle beaches is often high so that many sand-loving species may be found. However, other beaches may be largely sand-free, and in all the pebbles vary a good deal in size, from that of a walnut to the size of one's fist; sometimes, for example at the foot of cliffs, they may consist of much larger cobbles and boulders. Common species of plants include *Tripleurospermum maritimum*, *Silene maritima*, *Atriplex glabriuscula*, *Rumex crispus*, *Galium aparine*. *Potentilla anserina*, *Honkenya peploides* and *Sonchus arvensis* are locally common. An interesting plant on shingle is *Mertensia maritima*. Gimingham (whose 1964 source I have made considerable use of in this section) points out that the range of this plant has contracted northwards in the last half century. (In the early 1920s there was one plant on Blakeney Point, Norfolk.) On the other hand such common shingle species in England as *Suaeda fruticosa* and *Glaucium flavum* do not reach Scotland. Since shingle is often much shifted by the waves, vegetation may be absent or nearly so.

The salt marshes in Scotland are, generally, more like the west coast marshes of England and Wales. They are sandy with a relatively thin spread of mud or silt upon them. The Montrose Basin and some marshes in the Firth of Forth are rather more muddy. A number of marshes are discussed at some length in the body of this book so only an introductory statement is necessary here. On bare mud the algae are the first colonizers and include especially *Vaucheria*, *Rhizoclonium*, *Cladophora* and, less frequently, *Enteromorpha*. At a later stage *Zostera* appears, but is much less common than it was since it was largely decimated by disease about 1950. *Salicornia* spp. and *Suaeda maritima* follow, and *Puccinellia maritima* is often prolific. *Aster tripolium* and *Triglochin maritima* may accompany or follow these plants. Plants well-known in the south are also common in Scotland – *Plantago* spp., *Armeria maritima*, *Glaux maritima*, *Spergularia marina*, *Juncus* spp., *Festuca rubra*, *Agrostis stolonifera*, *Cochlearia officinalis*, and others. Salt marsh may grade into fresh marsh through a Phragmitetum. Chapman (1960) has shown that in Loch Creran the sequence of plant development is as follows: in what he calls the Lower General Salt Marsh, *Puccinellia* and *Armeria* are dominant, with *Triglochin*, *Glaux*, *Aster*, and salt marsh forms of *Fucus ceranoides*. Later, *Armeria*, *Glaux* and *Pucci-*

nellia are co-dominant and associated with *Plantago coronopus*, *Festuca rubra*, *Carex extensa*, *Hypochoeris radicata* and *Lotus corniculatus*. At still higher levels a Juncetum is present, with mosses. *Iris pseudacorus* may replace *Juncus*.

On the west and parts of the north coast, but not entirely limited to these areas, true salt marsh is found at the head of many lochs and often grades into stands of *Juncus articulatus*, *Carex demissa* and *C. flacca*. If there is a good deal of fresh water *Iris pseudacorus* or *Eriophorum angustifolium* and *Molinia caerulea* are often abundant.

In these sea-lochs the salt marsh often grades downwards into a lush growth of algae on stones and boulders which are covered at all, or nearly all, high waters. Fucaceae are usually dominant in the mid-shore region. At low water, and in good light, they give a strong golden-bronze colour to the upper part of the loch. The lower level flora depends much on the substratum and the amount of tidal movement. 'The flora is variously dominated by combinations of *Laminaria*, *Himanthalia*, *Halidrys*, *Codium*, *Fucus serratus*, *Zostera* and smaller brown, red and green algae which may form a compact turf or lightly attached floating masses' (Lewis, 1957). The play of light on the water, and on the lower marshes exposed at low water, and on the green of the higher marshes, the general setting of the loch, and the sky and clouds make these upper parts of the lochs most beautiful and fascinating.

Cliff vegetation varies with locality and especially in relation to distance from the waves and the amount of spray which reaches it. In this context the great extent of *old* cliffs – those which stand behind raised beaches – is important. Some of these are well sheltered, others may face open water and, in storms and strong winds, are thoroughly wetted by spray; others are well sheltered and carry a normal land vegetation.

On open coasts where the slopes plunge down into deep water there is often a great abundance of algae and lichens on the cliffs which add greatly to their colour and beauty. On exposed shores some belts occur with great regularity. Just above tidal limits there is often a broad band of *Verrucaria* giving a black colour; the height of the zone will vary with exposure. Above it a *Xanthoria-Lecanora* belt is often present and gives bright orange and grey colours. In many parts of the north-west coast, these belts on the cliffs and slopes of Lewisian Gneiss enhance the whole appearance of the coast – the grey cliffs, the background of high hills, and on a bright day in summer, the almost Mediterranean blue of the ocean.

On many cliffs, not least on the Torridonian and Old Red Sandstone, there are often patches and even large spreads of *Armeria* and *Cochlearia*. They usually appear first in cracks and crevices and small terrace-like features. They may be accompanied by bryophytes and lichens. Where there is a good deal of spray *Plantago* spp. are often common. When *Plantago* and other species have once taken hold, they in turn become a buffer to moving detritus and so enlarge to some small extent the place in which they began to grow. With increasing distance from the spray the vegetation becomes less and less maritime and more akin to that of the fields or country behind. If the cliffs are faced by much talus the nature of the vegetation must depend largely on the size of the fragments, the nature of the rock (this applies on all sorts of cliffs to some extent) and particularly on the amount of fine material within the coarser stuff.

Cliffs which are the resting place of numerous sea-birds are often very steep. On the ledges *Armeria* and *Cochlearia* are common. The *Armeria* on the cliffs along the west coast of the Mainland of Orkney is striking. Vevers (1936) noted on Ailsa Craig that where guano accumulated at low levels there was a growth of *Urtica dioica*, *Silene dioica* and *Poa annua*.

It will have been noticed in this brief account of the coastal vegetation that there is not any marked difference between east and west coasts, apart from the machair of the west. On the other hand, there are noticeable differences in the plants between north and south Britain. Northern species include *Carex maritima*, *Juncus balticus*, *Ligusticum scoticum* and *Mertensia maritima*. *Glaucium flavum*, *Crambe maritima*, *Limonium vulgare* and *L. humile* only reach southern Scotland. *Halimione portulacoides* and *Suaeda fruticosa* are absent.

In the chapters which follow there will be found several accounts of maritime vegetation of particular areas.

Chapter II

The nature and origin of the major inlets
of the coast of Scotland

In Chapters IV and VII an analysis of coastal features is made but no account is given of the nature and origin of the major inlets and indentations of the coast. In order to understand their origin we must leave the coast for the time being and discuss amongst other things the possible ways in which the river system of Scotland came into being. This will also involve some remarks on vertical movements, but in this chapter only those of significance for the present purpose are considered; the intricate Glacial and Post-glacial movements and their connection with raised beaches and carselands are best treated separately.

There have been several attempts to explain the rivers of Scotland. Gregory (1927) was responsible for the first comprehensive view, although Mackinder (1907) had already discussed the matter in more general terms. Gregory's views are based on assumed extensive tectonic movements. He gave no proof, however, that these movements had taken place. Moreover, one must also take for granted Gregory's geological time-table. Unfortunately his views, interesting though they are, are far too simplified. He supposed that in Eocene times Scotland suffered great denudation and only a few patches of the former cover of Cretaceous rocks survived. This was followed in the Oligocene and the first half of the Miocene by gentle folding along east and west axes, and in the Upper Miocene by subsidences on the Atlantic side along north-west and south-east lines. In the Lower Pliocene he invoked movements along lines at right-angles to these. All these movements he asserted produced torsion and fractures which 'preceded, accompanied and followed' the Upper Miocene vulcanicity. Eventually Scotland was worn down to a plain sloping south or south-east. He also believed that in the Middle and Upper Pliocene there was a general uplift affecting most of the British area, and it was at this time that the sea and inland loch basins were formed as a result of fracturing. There was also recurrence of movements along old fractures, causing them to widen. This, in turn, led to capture of the drainage from the higher land, and also allowed the sea to flow into the lower lands. While these Alpine movements were taking place, the east–west valleys north of the Great Glen were formed, and also the relatively straight south shores of both the Pentland and Moray Firth.

There is no doubt that extensive earth movements have taken place in Scotland, and that some may well have coincided with the initiation of the river system. Gregory's scheme is indeed hypothetical, and if it were true it is clear that rivers would flow primarily along fracture lines. This is not to deny that a number of lochs and rivers do follow lines of faulting; but this does not necessarily imply that the line of the river is directly caused by the faulting.

Davis (1895) wrote a well-known paper on the English rivers. He postulated that the Mesozoic rocks formerly extended far to the west and north-west of their present scarps. When, probably in the Late Cretaceous, these rocks emerged, an extensive coastal plain was formed, and streams rising in the western parts of the country (the parts which were not submerged) flowed south-eastwards across the plain. As time went on they cut down through the surface rocks, and strike streams developed at right angles to their initial courses. In other words, drainage became adjusted to structure; water gaps were formed and Davis maintained that 'the rivers of today, in the mature stage of the present cycle of denudation, appear to be the revived and matured successors of a well-adjusted system of consequent and sub-sequent drainage inherited from an earlier and far-advanced cycle of denudation'. This view was not applied by Davis to Scotland, but we shall see below how it has probably influenced the ideas of some later writers.

Mackinder's views were somewhat similar. He fully appreciated the lowering of the land by denudation to a base-level. This level he thought was

grained like sawn wood, because the plane of denudation intersects folds in the constituent rock-beds. If land thus denuded be raised afresh by crust movements, a series of streams is likely to be formed which will flow over the plateau following any initial tilt of the nearly flat surface but unaffected by the perhaps complicated structure and varying quality of the rocks beneath.

This is also, in essence, the view of Peach and Horne (1936). They assumed that, in the early Tertiary, Scotland was part of a continental mass continuous from Ireland to Scandinavia, and that this area had been reduced to a peneplain. Rivers flowed to the south-east across this surface to a Miocene sea. But by no means all Scottish rivers flow to the south-east. Mackinder fully realized that, e.g., the Dee, Don, Leven-Tummel and part of the Tweed are exceptions and, in his opinion, may follow re-excavated courses of some earlier system. In this way, too, he explains the Ness, Lochy, Findhorn, and Spey. He also makes the point, but does not explain it, that the east–west valleys are parallel to the north Sutherland–Caithness coast, the southern shore of the Moray Firth, and also to the Solway–Tyne line. Mackinder also suggested that the Tweed might be the lower part of a river, the upper part of which rose in the west highlands and followed the course of Lochs Goil and Gare. In a somewhat similar way the Nith may at one time have flowed eastwards along the line of the present Solway Firth to the Tyne. Bremner (1942) is rightly critical, and I fully agree with his statement that 'no hazy generalizations and qualifications about "easterly components" and "sympathetic curves" help the argument; a tilt is to the south-east or it is not'. What is more, whatever there may be of a peneplain in the Scottish highlands, it is far from perfect.

Bremner has added greatly to our knowledge of Scottish rivers. At the present time the rivers on the plateau flow over pre-Cambrian and older Palaeozoic rocks. But they did not originate on this surface. He maintained, as did other physiographers, that the plateau area was once covered by Mesozoic rocks, and of these the Upper Cretaceous covered the greatest area. It may well be that these rocks were deposited upon an earlier peneplain. Today the region of Tertiary vulcanicity is restricted to a narrow strip on the west coast and to the islands of the Inner Hebrides. These extrusive rocks were spread over dry land, and although they must have had a considerable effect on any rivers in that particular part of Scotland, they would have had little or no effect in the main region to the east. Bremner believed that the main uplift was at the end of the Cretaceous, but it was not a simple movement, and today only part of this uplifted surface survives in the dissected High Plateau of Scotland. He also maintained that the main axis of uplift was to the west of the present mainland coast. Skye and the Inner Hebrides (see also George, 1966) are part of the plateau. The problem of the Minch will be taken up later, but it is certainly a depressed area. It is more open to question whether The Long Island is, or was, part of the plateau.

If, then, we assume a main axis on or near to the west coast we may also assume eastward-flowing rivers, but there is no need to suppose that all were parallel one to another; there were, for example, irregularities around the Moray Firth (see below). It is not our purpose to follow the development of the rivers in any detail, but a brief summary of some of their probable early courses is necessary.

The Forth is an east-flowing stream; at one time it probably rose to the west of Loch Lomond and followed the line of the Inveruglas and Arklet valleys. The origin of Loch Lomond and of the possible Pre-glacial and earlier lines of drainage is discussed by Linton and Moisley (1960). The River Devon's earlier course has been reconstructed by Bremner, and this river also may have had an east–west course from beyond Uamh Bheag (north-east of Callander) to the Forth at Leven.

The Tay system presents interesting problems since it is usually assumed to afford good evidence of *south-easterly* drainage. The Earn flows obliquely to structural lines and has suffered some diversion, but in general follows an east–west line. The Almond and Bran join the Tay in its north–south course. The line of the Bran is continued eastward of the Garry–Tay into the Lunan Burn and the Newtyle gap. Bremner believed that these valleys, once continuous, had a similar relation to the Tay as had the Inveruglas–Arklet valleys to Loch Lomond. The Earn and Almond probably once ran via Newburgh and Dunbog to St Andrew's Bay. The Lyon–Tay line is continued in that of the Lunan Water between Forfar and Lunan Bay, and the Tummel is in line with the South Esk which reaches the sea at Montrose. The evidence for these former courses and the changes which have taken place is summed up by Bremner (1942) in the following lines.

The validity of the claim that the primary, consequent drainage was W.–E. and not N.W.–S.E. is suggested, if not established, by the following figures. The reconstructed Tummel (Corran to Montrose) in a course of 103 miles [166 km] (curves neglected) has 77 miles [124 km] represented by continuous W.–E. valleys; the figures for the Lyon–Strath Tay (source to Lunan Bay) are 80 [129 km] and 50 [81 km], and for the Earn (source to St. Andrews) 68 [109 km] and 68 [109 km], though for the last 16 miles [26 km] the high level valley is not particularly easy to trace.

Many changes as a result of river capture and of glaciation have taken place so that now these assumed former west–east courses are much broken up. The Tay flows south for part of its course; in some of the original west–east valleys, captures have led to reversals of flow.

Farther north, west–east courses still prevail. The present distribution of Old Red Sandstone around the shore of the Moray Firth, and its occurrence in outliers such as Ben Griam More and Ben Griam Beg imply a former much wider spread. It is also probable that Mesozoic rocks capped the Old Red. Streams originated on this cover, and cut their way down until today they run on the underlying schists. The Dee may at one time have risen in the far west and followed the line Loch Arkaig, Glen Spean, Glen Feshie. The Don's original course may well have been from Towie via the Tillifourie gap to Kemnay. North of these two rivers there is a greater variety in the direction of flow of the streams. Bremner's view is that (Fig. 2) the whole area within the curved line suffered a down-warp, probably in the early Tertiary, related to the movement along many pre-existing lines of fault. The axis of this down-warp followed, approximately, Strath Nairn. It was, therefore, towards this line that consequent drainage was directed. But Bremner also argues that north of the Great Glen the larger streams occupy only portions of through valleys, in the western parts of which there is usually a stream flowing to the west coast. It is also noticeable that these through valleys run to the south-east, north of the neighbourhood of Tain; between that place and the Beauly Firth the direction is eastwards, and then it becomes decidedly north-east. It is thought that the earlier drainage east of the Great Glen was on the cover of Old Red and later rocks. These streams were attacked by the headward cutting of rivers such as the Spey and Findhorn. Peach and Horne thought it possible that the Spey had intercepted drainage as far south-west as Loch Eil.

East of the Spey there are clear traces of a former west–east drainage. The most important of these is the line Rosarie Burn–Lower Isla–Middle Deveron–Ugie. This line cuts across all structural trends. There are traces of another valley close to the coast of the Moray Firth, from the Binn of Cullen to Boyndie Bay.

The hypothesis of easterly flowing rivers was developed further by Linton (1951). On the other hand he, too, makes an exception of the drainage to the Moray Firth. Although Linton is concerned more with an hypothesis of river development, his views on the rivers on the east coast of Scotland do not differ materially from those of Bremner, and moreover, he, Linton, is not concerned with tracing their detailed courses. However, by a reconstruction of the contours on which the assumed Cretaceous cover was laid down, he postulates a major divide. He follows somewhat closely Cloos's views (1939). Cloos, in a study of several

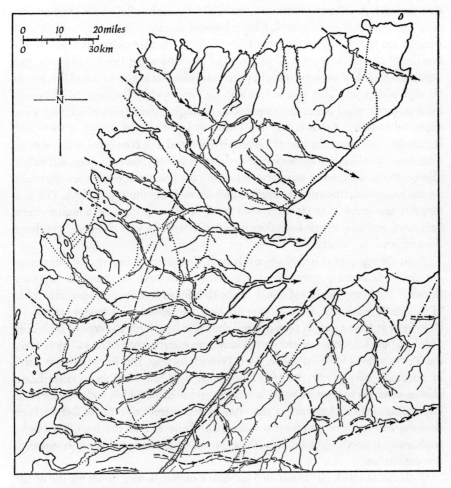

Fig. 2. Tertiary rivers in north Scotland: broken lines – original consequent lines of drainage; dotted lines – fault and lines of crush (after A. Bremner)

different regions, concluded that, after upward bulging of an area had reached a certain stage, collapse, probably associated with vulcanicity, was likely to occur. In Britain parts of Scotland and England form the eastern part of a dome; Ireland and the Hebrides the western part. To quote Linton

If the sub-Cenomanian surface is indeed represented by ... contours ... rising from 2,000 feet [610 m] about Stonehaven to some 4,500 feet [1,368 m] at Ben Nevis, then this same surface is depressed to about 1,500 feet [457 m] beneath the summit of Beinn Iadain in Morvern and descends to sea-level on the shores of the Sound of Mull. At Strollamus in Skye, at Gribun on the west coast of Mull, and extensively in Antrim, the Chalk and Greensand deposits outcrop on the coast. It would appear that the crown of Cloos's arch must have dropped down by an amount approaching 5,000 feet [1,524 m].

Bremner had, it will be remembered, spoken of through valleys between west and east, but he did not explain them. Linton believes that the main watershed followed roughly the line of high peaks from Cape Wrath to Ben Cruachan. It is across this line that Bremner's through valleys pass, and Mackinder and Peach and Horne had suggested that they were produced by streams rising far to the west and flowing to the then North Sea. This was, in fact, also the view held by Bremner. Sölch (1936) quite properly asked 'what has happened to the high ground from which they were supposed to flow?' Louis (1934) and Linton (1949) took the view that ice would accumulate first along the line of highest ground, and as it continued to increase its main mass would move somewhat to the east. It would, however, send out valley glaciers to the west, and it was these, following the low passages between the peaks on the watershed, that cut the valleys which we now call through valleys. This is a far more reasonable view; it overcomes the difficulty of presuming a high western land mass, and in its simple form does not conflict with the various views held about the origin of the Minch.

T. N. George (1965) is very sceptical about Cretaceous peneplanation. He points out that it is based largely on what is assumed to have happened in southern Britain. 'The Greensands and Chalk of the Hebridean area are deformed into folds of amplitude exceeding 2,000 ft. [609 m], they are faulted with abrupt changes in altitude of at least 1,400 ft. [427 m], and they nowhere form the caps of high hills which (where Cretaceous rocks occur) have a protective cover of lavas.' Moreover, the Tertiary lavas are involved in later-Tertiary folding, and in the western parts of Scotland, particularly the Hebridean area, it is not possible to co-ordinate plateau summits formed of lavas, intrusives and metamorphics, and reconstruct a Cretaceous surface. Any surface that may be reconstructed must, in fact, be later than the Tertiary igneous episode. George also believes that, in parts of the Highlands away from the igneous area, the evidence of geomorphology also supports this conclusion.

With this in mind, rivers initiated on such a surface flowed down the dip of the rocks, if sedimentary rocks persisted, or down the slope if only an eroded surface were present. George agrees that the Scottish rivers are superimposed, but that this is no evidence that they were superimposed from a Cretaceous surface. This is plain in some west-coast rivers, in places where Cretaceous rocks remain, not as parts of a surface from which the rivers were superimposed, but under thick masses of lavas. But the rivers, short though they may be, flow definitely across the grain of the country rocks, Pre-Cambrian, Mesozoic and Tertiary, and show as clear evidence of superimposition as do the much larger eastward-flowing rivers. The surface on which the rivers originated probably extended beyond the present coastal limits, and as a result of emergence the rivers are now fairly deeply incised into this surface. George then argues that if the present system of rivers indicates anything of the nature of the surface on which it originated, it seems likely that the uplift was of such a nature that a main watershed was formed following fairly closely the trend of the west coast, a feature that would account easily and simply for the contrasts between the short and relatively steep westerly flowing streams, and the longer and gentler ones flowing to the east. In any case there has been ample time for numerous modifications both by capture and by glaciation to have taken place since then. In this respect there is no essential difference between the views of George and other writers, the main difference is in the date and nature of the surface on which the rivers began. The date we have already discussed. The origin of the surface is usually assumed to be the uplift of a sea-floor. George, in the absence of any sediments to prove the emergence of a sea floor, thinks it possible that 'the emergent surface was mainly wave-eroded and sediment-free, the rivers being then superimposed directly on to the ancient geological structures beneath' (p. 42).

Sissons (1967) summarizes the views about the evolution of the Scottish rivers, and comes to the conclusion that they probably began on a cover of chalk and other Mesozoic rocks, and that they were well established before the advent of vulcanicity in western Scotland. The surface on which they are assumed to have risen has, of course, disappeared except perhaps for certain high parts of the Grampians. Sissons, however, is insistent on the effect of marine erosion. 'The remarkably close relationship between much of the present coastline and the known or inferred presence of the cover rocks implies that in relatively recent geological times the sea has been a very selective agent of erosion' (p. 28). Sissons thus agrees in many ways with Linton's views, but also in his references to marine erosion he follows George. We have much yet to learn about the power of marine erosion; the demonstrably little work that has been done by the sea, since the final retreat of the ice, on cliffs swathed in boulder clay, must be regarded as warning against a too ready assumption of wide wave-cut rock platforms, even when all allowance is made for the much greater periods of time.

How far can the sea cut into the land? Johnson (1919) thought that in unlimited time the sea could cut away a whole continent. This view is based on two assumptions: (1) the wave base is a horizontal surface approximately 656 ft (200 m) below the surface of the sea; and (2) the products of abrasion become sufficiently fine so that they can be taken into deep water. Zenkovich (1967) maintains that: '(1) the wave base, or limiting surface of abrasion is *inclined* and has a minimum gradient of approximately 0.01, (2) the particles are only broken down to the dimensions of fine sand . . . ' He also thinks that the lower limit of effective abrasion is about 328 ft (100 m) 'it follows that even under the most favourable conditions (a 100 m [328 ft] limit to abrasion and a gradient of 0.01 to the bench) the mainland with all its bays and accumulation forms cannot recede for more than 10 km [6 miles] *if the level remains constant*'. Zenkovich's views are based on many observations of the structure and nature of submarine slopes. Much more work is required on this problem. It is clear that with gradual submergence a much wider bench or platform may be cut, but enough has been said to warn the reader against accepting too readily an easy explanation of wide platforms, especially if they are associated with a falling sea level. We have a great deal to learn about the rates of, and conditions under which, subaerial and submarine planation take place.

Despite the attention we have given to the origin of the Scottish rivers, we are much more concerned with the ways in which they reach the present sea. There is general agreement amongst recent writers that a watershed, following approximately the present divide between west- and east-flowing rivers, came into being as a result of uplift. The time of the uplift has been placed earlier by Linton and others than by George. The careful analysis of the relation of the rivers to Cretaceous and other rocks undoubtedly suggests that the later date is more probable. However, from the present point of view that is of relatively small significance. We may assume that rivers ran east and west from the divide; those to the west were shorter and swifter. We may assume that some of these, wholly or in part, originally made use of faults, but this does not mean that such rivers are fault controlled, except in the sense that a fault is likely to mark a line of broken rock along which erosion takes place somewhat more easily than elsewhere. More emphatically, it does not mean that some of the rivers followed rifts as suggested by Gregory. The major accumulation of ice would probably be on and east of the watershed, and at certain periods of the ice age streams of water from the ice would not only course down both west- and east-flowing valleys, but would also, because of the position of the ice-cap relative to the divide, cut through the divide in suitable places and so produce the through-valleys we see today. At other times tongues of ice advancing down the valleys would transform them from ordinary river valleys into glacial valleys. With, perhaps even without, the subsequent rise of sea-level the lower parts of these valleys would be converted into sea-lochs. Once again the difference between east and west would be accentuated. The long eastern valleys were glaciated, but partly because of the much greater slope, partly because at times the Scottish ice ran up against the Scandinavian ice close to the Scottish coast, the effects of the ice in the valleys were different, and only in the upper part of the Dornoch Firth, and possibly in the Brora river are there any features on the east coast resembling the western sea-lochs. The Firth of Forth is wider and more funnel-sloped than any other Scottish inlet. This is partly because it lies in flooded lowland between the Pentlands and the hills of southern Fife. The narrow Firth of Tay is contained between the relatively high and resistant rocks of north Fife, and the Sidlaws to the north. At one time it must have occupied all that part now covered by the Carse of Gowrie. That both these firths are far more complicated than suggested here is made clear on pp. 261 ff, where the details of their shores are described.

The former mouth of the Lunan water must at one time have been much wider, and so too was that of the South Esk. In fact South Esk and North Esk both reach the sea through sands pushed up into the former wide bay between St Cyrus and Scurdie Ness. Bervie Bay and Stonehaven Bay are entered by small streams, and are quite unlike west coast inlets. The mouths of the Dee and Don are similar, but on a bigger scale. If all the recent and superficial deposits were removed, the coastal outline from Aberdeen to Newburgh and Hackley Head would be very different. Cruden Bay, the Ugie mouth and Strathbeg Bay are of much the same general character. In brief, in former times the eastern coast of Scotland south of Peterhead was much more irregular, and indented with wide, open bays and firths.

The Moray Firth is usually interpreted as a sunken area. It is probable that Cretaceous deposits rest on its floor. The disposition of the Mesozoic rocks around its shores emphatically implies this, and George suggests it may have extended in Tertiary times 'partly by inheritance of an ancient structure, partly by rapid marine encroachment on soft sediments preserved in a sagging belt of tectonic weakness'. Bremner put forward the idea of a 'basin-like downwarp' in the Moray Firth area. This may account for the radial pattern of drainage to it, as distinct from the west–east trends which prevail over most of Scotland. The northern coast of the firth is closely associated with the Great Glen Fault which runs north-eastwards along it and probably reappears in Orkney and Shetland. (See p. 209 for the possible effect of the Great Glen Fault on the shape of the Moray Firth.) The faulting which brings down Jurassic rocks near Golspie and Brora may be regarded as part of this system. The relatively straight east–west trend of the southern shore of the firth is not easy to explain. Peach and Horne (1936) wrote 'The apparently abnormal direction of the shore-line – from Fort George to Kinnaird Head – is probably due to the strike of the planes of unconformity at the base of the Trias and at the base of the Upper and Middle Old Red Sandstone'. Bremner dismisses this by pointing out that the plane is, in fact, a very uneven surface. If the whole of the Moray Firth is a warped-down surface, and if the surface was of the nature of a peneplain there is no particular reason why the resulting shoreline should not be fairly straight; it would depend much on the way in which the warp occurred. On the other hand there is no particular reason why any such movement should produce a straight coast. All that can be said is that since this coast came into being the waves have not appreciably altered its major characteristics.

The contours on the Admiralty Charts suggest the possibility of a major stream which formerly flowed eastward, close to, and parallel to, the present coast. The same suggestion is offered by the submarine contours following the north coast of the Moray Firth, and also along the east coast of Scotland between the Tay and Aberdeen. We are here, however, in the realms of speculation, and a great deal more work must be done before we can interpret these features with any certainty.

THE SEA-LOCHS OF NORTHERN AND WESTERN SCOTLAND

A great deal has been written about the west coast sea-lochs in the past. There is no need to discuss previous views in any detail. Gregory's well-known book (1913) and also one of his papers (1927) made a strong plea for their tectonic origin. Some he regarded as rift valleys; nearly all he associated with faulting, and certainly made this the primary cause. He did not neglect the work of ice, but it was of secondary significance. Most other writers have argued that ice has played a major part in the formation of the sea-lochs, and differences between views are those of degree – just how much did the ice do? There is no doubt that ice has played a major role in their shaping, but that does not mean that it is necessarily the main reason for their origin. I prefer to call these features sea-lochs; this does not imply any particular mode of origin.

In a general way the sea-lochs of north-western Scotland run approximately south-east to north-west; those in the mid-part of the west coast are more nearly aligned east and west, and those farther south trend north-east and south-west. Although this orientation is clearly apparent on any map, it has not always been emphasized. Ting (1937) drew careful attention to it; further reference to his views is made below. It would probably be accepted by everyone nowadays that faulting has played an important role in the trend and form of a number of sea-lochs, but is not present in all cases. A fault is usually a line of weakness and is often followed by a stream. Later, during the ice age, tongues of ice may have followed the same line and the river valley was deepened and widened to produce the outline with which we are now familiar. The steep descent to the western sea was favourable, both to rivers and ice.

There is much variety of form in the sea-lochs and at the risk of some little tedium it is worth while calling attention to the lochs individually and particularly to their bottom topography. On the north coast of Sutherland there are three, the Kyle of Tongue, Loch Eriboll and the Kyle of Durness. The Kyle of Tongue is shallow, seldom exceeding 6 fathoms (11 m) over much of the loch, but increasing to 8 and 15 fathoms (15 m; 27 m) along a line joining Eilean a Chaoil to the coast south of Eilean nan Ròn. South of Skullomie Point there is only a narrow channel, 1–2 fathoms (2–4 m), at low water. At the southern end there is much silting and a mud marsh is developing. The development of the marsh will repay study; it appears to be far more muddy than is usual in Scottish marshes.

Loch Eriboll is different. Depths increase from the head of the loch to a line running eastwards from A'Chléit (32–5 fathoms; 58–64 m). Farther north depths are less – in the 20s away from the sides, and in the open sea 28–40 fathoms (51–73 m) are commonly found. There is a broad ridge-like feature across the loch near A'Chléit, and the basin width extends to near Eilean Choraidh. Thus we may probably regard this ridge as a threshold. In both the Kyle of Tongue and Loch Eriboll there are many instances of raised beaches and old cliffs, and the spit running west from Tongue Lodge, and the tombolo tying Ard Neackie to the mainland in Loch Eriboll are both features formed at a slightly higher sea-level.

The Kyle of Durness is shallow, and apart from a winding channel in the sand banks, almost dries out at low springs. The Kyle should be separated from Balnakiel Bay which is enclosed by the dunes and the rocky promontory of Faraid

Head. The Kyle of Durness is in three sections; the middle one, trending roughly north-west and south-east, is almost certainly fault controlled, the trend being a direct continuation of the fault which extends to Kearvaig, near Cape Wrath. It is also likely that structure has played an important part in Loch Eriboll; it lies mainly in Cambrian rocks, and its eastern side is close to and roughly parallel with the lines of the great thrusts, especially the Moinian. Structural control is less evident in the Kyle of Tongue. It is worth noting that the Kyle of Durness, the smallest of the three, is associated with the Dionard river and its glacial valley. The streams now flowing into Eriboll and Tongue are minor ones.

What, however, is the significance of Loch Hope? It is connected with the sea by the Hope river, and was clearly a sea-loch when sea-level was but little higher than the present. Depths in the loch reach 175 feet (53 m), and a considerable drainage flows into its upper end. Most of the loch is in the Moinian rocks, and there is the possibility that it should be associated with the steep fault?-face of Ben Hope. But Ben Hope and the Loch are by no means contiguous. It is worth while to compare the loch with Loch Morar. The latter is much deeper, but even so here are two distinctly deep lochs which at one time ran to the sea and in both of which there is a pronounced bar or threshold. Why should the ice have, seemingly, eroded such deep hollows and then failed to cut down the threshold (cf. also Upper Loch Torridon)?

On the west coast Sandwood Loch needs more investigation. Strath Shinary, which drains into it, is a small glacial valley. The loch is not deep and is dammed back by a bar of sand. If this were removed, the inlet would be a fairly typical sea-loch.

In Loch Inchard depths increase from the head to about 30 + fathoms (55 + m) along a north–south line south of Loch Innis na Bà Buidhe; then they decrease, somewhat suddenly, to 12–13 fathoms (22–4 m) and remain about this figure until near the mouth where soundings of 18–20 fathoms (33–7 m) occur. In the wide outer part depths of 30 + fathoms (55 + m) are common. Thus there appears to be a marked bar. Glaciation has obviously played a major part in this loch, but its ground form owes, perhaps, more to faulting, especially the straight coast from Rhimichle to Rhiconich, and continued inland by lakes and rivers. Seawards, the same line may correspond with the southern harbour of Kinlochbervie, and the line of cliffs named Eilear na Molo.

In Loch Dùghaill depths generally increase outwards. In Loch Laxford, in the main channel, depths increase to the mouth, 37–39–31 fathoms (66–71–57 m), but south-west from Ardmore Point a line of 16–20–25–14 fathoms (29–37–46–26 m) can be traced, and suggests a small bar. Farther out, depths increase to 40 + fathoms (73 + m). In its northern tributary, Loch a'Chadh-Fi, depths increase to the main loch, and do not suggest any form of bar between them. In Loch Laxford there is probably a significant structural influence, and the general trend of the loch is continued far inland along the Laxford river and Loch Stack.

The irregular inlets to the south have undoubtedly been modified in some degree by ice, but are not typical sea lochs. In Scourie Bay depths increase outwards; Badcall Bay is beset with glaciated islets. The next major sea-loch is Loch Cairnbawn (A'Chàirn Bhàin) and the two inner lochs, Glendhu and Glencoul. In Glendhu there is a marked shallowing near to, and east of the ferry. Glencoul is deeper, reaching, in a small basin in common with Glendhu, 26 fathoms (48 m) close to the ferry. In Glencoul there is a marked shallowing between Unapool and Aird da Loch. The ferry marks the position of a pronounced bar; seawards depths increase to 50–60 fathoms (91–110 m), but there is another marked shallowing at Duartmore Point. Beyond, the depths increase, but the sea-floor is irregular between Oldany Island and Badcall Bay. The constriction in depth and width at the ferry in Loch Cairnbawn is similar to that between Inner and Outer Lochs Torridon. It is probable that tectonic influences have played their part in these three lochs. At the present time no streams of any magnitude drain into them.

Westwards and southwards from Loch Cairnbawn, typical sea-lochs are wanting until Loch Broom is reached. But the smaller inlets offer several problems. Loch Nedd is shallow and deepens regularly. Loch Roe is similar. Despite their shapes, there is no suggestion of thresholds. Loch Inver is a considerable inlet. Depths increase to 19–21 fathoms (35–8 m) and there is a slight suggestion of shallowing along a north–south line at the inner end of Kirkaig point; but even so depths of 13–15–17 fathoms (24–27–31 m) occur in mid-channel. Farther outwards depths increase fairly regularly. It is, therefore, difficult to picture a true bar in this loch; at best it is a very gentle ridge. Two considerable streams drain into the north-east and south-east corners of the bay. The part between them, on which Lochinver village is situated, is fairly straight, and it is difficult to envisage precisely how ice has played its part in this loch. A local investigation might well be rewarding. Loch Kirkaig is in some ways a smaller version of Loch Inver, but it is shallow, and depths increase outwards, though only to 4–5–6 fathoms (7–9–11 m) north of Sgeir Mhór.

Loch Broom is a major feature. There is a ridge of shallows, 6–9 fathoms (11–17 m) at Corry Point (above Ullapool), and depths fall to 20 + fathoms (37 + m) farther in. The loch remains shallow between Corry Point and Ullapool, but deepens fairly regularly below Morefield. There is a narrow channel, 12–15 fathoms (20–7 m), at Ullapool Point, and a slightly deeper basin between there and Corry Point. Inland the loch is prolonged into Strath More, a broad valley reaching as far as Braemore where two streams join. Little Loch Broom deepens to 50 + fathoms (91 + m) (one sounding of 60 fathoms: 110 m) between Badrallach and Durnamuck, then shallows to 20–30 fathoms (37–55 m) along a line drawn north from Durnamuck. Beyond it is somewhat deeper, reaching 30 fathoms (55 m), but at and near the mouth a line can be drawn not exceeding 22 fathoms (40 m). It deepens farther seaward. This valley is continued about two miles (3.2 km) inland by Strath Beag and receives the waters of the Dundonnell river. Both lochs are glacial features, and follow lines of former river valleys. There is no obvious evidence of structural control.

Gruinard Bay and Loch Ewe, which is separated from the open sea by a distinct sill, are discussed on pp. 70 and 71. All that need be emphasized here is the strong structural control of Loch Maree and the continuation of the fault towards Rubha Réidh. Loch Gairloch is of unusual form. Although there are some slightly lesser depths, 27–9 fathoms (49–53 m), near the entrance, there is no real suggestion of a bar. Inside the loch depths are fairly consistent, 25–34 fathoms (46–62 m), and 40 fathoms (91 m) is soon found in the open sea to the west. Caolas Beag, the channel between Longa Island and the north shore, is distinctly shallow. The north side and the south side as far as, approximately, Badachro, are in Torridonian rocks; the crystallines form the south-east and east. In that sense there is some structural control of shape. The east shore, above Gairloch, is steep, and the streams reaching the loch are small. In some ways it may be compared with Loch Inver and perhaps Loch Kirkaig, but in neither of those is there any Torridonian.

Loch Torridon is a particularly interesting loch. The upper loch deepens to 48–9 fathoms (106–8 m) along a north–south line just west of Alligin Point. In the Kyle, i.e. the narrows, 12–13 fathoms (22–4 m) is the greatest depth. But in outer Loch Shieldaig and within Diabaig Point 70–80 fathoms (128–46 m) occur. Again, in the narrows between Diabaig Point and Ru Ardheslaig 55–7 fathoms (101–4 m) are found, but shallowing to 43–25–26–27 (79–46–48–49 m) a little to the west, then deepening, but again shallowing at the mouth. On a line from Red Point to Rubha na Fearn most soundings are between 30 and 40 fathoms (55 m; 73 m), but there are one or two at 50+ (91+ m). Then it deepens westward. The inner loch was a collecting ground for ice from the high ground surrounding it. The constriction between the two lochs is a ridge of Lewisian Gneiss. Doubtless this was resistant and represents one of the many major irregularities on which the Torridonian accumulated. Nevertheless it is not easy to see just how and why such a narrow ridge and small sea channel were spared between the Upper Loch and Loch Shieldaig. There is a rather similar feature between Loch Shieldaig and the Outer Loch.

Applecross Bay is a relatively small feature and may be related to the major fault which continues up the valley of the Applecross river. The picturesque, but usually north–south trending inlets at Camasterach and Toscaig may probably be associated with faulting.

In Loch Kishorn depths increase fairly consistently outwards, but there is some suggestion of a broad ridge. The existing chart is dated 1850, and new work may possibly modify this. The Loch Carron chart is also more than a century old, but the features of this loch are pronounced. There is a marked basin above Strome Ferry where depths reach 50–60 fathoms (91–110 m). At the ferry depths are 3–6 fathoms (6–11 m), and the loch narrows considerably. Westwards, towards Plockton, Lochs Carron and Kishorn unite. The upper parts of both must have been much more extensive at higher sea-levels. In Loch Carron, the flats reach back almost to Loch Dughaill, and the valley is strongly glaciated.

It will be convenient here to comment on Inner Sound, the Sound of Raasay and some of the sea-lochs of Skye. Inner Sound is deep, 70–100 fathoms (128–83 m), and one sounding at 133 fathoms (243 m) (Robinson, 1949). The Sound of Raasay is usually 50–70 fathoms (91–128 m) deep, but at the southern end only 16 fathoms (29 m) and but 10 fathoms (18 m) between Raasay and Scalpay, and 4–8 fathoms (7–15 m) between Scalpay and Skye. In Skye, there are no sills in Lochs Snizort, Dunvegan, Slapin and Eishort. There is a slight sill at the mouth of Loch Harport before it turns westwards. Loch Scresort, in Rhum, which opens somewhat anomalously to the east, has no sill, and depths at its mouth fall quickly to 28+ fathoms (51 + m).

Loch Alsh with Lochs Long and Duich is reminiscent of Loch Cairnbawn. Loch Long is narrow and shallow, but attains 17 fathoms (31 m) a little above Dornie. It makes a shallow junction with Loch Duich which has depths of more than 60 fathoms (110 m), but only of about 10 fathoms (18 m) between Totaig and Eilean Donnan. After their junction to form Loch Alsh, the major loch shows soundings down to 60 fathoms (110 m), but near Kyleakin it shallows markedly to about 10–13 fathoms (18–24 m). On its south side there is the remarkable Kyle Rhea, a river-like feature, with depths between 7 and 13 fathoms (13 and 24 m), deepening to 18–19 fathoms (33–6 m) near the Sound of Sleat (see Plate 43). In the Sound of Sleat 30 to 60 + fathoms (55–110 + m) are common, and into it drain two of the most beautiful sea-lochs, Hourn and Nevis. In Loch Hourn the innermost basin which extends to Eilean Mhogh-sgeir has depths down to 13 or 14 fathoms (24 or 26 m); the second basin deepens to 18 fathoms (33 m) and the second bar is at Caolas Mór at 2 to 5 fathoms (4 to 9 m). There follows a small basin and then another bar at Eilean à Gharb-làin with soundings reaching 8 fathoms (15 m). Then follows the main loch with soundings reaching 80–90 fathoms (146–66 m), and a gradual shallowing to the Sound of Sleat which hereabouts ranges between 40 and 50 fathoms (73 and 91 m). Loch Nevis is wider at its inner end, and the first basin, with soundings down to 50 + fathoms (91 + m), runs as far as the narrows at Kyles-knoydart where depths of 2–4 fathoms (4–7 m) occur. Then the loch deepens and widens. There are many soundings between 40 and 50 fathoms (73 and 91 m), and occasionally more than 80 fathoms (146 m). There is a distinct ridge at its mouth where depths of 13–20 fathoms (24–37 m) are common, and also shallower patches. Then the bottom falls to the Sound of Sleat.

A mile or two south of Loch Nevis is Loch Morar, the deepest loch in Scotland, it is more than 1,000 feet (305 m) deep in its mid part. There are no sills separating different basins as in Hourn and Nevis, and today it is cut off from the sea by a low neck of land. In higher raised beach times it must have been open to the sea; it is considered here with the present sea-lochs. Towards the seaward end the bottom slopes fairly steeply up to and above the 100-foot (31 m) contour. The bar is comparable to that between the two lochs Torridon, but is even more pronounced. It is a little difficult to explain this sill completely. There is no doubt that Loch Morar is a glaciated basin, although no valleys of any size reach it. Why should erosion, of what must have been a very powerful glacier, have ceased so abruptly? The col between South Tarbet Bay on Loch Morar and Tarbet Bay on Loch Nevis is an admirable example of a channel produced by an overflow of a small tongue of ice to the north.

Loch nan Ceall is hemmed in by a large number of skerries and small islands; in this sense it is unique. Loch Ailort falls to 26 fathoms (48 m) about 1½ miles (2.4 km) from its head; in the narrower part it is shallow, 2 to 11 fathoms (4–20 m), and even 1–3 fathoms (2–6 m) west of Eilean nan Trom. Then it deepens to south of Arisaig. Loch nan Uamh has a somewhat irregular floor, but there is no marked sill.

Loch Shiel may also be regarded as a sea-loch since, like Loch Morar, it had a connection with the sea in higher raised beach times. It is a long narrow loch trending north-east and south-west, and so somewhat athwart the east–west lochs of this part of the coast. It reaches depths of 300 + feet (91 + m) in its central area, and there is a smaller basin at 400 + feet (122 + m) somewhat higher up the loch. It deepens fairly quickly from Glenfinnan. At its lower end it shallows more gradually to the narrows (c. 50 feet: 15 m) at Eilean Fhinaian, and eventually drains by the Shiel river to Loch Moidart.

Loch Sunart (see Plate 48) is the longest of the east–west sea-lochs. In the upper part soundings of 3 to 30 + fathoms (6 to 55 + m) are found, but only about 13 fathoms (24 m) at the shallows at Laudale. Then it deepens to 20, or even 40, fathoms (37 or 73 m), but falls to 21 (38 m) in the narrow strait north of Carna island. The bottom then falls to 40 or even 50 + fathoms (73 or 91 + m), and there is some evidence of a slight ridge at the mouth. Loch Teacuis, on the southern shore of Loch Sunart, has an upper basin at 15 or 16 fathoms (27 or 29 m), and a marked bar at the narrows (1–2 fathoms: 2–4 m).

In a sense the Sound of Mull is one with Loch Sunart. In the sound, apart from a narrow channel at 50–70 fathoms (91–128 m) on the east side at the south end, depths are less both at the northern and southern entrances. In the main part of the sound they range from 20 to 50 fathoms (37 to 91 m), and from about 32 to 60 fathoms (59 to 110 m) near Auliston in the deep channel leading to Loch Sunart. The significance of the sound is discussed on p. 38. On the island of Mull, Loch na Keal shows a distinct sill cut by a narrow channel at 20–35 fathoms (37–64 m); otherwise it deepens gradually seawards. Loch Tuath shows scarcely any sign of a sill; the same is true of Loch Scridain and Loch Buie. Loch Spelve has an inner basin reaching 20 fathoms (37 m), but the narrow entrance, apart from a channel at 3–4 fathoms (6–7 m) largely dries out at low water.

Loch Eil, Loch Linnhe and the Firth of Lorne make the greatest indentation on the west coast of Scotland. Loch Eil is separated by a narrow and shallow sill from Loch Linnhe. Depths may reach 30 fathoms (55 m) within, but only 2–4 (4–6 m) in the narrows. Then they fall to about 70 fathoms (128 m), but there is a second narrowing and shallowing (to 6–10 fathoms: 11–18 m) at Corran. The water remains fairly shallow as far as Sallachan Point and then the bottom falls to between 30 and 40 fathoms (55–73 m) as far south as Balnagowan island. In the outer part of the main loch depths of more than 110 fathoms (201 m) occur on the north side of Lismore island, but then the bottom rises somewhat to the Sound of Mull. The channel on the south side of Lismore island is often narrow and shallow, only 4 to 8 fathoms (7–15 m) at the northern entrance. Loch Leven joins Loch Linnhe just below Corran narrows. In the upper basin depths of rather more than 20 fathoms (37 m) are found except near islands, but only 2–4 fathoms (4–7 m) in the narrows and also at Rubha Charnuis. Between these two places there is a small basin (*c.* 17 fathoms: 31 m), and there is also a shallow dredged channel at Caolasnacoan. Thus there are two marked shallows and narrows.

Loch Creran, a few miles farther south, has a shallow entrance (2–5 fathoms: 4–9 m) and depths reach 20 fathoms (37 m) within. There is another narrow at the old railway bridge (Creagan) and a small inner basin. Loch Etive is comparable with Lochs Morar, Shiel and Hope; it makes a connection with the sea only at high water. The Falls of Lora are reversing, falling to the lake with a rising tide, and to seaward at the ebb. There is another marked constriction at Bonawe Ferry. The Firth of Lorne is relatively shallow between Oban and Lismore, usually less than 20 fathoms (37 m), but 40 fathoms (55 m) are found between Lismore and Duart Point. Farther south depths increase to 80 or even 100 + fathoms (146 or 183 + m); although there are lesser depths south-west of Kerrera island. In Kerrera Sound 20 + fathoms (37 + m) occur, but less in the narrows north of Oban and also near the south entrance of the sound.

Loch Feochan, in the Firth of Lorne, dries at low water at the narrow mouth; there is a basin to about 10 fathoms (18 m) within. Seil Sound shows 2–3 fathoms (4–6 m) above the bridge, and is even shallower as far as about half way along the island, where depths increase and briefly reach 30 fathoms (55 m). Similar depths are found inside Luing. Loch Melfort has traces of a sill and depths to perhaps 20 fathoms (37 m) within. On the west of Luing the water deepens to the south; there are several islands at the north end of this sound. Shuna Sound is slightly deeper (19 + fathoms: 35 + m) in the north than in the south (11–13 fathoms: 20–4 m). The Sound of Jura varies a good deal in depth, but 90–100 fathoms (165–83 m) are found along a fairly wide channel from Loch Crinan to the entrance of Loch Caolisport. Loch Craignish, in a sense the head of the Sound of Jura, shows 8–9 fathoms (15–17 m) in its upper part, and 4–5 (8–10 m) at its southern end; there is a narrow and deep channel (reaching 20 fathoms: 37 m) between the long islands of Eilean Righ and Macaskin and the mainland. Loch Crinan is shallow (*c.* 5 fathoms: 9 m), but deepens quickly to the Sound of Jura along a line from, approximately, Ardnoe Point to Scodaig. Loch Sween shows two small basins separated by sills, one at Kilbride (Kilbryd) and the other a little above the entrance.

On the other hand Loch Caolisport deepens gradually and shows no sign of a sill. West Loch Tarbert (i.e. the loch separating Knapdale from Kintyre) deepens gradually from 1 to 11 fathoms (2 to 20 m). There is a sill at 4–5 fathoms (7–9 m) at the entrance near Eilean Traighe. In the mid part of the loch there is a channel 11–12 fathoms (20–2 m) deep, with one sounding of 18 fathoms (33 m).

West Loch Tarbert (Jura) deepens fairly regularly seawards, and the Sound of Islay is deeper, about 20–30 fathoms (37–55 m), in its mid part than it is at the north end (6–8 fathoms: 11–15 m). In the south depths are a little less than in the mid-part.

Loch Fyne shows distinct basins. The upper loch increases in depth fairly regularly to 70 + fathoms (128 + m), and then there is a ridge, c. 30 + fathoms (55 + m), along a line south from Furnace. Depths of this order prevail as far as the islands near Minard Castle (15–21 fathoms: 27–38 m). Then a basin, to 30 fathoms (55 m), follows as far as Glas Eilean, where there is a sill which, however, is cut by a narrow channel to 20 + fathoms (37 + m). In the outer loch depths fall to 90 + fathoms (166 + m), but there is a shelf near Ardlamont Point, and there is a deep channel. Loch Gilp is shallow and deepens gently to Loch Fyne.

Loch Riddon is shallow near its head. It deepens to the Kyles of Bute. The western kyle is perhaps a little shallower, and there is a slight sill near Tighna-bruaich; the eastern kyle is shallow in the north, but deepens southwards. Loch Striven is deeper; there is a sill or ridge, not a marked one, between Ardbeg Point and the Ardyne burn. Loch Long affords some evidence of basins. From its head it deepens to 20–30 fathoms (30–55 m), but is somewhat shallower on a line running north from Finnart jetty, south of which it depends to its junction with Loch Goil. This smaller loch has depths of 40 + fathoms (73 + m) in its main basin, but only of 8–10 fathoms (15–18 m) in the narrow channel about half-a-mile above the junction with Loch Long. The lower part of the main loch has soundings of 40 + fathoms (73 + m) but is somewhat shallower near Strone Point. Holy Loch varies but little in depth; it falls to 20 + fathoms (37 + m) at and near its junction with Loch Long. Gare Loch has a marked inner basin, succeeded by a narrow and shallow sill of 5 to 6 fathoms (9 to 11 m), through which is a channel dredged to 36 feet (11 m) (1966). Lower down natural depths are greater but the Helensburgh shore is shallow.

THE LONG ISLAND

In the Long Island there are innumerable sea-lochs, but comment will be restricted to relatively few. Nearly all open to the east and are presumably partly associated with a local glaciation. Loch Erisort shows no sill. Loch Shell deepens from 5 to 20 + fathoms (9–37 + m) and there is a slight ridge at the mouth. There is possibly a small sill in Loch Claidh. The largest of the Hebridean lochs, Loch Seaforth, is more complicated. There is a pronounced sill between the innermost, east–west trending, part and the middle section, the trend of which is approximately north-east and south-west. Seaforth island stands at the junction of this part with the main south-eastern trending arm. There is a sill in the channel to the east of Seaforth island, but not in the westward channel. There is also a marked sill at the entrance to the loch. In the innermost part depths scarcely exceed 10 fathoms (18 m), and the sill is at 1 fathom (2 m) (see Plate 36). The middle part is somewhat deeper, and in the outer loch soundings locally exceed 50 fathoms (91 m) just inside the sill on which they are less than 20 fathoms (37 m). In East Loch Tarbert there is no true sill.

On the west coast of Lewis–Harris sills are absent in Lochs Resort and West Loch Tarbert. The Sound of Harris in general deepens to the east, but it is much encumbered with islands.

Loch Maddy in North Uist has some shallows at its mouth, but a true sill is absent. The strait between Ronay island and Rossinish deepens steadily, but Loch Flodday, to the north of Ronay island, is distinctly shallower between the islands at its mouth than it is within. In Benbecula Loch Uskavaig (Uskavagh) is without a sill; there is a small one in Loch a'Laip and also in Loch Keiravagh.

The strait between Benbecula and North Uist deepens to the east, and in Lochs Carnan and Skiport and also in Lochs Eynort and Boisdale sills are absent although islands may obstruct their mouths. The sounds of Eriskay and Barra also deepen a little eastwards.

This long description of the sea-lochs is worth while in that it serves to emphasize their differences as well as their similarities. There is no doubt that ice has played a prominent part in their shaping, and that the glaciers followed pre-existing valleys, some of which are clearly associated with faulting. If now we turn briefly to a discussion of the contours of the sea-floor between The Long Island and the mainland we shall see that there is at least a strong suggestion that the valleys of the sea-lochs were at one time the probable headwaters of a major system of Hebridean rivers. This idea was put forward by Ting (1937). Ting concerned himself mainly with rivers north of the Clyde, and made a careful analysis of the charts covering the whole area, not only of the waters within the archipelago but also as far seawards as the edge of the continental shelf. Reference to Fig. 3 will show his reconstruction. He interpolated contours on the charts and found clear evidence of major trenches, which he assumes to be former river courses, cutting the sea-floor. It will be seen that to the south of Skye the contours suggested south- and west-flowing streams, and that north of Skye the flow was northerly. He then proceeded to prolong the rivers, represented now by the sea-lochs and the streams flowing into them, and so built up a consistent system of major streams. It would doubtless be possible to make other interpretations of these ancient courses, but we may accept Ting's view as a most reasonable one. (On p. 27 reference is made to the possible association of east-coast rivers with sea-bottom features.)

But how was the sea-floor formed on which these rivers may once have flowed? If we examine the cliffs of much of Skye, Canna, Rhum and some other islands we

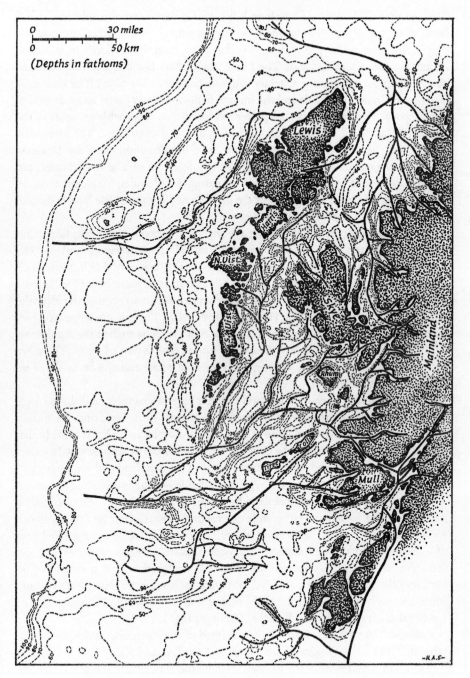

Fig. 3. The submarine topography off western Scotland (after S. Ting)

shall see that thick flows of basalt and dykes of dolerite end abruptly in the cliff faces, clearly implying that at one time they extended much farther. Here is one simple and spectacular proof that there has been extensive collapse in this area. Alongside the eastern shore of The Long Island there is a long, continuous submarine scarp. Its appearance certainly suggests faulting. (See p. 8.) The continuation of the Great Glen Fault and the known faults in many coastal areas, as well as others, possibly of major size, suspected on more than reasonable grounds in the Sound of Raasay, in Morvern and elsewhere, all point to collapse. The same conclusion may be drawn from a consideration of the distribution of the Lewisian Gneiss on the mainland and throughout the whole length of The Long Island, and also in Coll and Tiree. There are also numerous small faults, including those of mid-Tertiary age in Skye, which cut the granites and basalts of the Red Hills, and are also abundant in north-western Skye.

But despite the evidence of faulting the sea basin between The Long Island and the mainland is not a true rift. George (1966), in an important paper, thinks of The Long Island fault 'not as a graben but, much as the depths of Loch Ness, Loch Lochy and Loch Linnhe delineate a Great Glen fault, as a "subsequent" valley: the valley was differentially eroded in fluviatile adjustment to structure long after the initial feature was mobilised by Hercynian stress . . . ' There is no suggestion of a corresponding fault on the mainland side; the oblique faults such as the Applecross Fault are ancient, and are in no sense to be interpreted as part of a rift. The more modern faults such as those cutting the Mesozoic rocks and the lavas 'have no regional effects' (George).

Skye is separated from the mainland only by narrow channels, and the geology of the Sleat peninsula is really a continuation of that of the mainland. In Raasay basalts rest on Mesozoics and these in turn on Pre-Cambrian rocks. It is likely that the Jurassic beds of Raasay once covered the Liassic beds at Applecross. It seems, therefore, that Raasay and Skye are really parts of the mainland and were separated from it by erosion and some submergence which led to the intervening straits.

Mull is separated from the mainland by a long strait which is in one place only a mile wide. The basalts on either side correspond closely. George (1966), unlike Linton, does not regard the Inninmore Fault on the mainland as the marginal fault of a Hebridean rift. The basalts in Morvern indicate a former wider extent, and since the basalts usually rest on Mesozoics, the suggestion is that structures on Mull and the mainland are in accord. Mull was isolated in the late Tertiary, 'and south-easternmost Mull is structurally in its foundations on the Grampian side of the Great Glen fault, in Argyll facies' (George, 1966).

Coll and Tiree are, apart from dykes, formed of Pre-Cambrian rocks. Colonsay, Islay and Jura are geologically analogous to Knapdale and Kintyre.

The Long Island is far to the west and beyond the Minch fault. It is almost wholly Pre-Cambrian, and closely related to the same rocks on the mainland.

There is no doubt that the Hebridean area has sunk. The sea-floor today varies in depth. It is usually less than 100 fathoms (183 m), except for the deeper area between north-western Skye and North Uist, and again in two elongated deeps of more than 100 fathoms (183 m) between Coll and Tiree and Barra, the greater one being much nearer Barra. In other words the floor of the sea is, apart from particular places, for the most part some 300 to 400 feet (91–122 m) below the water surface. If the interpretation put upon the submarine features by Ting is approximately right, it is interesting to consider at what height above sea-level the present submarine valleys were cut. The present sea-lochs are heads of the submarine valleys, so the change of level may not have been unduly large. George suggests (1965) that some of the Hebridean Mesozoic basin never stood at high altitudes but in other parts

the Mesozoic rocks were subaerially degraded by removal of not less than an estimated 3000 ft. [914 m] before the outpourings of the Palaeogene lavas. It was presumably on these lavas that part of the west coast river system originated. The west coast Highland rivers, in country where Cretaceous rocks still remain and form not a cover but a floor to thick lava flows, are very short and occupy deeply glaciated valleys; but, allowance being made for local and secondary adjustments, they are predominantly west-flowing, across the geological grain not only of the pre-Cambrian rocks but also of the Mesozoic and Tertiary rocks. They also appear to be as completely superposed on all elements of the foundation just as are their longer counterparts in the east. In the south-west Highlands and in the Southern Uplands a similar discordance includes lava free valleys incised across the Tertiary dykes, and implies correspondingly a late-Tertiary superimposition.

At what height this surface stood is a matter of conjecture. In the area of the Minch, assuming that the sea-contours have been reasonably interpreted, it may not have been very high; a height to be measured in hundreds rather than thousands of feet. This does not exclude the possibility that its margins, especially on the mainland, may have been a good deal higher, but the apparent accordance of sea-lochs and submarine features, except in places such as the Minch fault, seems to imply a gentle warping of a relatively low area. The subsequent glaciation modified the valleys both terrestrial and submarine, and may well have accentuated the declivity of the Minch fault.

Before dealing with southern Scotland it will be rewarding to glance at some of the firths on the east coast. Mention was made on p. 26 of the Brora river. This does not form an estuary, in fact its mouth resembles a small delta. But a few miles above the mouth is Loch Brora in which there are three distinct basins, the inner two of which reach 60 fathoms (110 m). They are separated by a narrow channel and a low, but broad, spur. The middle basin is separated from the third one by another narrow strait. All the loch is bounded by relatively steep slopes which rise to 700 or more feet (213 + m). If the low ground behind Brora were eliminated, it is mainly alluvial, Loch Brora would resemble the west-coast lochs, and in highest raised beach times it must have been penetrated by the sea.

Loch Fleet (see p. 205) is enclosed to seaward by recent accumulations, and the embankment is an artificial barrier at its upper end. But Strath Fleet continues inland and would have been an open sea-loch at least as far as Rogart in higher raised beach times. The slopes on either side of the strath are fairly steep, and when the floor was flooded it would have been a small sea-loch.

Dornoch Firth is still relatively open. The accumulations at its seaward end are described on p. 206. It is tidal to Bonar Bridge and in earlier times the sea would have penetrated the Kyle of Sutherland. Although it has resemblances to west-coast lochs, it is, for most of its length, very properly called a firth. The Cromarty, Beauly and Inverness firths are discussed in Chapter VII and need little comment here.

THE FIRTH OF CLYDE, THE NORTH CHANNEL AND THE SOLWAY FIRTH

The general configuration of south-western Scotland and its coastal evolution are not easily explained. That it is submerged area may be accepted, but it is not so clear why, for example, Arran remains as a high mass and why the bays of Galloway, Luce, Wigtown and Kirkcudbright lessen both in width and depth eastwards.

It will be helpful to refer to some early ideas on river development in this region; in this way the change from older to more recent views is made clear. We have seen that Gregory suggested that the rivers of all Scotland have no general community of direction, but that their courses were determined by his hypothetical system of folds and fractures. In the Clyde area he suggested streams flowing to the south-east, across Lochs Fyne and Long to a Clyde also flowing south-eastward. McCallien followed this view to some extent, and assumed that what is now the Clyde estuary was part of a land area sloping from Loch Fyne to the Firth of Forth. Later, he supposed that the sea-lochs draining into the Clyde were cut in a relatively flat surface at about 1,000 feet (305 m). Parts of this now remain in Arran and around Helensburgh. Peach and Horne (1936) followed Mackinder (1907) in the view that at one time Scotland was part of a continuous land area stretching from Ireland to Scandinavia. This land had been peneplaned and in the area with which we are now concerned it was assumed that drainage extended from Argyll to the Solway and Tweed. Mort (1918) also speculated freely in this matter, and wondered if the valleys which cut across the Kintyre peninsula, and the Sound of Islay were parts of some ancient river system which reached, or even crossed, the Solway Firth. Views such as this imply that the present Clyde is flowing in reverse of its old course. The idea is connected with the Biggar Gap, a marked valley feature near the town of that name. Linton (1933, 1934) suggested that what is now the Upper Clyde formerly flowed through this gap and so into the Tweed system. Examination of a map shows that the Upper Clyde flows directly towards the gap, and suddenly turns through a right angle and flows to the north-west. The Tweed crosses the Southern Uplands Boundary Fault a little to the east of Biggar, and at one time the river may have risen where now is the source of the Clyde. The

Clyde valley can be divided into three different parts: the Upper Clyde above Roberton is wholly in the Southern Uplands plateau, and is a transverse and consequent stream; the Middle Clyde between Roberton and Lanark consists of disjoined reaches, some longitudinal some transverse; the Lower Clyde, below the falls, is a subsequent and longitudinal stream along the axis of the Lanark coalfield. By headward cutting in these relatively less resistant rocks the Clyde, through its tributary the Douglas, cut back into the Tweed headstreams, and eventually led to the final diversion of the Upper Tweed into the Clyde, so that what was once the upper waters of the Tweed reached the Firth of Clyde instead of the North Sea (Linton, 1933, 1934).

Linton argued that these changes were induced by superimposition from a Cretaceous cover. There is general agreement that superimposition has taken place, but George (1954–5, 1958) shows that it is more probable that it was of later date. It is more than likely that the upland surface on which the rivers originally ran was post-dyke in time – long after the Cretaceous. The dykes are most significant because truncated dykes 'from which no lava spills on either side [were] truncated *after* intrusion' (George, 1958).

The Clyde crosses major structures indifferently. These include the Southern Uplands Fault and the central coalfield. The firth of Clyde is an extension of Loch Long which also traverses outcrops, dykes and faults. George, following Hollingworth, maintains that in the Clyde basin several cycles of erosion, represented by benches, can be demonstrated. This is also true of several other river basins where not remoulded by ice work. Knick-points are common, and the Falls of Lanark are a magnificent example. For about 20 miles (32 km) above the falls the gradient is gentle, and graded to a base-level of about 600 feet (183 m). Below the falls the river is in a gorge through which it drops about 200 feet (61 m) in little more than a mile.

A gravitational survey of the Kelvin valley showed that a deep Pre-glacial valley exists. It seems that the Pre-glacial Clyde was a hanging tributary of the Kelvin which was graded to a sea-level about 300 feet (91 m) below the present level. Today the Clyde basin is eight to ten miles (13–15 km) wide in the region around Paisley and Glasgow, and narrows to about one mile (1.6 km) at Dumbarton. Ting (1937) referred briefly to the Clyde. In that Firth there are long and narrow troughs of deep water, the deepest reaches 92 fathoms (168 m). Ting argues that these troughs seem to have developed on a surface between 30 and 35 fathoms (55 to 64 m), a depth not consistent with that suggested above for the Pre-glacial Kelvin. If, as is probable, a river followed Loch Fyne, it may well have continued along the eastern, or even the western, side of Arran. Other streams may well have followed along the lines of Loch Riddon and the Kyles of Bute, and also, as already mentioned, down Loch Long. All of these may well have joined a main stream which passed through the well-marked deeps in the North Channel. There are difficulties in assuming this view too readily; south of Arran the contours of the sea-floor do not suggest a continuation of these postulated rivers. The deeps on either side of Arran give way to a shallower plateau which separates the northern troughs from that in the North Channel. On the other hand, it may be that this shallow area has been built up partly of glacial material. In general it is reasonable to accept an hypothesis which led the drainage of Lochs Fyne and Long and smaller lochs to the North Channel; it is however proper to realize that it is not yet capable of direct proof.

George (1954–5) and Jardine (1959) follow Linton in thinking that the rivers of the Southern Uplands are superimposed, but they think that superimposition took place at a later time than does Linton. Neither George nor Jardine suppose that these rivers crossed the Uplands from some distant source. Jardine is definite in his opinion that the drainage originated in the mid-Tertiary along the northern anti-clinorium of Ordovician strata. Captures and other changes have taken place since, but the rivers are autochthonous. Both George and Jardine recognize several levels in this part of Scotland, levels which represent former stands of the sea. Whether these levels are marine or subaerial is from the present point of view relatively unimportant, but as Jardine remarks the 'stepped morphology of the land has played an important part in the evolution of the present drainage system'.

These rivers now drain out into Wigtown Bay, Kirkcudbright Bay, Rough Firth and the Nith estuary. Although small streams reach Luce Bay, this bay must be regarded as different from the others. It was at no distant date part of a strait; the raised beach deposits have filled up the mid-parts. That it may, at one time, have been followed by a through river is possible, but there is no proof of this. The Galloway coast has every appearance of a drowned area, but it is doubtful if Mort's view that the subsidence was greater in the west than in the east is correct. It is not unreasonable to suppose that in earlier times the Cree, Fleet, Urr, Nith and Annan once extended farther southwards and that they joined a Solway river, an extension of Esk and Eden. (Linton (1933) regarded the Tweed as a main east-flowing consequent river. Its course is independent of structure, and it is not a strike stream. He also suggested that at one time it extended westwards of its present source, possibly even as far as Arran. The Clyde, in his view, cut back and captured the headwaters of the Tweed. Bremner postulated a Solway-Tyne river. The collapse of the Solway area reversed the flow of the western half of this river. The Luce, Cree, Nith, Annan and Esk were assumed to be north bank tributaries of this river.) The contours of the Solway, in so far as they are not obliterated by sands, are in conformity with these views, and the combined river presumably reached that flowing southwards via the North Channel to form an Irish Sea river. The minor changes of level since the ice age, the accumulation of marshes and sands, and the somewhat limited amount of erosion have produced the present coastline. Nevertheless, a considerable amount of detailed work remains to be done, not least, as in other parts of Scotland, on the relative severity of erosion in pre- and post-'25-ft' beach times.

Chapter III

Recent vertical movements of the coastline

A great deal has been written about the raised beaches around the coast of Scotland. It is not intended that this chapter shall analyse the problem in detail, but it is essential to appreciate the nature and intricacies of the subject in order to understand many features of the coast.

In the considerable number of sheet and regional Geological Survey memoirs, some of which are now very old, there are numerous references to raised beaches in particular localities. Usually the descriptions and details are clearly given, but since it is only in quite recent years that the need for precise measurements of their heights has been realized, many of the heights given are somewhat approximate. This is not necessarily the result of inaccurate work, but rather that it was not fully appreciated what heights should be measured. Moreover, a beach and a wave-cut bench are not the same thing. A beach rests on a bench, and the true beach material may be piled up well above bench level by the action of storm waves, and much will depend upon the degree of exposure. On the other hand the bench is seldom, if ever, a level surface. It slopes downwards from land to sea; its surface may be irregular on account of the varying resistances of the rocks of which it is composed. Moreover, it may be tilted, so that if it is possible to follow it for any distance a definite slope may be detected. This tilting is usually the result of isostatic movements caused by the waxing and waning of the Quaternary ice-caps. There may also be local reasons which will lead to some irregularity in the surface of a bench.

Which part, then, of a bench should be measured to obtain its height? It is now generally agreed that the best place is the angle where the bench gives way to the cliff behind it. Unfortunately this angle is often obscured by talus, but this can be moved if not too thick. But there is another problem. Many raised beaches are in remote places, and it is no easy job to tie their heights to a fixed mark, a bench-mark or some other carefully surveyed point. Parts of Scotland present serious difficulties of this sort; in some countries it is impossible to use this method since detailed and accurate levelling does not exist. Even in Scotland some local datum may have to be used, the higher limit of barnacles, or a marked line of algal growth. In Queensland I used the upper line of range of the rock oyster, *Ostrea mordax*. It would be easy to expand this theme, but if anyone interested in the problem will consider it in the field there is little likelihood of the difficulties being minimized.

As a result of the many observations made by geologists and others in the nineteenth century and first half of this one, it came to be realized that beach or bench remains in Scotland grouped themselves around certain levels. This led to the naming of the beaches by their approximate height, a method that has not only lasted until today, but has given rise to much misunderstanding and imprecision. Unfortunately, however, there remains so much detailed work to do, so that even now we cannot be exact. We must, in fact, still use the old height names, but in order to emphasize their vagueness, they will, in this book, be given within inverted commas. It was soon realized that many remains were about one hundred feet above present sea-level; others varied between about thirty and sixty feet; there are also numerous benches at about twenty-five and fifteen feet. Hence the names '100-ft', '50-ft', '25-ft', and '15-ft' beaches.* Although it was soon realized that, for example, the '100-ft' beach might vary perhaps twenty or more feet in height from one place to another, the method of nomenclature inevitably forced the view that any beach remains within a reasonable range of the named height must all be part of that same beach or bench. The so-called '50-ft' beach was soon seen to present serious difficulties. There are numerous traces of beaches, the average height of which is not far from fifty feet. But this in itself means nothing. A bench is cut by waves; a beach is built by waves. It is obvious that, whatever the level of the beach or bench may be, at one time the waves were working on the land at about this level. Some of the benches are cut in solid rock; many others are in boulder clay or drift deposits. The most prominent bench in Scotland is that associated with the so-called '25-ft' beach. There are many miles of coast where it can be followed; several coastal roads are built on it, and in many places it affords areas of cultivated land. Behind it there is usually a well-marked cliff, now dead and commonly covered with grass and other vegetation. At first sight it looks as if beach, bench and cliff are of comparatively recent origin. In fact, the '25-ft' sea has been called the cliff former *par excellence*. This is not a view that can be held in simple form today. It is more than probable that the bench and cliff are much older, and were re-touched in '25-ft' times. We shall see later that it may be possible to eliminate the '50-ft' level: the remains of beaches that once were thought to have been formed by a 50-ft sea are probably parts of higher or lower levels; in other words they belong to the '100-ft' or '25-ft' groups. It is better to designate them as Late-glacial and Post-glacial beaches.

We must now consider the order in which those beaches were formed. The first great step forward was made by Jamieson (1865). He showed conclusively that the '25-ft' beach overlies the highest peat beds in the River Ythan (Aberdeenshire) and other places. Later it was realized that the '100-ft' beach was contemporaneous with large glaciers. Its exclusion from the upper parts of certain sea-lochs has been explained in this way, and it is associated with clays, containing an arctic fauna, laid down in deeper water off-shore. There are also off-shore clays associated with the '25-ft' beach, but these contain a fauna, as does the beach itself, comparable to that

* 100 ft = approx. 31 m; 50 ft = approx. 15 m; 25 ft = approx. 8 m; 15 ft = approx. 5 m.

of the present day. Features at an intermediate level still presented difficulties – did they represent stages in a fall of sea-level from the 'roo-ft', or did they indicate a subsequent rise of sea-level? It was clear that much later sea-level continued to fall until it reached a level about 200 feet (61 m) below the present. We know this from the occurrence of peats dredged up from the Dogger Bank, and also from borings made alongside the mouths of rivers. These records clearly indicate deeply-cut channels which have subsequently been filled up by deposits during the ensuing rise of sea-level. Then, as Jamieson proved, the level of the sea rose above that of the present and the '25-ft' beach was formed. Since then there have been fluctuations of sea-level but, relative to the land in Scotland, it has fallen.

Let us turn for a moment to consider the mechanism of this rise and fall of sea-level *relative* to the land. The formation of the great ice-caps and glaciers of the Quaternary implied a fall in sea-level, since the oceans were the only major source of moisture of which the ice-caps were formed. It is estimated that at the maximum of glaciation sea-level may have fallen 300–400 feet (91–122 m). But the growing ice-caps necessarily meant that those parts of the continents on which they formed were very heavily loaded. In consequence those parts were depressed. Since we are not here concerned with the structure of the earth, it will be sufficient to say that the overweighting by ice of parts of the surface caused a sub-surface translocation of material so that the parts peripheral to the ice-caps were forced to bulge outwards. In other words, relative to the centre of the earth, the overloaded parts sink, and the adjacent parts rise. These movements are very slow and long drawn out and are entirely independent of shifts of sea-level but the sea-level and the land movements work together, and it is their interaction which causes so many of the complications associated with research on raised beaches. It will be clear from what has been said that places far distant from ice-caps will, within the limits of our argument, only show the effects of a fall and a rise of sea-level. Places under the ice may have been depressed as much as, or even more than, the fall in sea-level. At other places the two movements may have been so balanced that no apparent movement of land relative to sea-level took place. In general, however, the isostatic movements have caused the beaches to be tilted, although the effects are only noticeable over long distances. But these movements are not rapid, and they are still going on. There is at the present time a very slow and slight rise of sea-level caused by the melting of the polar ice. This eustatic movement is to be measured in millimetres per century. But isostatic movements of the land caused by the growth and melting of the ice-caps are still relatively active. The Baltic area, for example, is still rising; the lands around the southern part of the North Sea, the Flemish Bight, are sinking. Precisely similar movements are taking place in America and elsewhere.

This brief discussion will make it clear that, since these movements have been in progress ever since the incidence of the ice age, and are still in progress, the heights of all raised beaches are likely to be affected. Some will be tilted one way, some another; locally they may remain horizontal, but only because the processes affecting them have balanced, and not because there has been complete stability in the places where they occur.

Jamieson had realized this, but it was Wright (1937) who made us familiar with the problem. He fully realized that the Scottish beaches had been affected in this way, and that a height name had little real significance. But progress was slow because at that time (i.e. the first quarter of this century) so little was known of the true heights of the many beach remains. Donner (1959) was among the first to apply careful height measurements in Scotland (see note at end of this chapter). Unfortunately he only investigated thirty sites around the whole coast of Scotland, and came to the conclusion that there were four main shorelines at 100 ft (tilted), 50 ft, 25 ft, and 12 ft, the three lower ones all being horizontal. This conclusion was at variance with that of Wright and others. A few years later two other investigators, McCann (1961) and Sissons (1962) began some detailed work, the former on the west coast, the latter on the east coast, particularly in the Firths of Tay and Forth. McCann's major contribution is contained in a thesis for the degree of Doctor of Philosophy at Cambridge, and can be consulted in the University Library; fortunately he has also published certain papers, and some of the more important results of his work and also that of Sissons and his colleagues will be found in the special number of the *Transactions of the Institute of British Geographers* (Oct. 1966).

McCann studied three parts of the west coast in great detail: Loch Broom to Arisaig, Loch Linnhe and the mainland coast of the Firth of Lorne, and the islands of Islay and Jura. In these three areas he made carefully checked measurements at 191 sites. All were plotted and, as shown on Fig. 4 (b and c), the heights fall clearly into two groups, or bands, both of which are tilted. The upper band (see (c)) which in a general way corresponds to what in the past has been called the '100-ft' beach, declines in height in all directions away from Callander. 'The shoreline curve indicates that there is an initial decline in height in all directions away from Callander at the rate of 0.5 feet per mile, from 115 feet [35 m] at 45 miles [72 km] to 90 feet [27 m] at 95 miles [153 km]. At this distance the tilting . . . becomes more pronounced, the rate of decline . . . being 1.5 feet [0.45 m] per mile, from 90 feet [27 m] at 95 miles [153 km] to 45 feet [14 m] at 125 miles [201 km]'. McCann admits that these figures may be modified by later work, especially on the east coast. The lower ('25-ft') band also shows a tilt which varies with direction. Approximately west-south-west from Callander this shoreline falls at the rate of 0.12 feet per mile, from 28 feet (9 m) at 32 miles (52 km) along Loch Fyne, to 20 feet (6 m) at 97 miles (156 km) in Islay. In the rest of south-western Scotland there is no appreciable fall in height away from Callander. To the north-west the slope is more pronounced. 0.26 feet (0.007 m) per mile from 31 feet (9.5 m) at 45 miles (72 km) along Loch Linnhe, to 10 feet (3 m) at 124 miles (200 km) in Loch Broom. These figures are taken from the thesis.

If we accept McCann's interpretation there can be no '50-ft' beach. The higher, Late-glacial, and the lower, Post-glacial beaches are both tilted, so that what is regarded as the same beach may be at very different heights in different places. This certainly simplifies the problem, but we need a much greater number of measurements before we can be convinced that the problem is fully solved. McCann has demonstrated most successfully that some features, formerly taken as beaches, are deltas or glacial outwash fans. They may be related to a specific sea-level, but cannot be regarded as indicators of sea-level in the same way as raised beaches. In western Scotland there are several other puzzling features connected with raised beaches and benches. Around the island of Great Cumbrae (see Plate 56) there is on all sides a beautifully clear-cut bench; the same is true of much of Arran. But how was such a bench cut in the enclosed waters of the Firth of Clyde? Even at a time of slightly higher sea-level and possibly of far more stormy conditions it is difficult to explain this bench, especially on the eastern side of Cumbrae and the western side of Arran (see Plate 57). In both places the amount of open water is limited. The benches in the enclosed waters between the islands of Seil and Luing and the mainland present this problem in a much more acute form (Plate 52). It is emphasized in several places in this book that a rock-cut platform is not necessarily of one specific age. The fact that there is little evidence of a modern platform in process of formation at the outer margin of the '25-ft' bench is in itself suggestive that the fine development of bench and cliff associated with the '25-ft' beach is multiple in origin. The same holds for the platforms on Cumbrae, Arran and elsewhere.

In the island of Islay there is a magnificent development of raised coastal forms. The north-east part of the island is built of massive and well-jointed quartzite, a rock which breaks away in large angular blocks and produces nearly vertical cliffs which remain 'fresh' for long periods of time. For seven miles (11 km) south from Rubha À Mhàil there is a remarkable rock platform which is overlain by gravels belonging to the '100-ft' sea. Wright (1936) regarded it as a Pre-glacial structure. In places it is backed by high cliffs which, near Aonan na Uamh Mhór, are 200 feet (61 m) high. McCann determined the height of the inner edge of this platform as 105 feet (32 m) at two places. The height of the seaward margin depends upon the amount of erosion in later times.

Since the formation of the old marine platform in north-eastern Islay there is, therefore, evidence from the deposits which now rest on the platform, of a period of general glaciation (i.e. the boulder clay), followed by a period of high sea-level (i.e. the high level marine gravels), followed in turn by a period of local glaciation (i.e. the Coir Odhar moraine and the solifluction deposits at the foot of the old cliff line). The local glaciation, and the deposition of the high level marine gravels can be related to the general chronology of Late-glacial events in Scotland, but the period of the formation of the rock platform itself remains problematical. (McCann thesis)

In short, we have here a platform older than any other on the west coast. It may be Inter-glacial, or conceivably (Wright, 1936) Pre-glacial.

In recognizing these three main platforms, McCann is in general agreement with

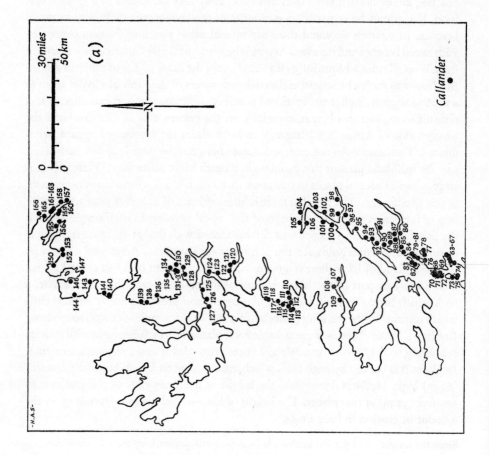

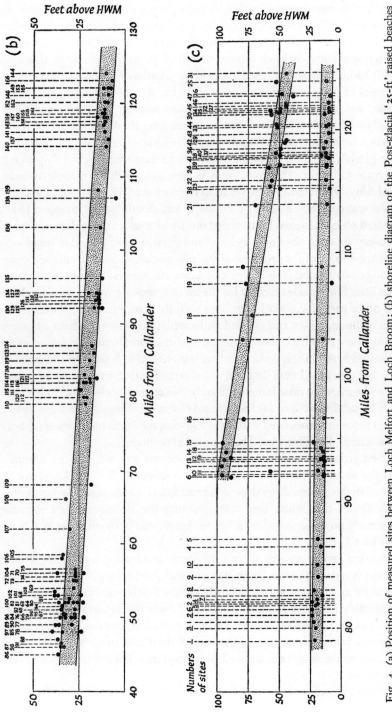

Fig. 4. (a) Position of measured sites between Loch Melfort and Loch Broom; (b) shoreline diagram of the Post-glacial '25-ft' raised beaches of western Scotland, north of Loch Melfort; (c) shoreline diagram of the raised beaches of north-west Scotland from Loch Broom to Loch Hourn (all after S. B. McCann)

Wright, but he also recognizes that in some respects his own views may need modification. It follows from what has been said that there must be a close relation between the extent of the inland ice and the platforms. McCann thinks that the higher terraces between Loch Broom and Red Point all belong to this higher ('100-ft') band. If, however, the view held by Charlesworth (1955) is correct, i.e. of the extent of the Highland re-advance of the ice, then McCann's hypothesis must disappear, and the Loch Broom to Red Point terraces must be regarded as a separate '50-ft' level. If McCann's views concerning the Loch Broom–Red Point terraces is accepted it is also probable that there were several halts in the retreat of the Late-glacial (i.e. '100-ft') sea, and that some of them were sufficiently prolonged for a terrace to be cut. This hypothesis is consistent with the shingle spreads in Jura and Islay, which indicate two halts in the retreat of the Late-glacial sea.

On the east coast some remarkably careful and detailed work on raised beaches and kindred phenomena has been carried out by Sissons and his collaborators. The work mainly concerns the Forth and Tay and the county of Fife. It is based on an intimate knowledge of the ground and the information obtained from the measurements of 10,000 heights, the analysis of 2,000 commercial bore-holes, and more than 700 hand bores made specially for the investigation. It may, I think, be claimed that no other part of the world has been so fully investigated from this point of view. In addition, pollen analysis and radiocarbon methods have been extensively used, and where possible a close correlation with archaeological evidence has been made. Raised beach in this context covers not only what is usually understood by the term but also raised mud flats and raised estuarine deposits; moreover, some of the raised beaches are now buried and only detected in bore-holes. These shorelines can be subdivided into (a) Late-glacial (formed before the Perth re-advance); (b) raised shorelines associated with and following the Perth re-advance; (c) buried shorelines; and (d) visible Post-glacial raised shorelines.

The first group is found in eastern Fife, since this area was first freed from the ice after the so-called Aberdeen–Lammermuir re-advance. Six shorelines have been recognized in this group. They slope downwards to the east; their relation to glacial outwash shows that when they were forming the ice margin was retreating. Unfortunately no direct correlation has yet been made between those in Fife with those in East Lothian.

The second group, associated with the Perth re-advance, are striking beaches; the highest is called the Main Perth Raised Beach, and is locally conspicuous, and reaches as far as Burntisland on the north and the Forth road bridge on the south; it may possibly be traced as far as Aberlady. Lower beaches in the Earn and Tay were built as the ice melted; they also decline eastwards, one can be followed until it disappears below the surface to the east of the Carse of Gowrie. Two of the beaches, below the main one, are well developed near Falkirk in the former delta of the Carron.

The buried shorelines have been found beneath carse clay and peat. The highest

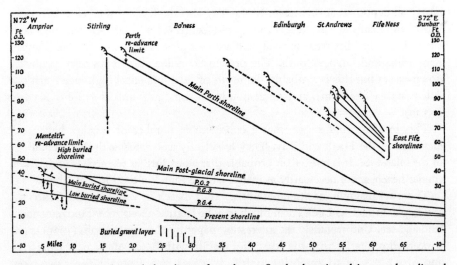

Fig. 5. The principal raised shorelines of south-east Scotland projected into a plan aligned approximately at right angles to the isobases so far as these are known at present (after Brian Sissons)

one is associated with the Menteith moraine and does not occur inside it. It slopes down (see Fig. 5) to the Main buried beach which is now about 9,500 years old. It also falls to the east. The Lower buried beach, about 8,800 years old, occurs in much the same area. Similar beaches occur in the Earn valley, but more evidence is required about them.

Around Grangemouth a large number of commercial borings have been put down, and their analysis has established a well-marked gravel layer, two to five feet thick (0.3 to 1.5 m). It slopes to the Forth. It has been definitely recognized as a former raised beach. The material of which it is formed is partly of stream origin, but mainly derived from the erosion of Late-glacial sediments.

It is relevant to note that the present coast of the Forth eastward from the vicinity of Bo'ness has suffered considerable erosion in relatively recent times where it is composed of drift: in places erosion is in progress today. The result of this erosion is that the present shore is often littered with a residue of stones and boulders of all sizes mixed with sand, mud and shells, this beach deposit closely resembling the buried gravel layer of the Grangemouth area (Sissons et al., 1966).

After the formation of this lower buried beach it seems that sea-level fell somewhat so that this beach was exposed, but soon after it rose again, at first rapidly, to form the Post-glacial transgression. This caused the buried beaches to be covered with carse clay, except locally where peat growth kept pace with the rise in sea-level. The transgression ended about 5,500 years BP. The main shoreline formed now slopes eastwards from about 49 feet (15 m) in the far west of the Firth of Forth to 19–20 feet (c. 6 m) at Dunbar. In the Tay, the clay of the Carse of Gowrie was nearly all deposited at this stage. In the Tay also the level of the beach falls east-wards, and finally at Fife Ness is about 21–2 feet (*c.* 6.5 m) high. At somewhat later stages lower beaches were formed, and are best preserved in the carselands above Kincardine and Grangemouth. The present shoreline also slopes very gradually downstream but this is probably the result of tidal influence; high-water mark at Alloa is 10.3 feet (4 m), at Grangemouth 9.6 feet (3 m) and at Dunbar 8.3 feet (2.5 m).

This brief account shows that a considerable number of beaches have been identified in the Forth and Tay. They are clearly associated with the fluctuations of the inland ice. It is more than probable that when further research is undertaken similar beaches, not necessarily as many, will be found in the Firth of Clyde.

There are magnificent stretches of the '25-ft' beach in Ayrshire and Galloway, and in places such as Loch Ryan much may be learned about recent movements of land and sea. Unfortunately the interesting paper by Harvey Nicholls (1967) is not concerned primarily with the coast, but mainly with Racks Moss, near Dumfries, and Aros Moss, near Campbeltown. It is, however, clear from the paper that more detailed work on the Kintyre and Galloway coasts is likely to be rewarding.

The raised beaches in the extreme north of Scotland have been recently investi-gated by King and Wheeler (1963). Little was known about these beaches and all too often it was assumed that beaches were poorly represented, or even absent, in that area. King and Wheeler made some careful measurements between Durness and Melvich. They were not able to make as many observations as did McCann and Sissons, but their work is of great significance, and it is to be hoped that a detailed correlation can eventually be made with that farther south. Four distinct levels were recognized, and they are tentatively correlated to the '100-ft', '50-ft', '25-ft' and '15-ft' beaches. The upper two are tilted; the two lower are horizontal.

The coast of north Sutherland is not only beautiful, but also varied in rock type and structure. The rocks on the coast are nearly all resistant, but often show the effects of glaciation. Many details of the coast are related to dykes. The spring tidal range is a little more than 13 feet (4 m) and, quite apart from the practical difficulty of obtaining heights of features and relating them to Ordnance Datum, there is the further difficulty of knowing at what stage of the tide certain features were formed, the more so since we know nothing of past tidal conditions. Four main types of features were recognized:

(1) Rock shelves or platforms backed by cliffs, usually much degraded.

(2) Flat features produced by wave action, but not backed by cliffs. These are usually irregular in detail.

(3) Deposition features, found for the most part in more sheltered places. These are out of reach of modern wave action, and are often formed of shingle or sand.

(4) Terraces cut in glacial drifts, and often associated with extensive outwash deposits in the mouths of large rivers.

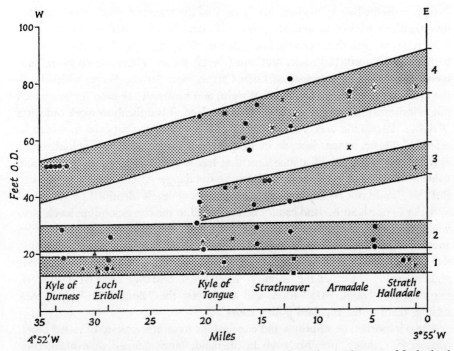

Fig. 6. The height range of the four main raised levels along the north coast of Sutherland, arranged from west to east: ●, rock platform backed by a cliff; ×, drift terrace; ▲, shingle or sand deposit (beach); ■, rock skerry (after King and Wheeler)

Sixty localities were investigated, and when, after taking all due precautions, their heights were plotted, they appeared to fall definitely into four groups and, as shown in Fig. 6, it seems that the two higher levels are at a greater elevation in the east than in the west. If this is a genuine difference of level, the authors think it is the result of a greater isostatic recovery in the east following from the melting back of the great mass of the Scandinavian ice which for some time covered the sea floor thereabouts. It is also probable that the rate of isostatic recovery varied from about 28 feet (8.5 m) in a thousand years 12,000 years ago, to 2¼ feet (0.7 m) in a thousand years in the last 4,000 years.

In the uppermost level glacial and fluvio-glacial deposits play a significant role, and since they may also slope downwards to the sea, heights are not always easy to measure. There are a few rock benches at this level. The next lower level is characterized by many wave-cut shelves in rock; it also includes a cuspate delta at Tongue. The next lower level is better preserved and there are numerous erosional features; in more sheltered places there are beach deposits on the benches. The lowest level affords many shingle spreads, rock benches and skerries. King and Wheeler make some general comparisons with beaches farther south, but this is based largely on Donner's work and they make only one reference to Sissons

(1962). It seems best to neglect this in view of the later and much more detailed investigations of Sissons and McCann. This may be illustrated in another way: reference is made to the gravels in Loch Carron (Wright, 1936). They are compared with those in Strath Halladale. McCann (thesis) writes 'There are no gravel embankments along the entrance of Loch Carron, near Strome Ferry, which reach the 138 feet (42 m) noted by W. B. Wright, and no steeply sloping surface which might indicate the form of an outwash fan. Indeed, it is difficult to work out from Wright's description and . . . map the exact location of the deposits to which he referred.' This comment is made in no querulous manner, but to emphasize how easy it is to make use of a statement that has found its way in a standard and valuable work, and even more to illustrate the danger of correlation in general. But this should not be taken to mean that the four levels identified by King and Wheeler in north Sutherland cannot be compared to the corresponding levels elsewhere. It does mean that a great deal more detailed investigation is necessary. With another suggestion made by King and Wheeler there should be much sympathy; that the beaches should no longer be referred to in terms of height but that 'the highest level . . . be called the "Zone I" level, the second the "Zone III–IV" level, the third the "Zone VIIa" level, and the lowest the "Zone VIIb" level, thus relating them to the standard pollen zones'.

These important descriptions and conclusions make the scarcity of raised beach features in Orkney, possibly even in Shetland, more difficult to explain. (See Chapters VIII and IX. There are several features in Orkney, e.g. the bench and old cliffs on the north side of Scapa Flow, which, in my opinion, are raised features. Certain wave-cut platforms in Shetland are also suggestive.) My own view is that a great deal more attention should be paid to remnants of benches and platforms found in various places in both groups of islands. There is perhaps no reason why isostatic movements should not have carried Orkney downwards relative to Caithness, but it is interesting to recall that King and Wheeler found good evidence of an east to west tilt. It is true that their observations did not include Caithness, but they serve to emphasize the complexities of the whole matter, and suggest that a careful analysis should be made of the Caithness coast. There seem to be few traces of raised beaches, and those few at a low level in eastern Caithness. If this is correct, and if King and Wheeler's view of a westward tilt is right, how does this abrupt change come about?

In the Outer Hebrides there is plenty of evidence of submergence, but little of any raised beaches. Baden-Powell and Elton (1936–7) call attention to a raised beach at Galson on the north-west of Lewis. It is in a notably exposed situation, and is 10–25 feet (3–8 m) above high-water mark. It is certainly earlier than a midden on it which can be dated to the last part of the pre-Christian era. The beach is also probably earlier than an Iron Age house at the same site. Along this part of the coast of Lewis there are traces of an ancient line of cliffs. There is some uncertainty about the true interpretation of a bench cut in the Stornoway conglomerate on the Eye peninsula. It has been suggested that it is a continuation of a raised beach at Loch Branahine. Baden-Powell and Elton investigated the site. The bench certainly resembled a marine-cut platform, but they were not sure if it represented any change of level.

More recent work by Ritchie (1966) in North and South Uist and Benbecula shows clear evidence of subsidence, but none of elevation. He does not refer to Baden-Powell and Elton, nor to a remark of Gregory's that there is a raised beach 10–12 feet (*c.* 3 m) above OD in Loch Maddy. There is, however, no doubt that submergence has been all important in all these islands, and that it is Post-glacial in age. The rise in sea-level has been of about 200 to 240 feet (61 to 73 m), and most of this change took place before 5,700 BP. 'After 5,700 B.P. it must be concluded that on the basis of the evidence so far available, any positive or negative changes in sea-level are, in keeping with the peripheral position of the islands, too small to be detected or distinguished from either the results of continuing coastal modification or present marine processes.'

APPENDIX

In a chapter in D. Walker and R. G. West, *Studies in the Vegetational History of the British Isles, Essays in honour of Harry Godwin* (Cambridge, 1970), J. J. Donner discusses land- and sea-level changes in Scotland. The main significance of the work is that he was able to date seventeen occurrences of submerged peats and associated deposits by means of radiocarbon methods and pollen analyses. The sites are scattered (see below) and vary in height. The marine sediments can be divided into Late-Weichselian and Flandrian. 'The Late-Weichselian culmination of sea-level was in Zone I, before 10,000 B.C., whereas the Flandrian transgression reached its maximum sometime between 6000 B.C. and 4000 B.C. after the early Flandrian regression, during which the peat beds were formed between 8000 B.C. and 6000 B.C.' Curves of movements can be drawn from these findings. At present their value is limited by the small number of stations; the probability that the work will be extended in future suggests great possibilities.

List of stations: Eastfield of Dunbarney (Bridge of Earn); Broombarns (Forgandenny); Airth Colliery; Flanders Moss; Kippin; West Flanders Moss; Garscadden Mains; Irvine; Girvan; Girvan railway station; Newton Stewart; Gatehouse of Fleet; Lochar Moss; Redkirk Point; Loch Mór (Soay); Isle of Calvay (S. Uist); Borve (Benbecula); Symbister (Whalsay, Shetland).

Chapter IV

The mainland coast: Melvich Bay to the Solway

MELVICH BAY TO CAPE WRATH

To choose Melvich Bay as the dividing place between the two parts of this volume which deal with the mainland of Scotland may seem a little arbitrary. The bay, however, marks the western end of the continuous outcrop of Old Red Sandstone which forms almost all the coast of Caithness. There are outcrops farther west on the coast, but these are disconnected from the main mass. In Melvich Bay also we find the first appearance of the Moinian rocks on the north coast of Scotland.

The coast from Melvich Bay to the Kyle of Tongue is very beautiful. It is not the only coast in Scotland in which these rocks outcrop but it represents the only place where, because of the earth movements to which they have been subjected, they run out to sea in narrow bands. These, partly as a result of differential erosion, produce a jagged coast in which two or three relict patches of Old Red Sandstone and coverings of boulder clay add to the variety of the cliffs. Melvich Bay is the end of Strath Halladale, a major valley which is broad and open in its lower parts. There is a good beach in the bay, and it is enclosed by granitic rocks on the west and by Old Red flagstones covered in boulder clay on the east. Along the western side the sand in the inner part of the bay gives place to a coarse boulder beach and then to a rock platform which is associated with the raised beaches (see p. 52) of this part of the coast.

The coast at and near Port Skerra is much indented by narrow inlets, called geos, and these are cut at Rubha Bhrà in a small outlier of Old Red, but in the bay to the west and at Rubha na Claiche gneisses and granites run out to sea and differential erosion along joints and other lines of weakness has given rise to the present coastal outline. All along this part there is a fossil cliff which can be followed to Strathy Bay where there is another considerable outlier of Old Red Sandstone. On the east side of the bay Caithness Flags of the Old Red, covered by boulder clay, give an outline and profile comparable to the corresponding part of Melvich Bay. The Strathy river reaches the bay in a valley smaller than, but similar to, Strath Halladale. This valley follows the contact between the sedimentary and crystalline rocks. The western side of the bay is prolonged into Strathy Point which projects about two-and-a-half miles to the north. The point itself is an intrusion of epidiorite, but the peninsula which is more than a mile wide is a complex of granite and injected rocks of granulites and granites with a marked northerly to north-north-west strike.

In Strathy Bay there is a beach which is supplied from several sources, erosion of the sandstone cliffs, glacial material off-shore, and the Strathy river at the mouth of which there is coarse shingle which has probably been brought down by the river. There are also dunes in two separate ridges, the inland one of which is the older. The outer is larger and is still in process of evolution. On the western side of the beach and bay a rock platform stands in front of fossil cliffs, and can be traced round some of the small indentations, such as Port an Uillt Ruaidh, which characterize the coast right out to the point. The northern part of the peninsula shows similar features together with stacks and arches cut out along lines of weakness in the rocks. This is equally true of the coast as far as Armadale Bay. This bay is cut in the Moine Schists and epidiorites form the eastern, and the southern part of the western, shores of the bay. The Armadale burn which reaches the sea at this place is in a much smaller valley than those farther east. The bay and valley both follow the strike of the rocks. There is little evidence of erosion in the bay and the beach is not supplied from this source. Ritchie and Mather (1969) are of the opinion that most of the beach material is derived from fluvio-glacial deposits offshore. The beach is backed by dunes.

Between Armadale and Farr Bays the coast has not been studied in detail from a geomorphological point of view. The general structure is like that to the east of Armadale. The trend of the rocks is more to the north-west. In Geodh' Ghamhainn and at Kirtomy there are small outliers of Old Red Sandstone. Port a'Chinn and Port Mór are north-facing bays like Armadale, but without sands. The fossil cliffs and rock benches are present, and it may reasonably be assumed that they and smaller inlets have been cut along lines of weakness. Numerous dykes and several small streams reach the shore between Armadale and Kirtomy Point. Kirtomy Bay on its western side is enclosed by an elongated skerry. The One Inch Geological Map brings out extremely well the general relation of structure to coast and indicates how closely the cliff scenery depends upon the strike of the rocks and on intrusions. This is true also of all the intricate coast including the Bay of Swordly, Borve Castle, Farr Point and Glaisgeo. The old cliff can be followed round most of this stretch. Farr Bay faces north-west, the strike of the rocks having turned in that direction. In this district there are also important depressions which run at right angles to the strike, the chief of which is the shallow hollow in which Clerkhill is situated, the other is farther inland and is followed by the main road. The bay contains a good beach at the back of which there are dunes and machair. Three distinct levels are found; at 25 feet (8 m) where there is an old cliff line, at 50–60 feet (15–18 m), and another platform at 70–90 feet (21–7 m). On the west of the bay similar rocks and trends prevail in the small peninsula that culminates in Creag Ruadh.

Torrisdale Bay has many features of interest. The Borgie and the Naver run into it, the former on the west side, the latter on the east. Between the two is the glacially-scoured ridge of Druim Chuibhe which is formed of strongly foliated

Moine Schists. The top is generally flat and boulder covered. The boulders are often very large. The outer part of the bay is enclosed by Creag Ruadh on the east and Aird Torrisdale on the west; both are in resistant Moinian metamorphics. The lower parts of the river valleys, Borgie and Naver, are filled with fluvio-glacial deposits. On the east side of the Naver three terrace levels are easily traced; two sand and gravel terraces are seen on the east side of the Borgie and west of the Naver. The terraces and similar features indicate the great amount of material brought down by these rivers in the glacial period, and this has given rise to the wide expanse of inter-tidal sands. Ritchie and Mather (1969) note the absence of shingle on the beaches and its great abundance in the outwash areas. The wide expanse of sand uncovered at low water is dried and much sand is driven inwards by strong northerly winds. On the sides of the ridge, Druim Chuibhe, it forms dunes and machair. Parts of the flats are covered by salt marsh; the best development is at Torrisdale. The marshes appear to be rather old, and where they are eroded often show a shingle layer beneath the beach sand. This suggests that shingle may be found elsewhere below the present sand surface. There are extensive dunes to the west of Druim Chuibhe; sand has spread inwards to form hummocky ground, but after about 50 yards (46 m) gives place to machair. There are also dunes climbing the north and west faces of Druim Chuibhe. The corresponding dune area on the Naver side of the ridge is much smaller. Sanding reaches 360 feet (40 m) on the east of the ridge and

Erosion scars and terraces are features of the ridge top. This is ecologically extremely important as it is one of the main factors giving floristic richness to the Nature Reserve, which in this area includes associations of *Dryas, Calluna, Empetrum, Carex* and *Juniper* along with rarer individual species of which *Primula scotica* is best known (Ritchie and Mather, 1969).

The terrace which is situated on the east of Druim Chuibhe and almost opposite to Bettyhill has many features of archaeological interest – cairns, hut circles and cist burials. There is also a broch at a rather higher level and just north of the small but marked valley which drains from Lochan Druim an Dùin. This is one of several lines of structural weakness, and this particular valley channels much blown sand up to the top of the ridge.

Westwards from Torrisdale Bay the coast is irregular; the fossil cliff can be traced; as far as Port an-t Strathain the main rock is epidiorite and hornblende schist. The north-westerly strike continues as far as Caol Beag, the strait separating the mainland from Neave or Coomb island which consists wholly of Moine Schists. A mile or more to the north-west is the bigger island of Eilean nan Ròn and its smaller western neighbours. All these are formed of Old Red Sandstone much cut by faults. The faulting is responsible for the lines of weakness which have been eroded into narrow channels between the main island, Eilean Iosal, and Meall Halm. The prominent indentations on either side of the main island have a similar origin.

The road from Torrisdale to Clasheddy follows a marked valley which reaches the sea in Skerray Bay. It follows the northwest strike, and coincides with an outcrop of Moine Schist between two areas of epidiorite and hornblende schist. The two small bays farther west, Lamigo Bay and Port an't Strathain, are also at the mouths of two deeply cut valleys containing small streams. The fossil cliff can be traced all along this part of the coast and on to Tongue Bay and the Kyle of Tongue. The Moine Schists reach the shore again beyond Strathan Skerray. The coastal outline is relatively simple as far as Coldbackie where, as far as the mainland coast is concerned, is the last outcrop of the Old Red Sandstone. There is a small beach at Coldbackie; on its west it is contained by a long sandspit associated perhaps with the outflow from the Kyle of Tongue, from which a strong ebb flows to the north-east. The old cliffs in the bay are cut in conglomerate and there are also ancient caves. Traced northwards towards Skullomie a series of benches can be seen both in the conglomerates and the succeeding schists. The stream which reaches the sea in the north-east corner of the bay cuts a deep valley in drift and finally reaches the dunes at the back of the bay over a waterfall over the old cliff.

The Kyle of Tongue is shallow, and there are traces of higher shorelines and old cliffs around its shore. The spit of land extending westwards from Tongue Lodge is a raised feature, and on the western side there are two small triangular forelands, near Melness House and the Achuvoldrach burn, which also may be raised features. At the head of the Kyle there is a considerable development of salt marsh. The setting of the Kyle is greatly enhanced by the outline of Ben Loyal as seen from the main road on the west, and the minor road to the north. The extensive sand flats exposed at low water are locally fringed by coarse boulder beaches, and the sand, submerged at high water, which joins the Rabbit Islands to the mainland is the western equivalent to that which encloses Coldbackie Bay. The two larger Rabbit Islands are joined by a spit (? raised); the smaller islands to the north are separated from the main group by deeper water. They are all built of Moine Schists. The coast is easily accessible as far as Achininver where there is a deep inlet the inner part of which is filled with sand. A stream with ramifying headwaters draining the bleak A'Mhòine peninsula drains into it. The lower valley is open and where it joins the sea mica schist and epidiorite replace the Moine Schist which forms the coast to the Kyle of Tongue. This part is more irregular than the outer part of the A'Mhòine peninsula which is remote. The cliffs, not wholly of present-day origin, are high. Along the coast a strip of mica schist runs from the north-western corner of Achininver Bay as far as Loch nan Aigheann. Westwards of this place the coastal rocks are more complex and cut by the Moine and Arnaboll thrust-planes which are visible in the cliffs. The most northerly point is Whiten Head (Ceann Geal Mór). There the cliffs are lofty, and the unconformable junction of the Cambrian quartzite and the Lewisian Gneiss is visible in the cliff about a quarter of a mile south-west of Whiten Head, and also on a stack offshore. The head itself and the cliffs for about a mile to the east are in the Lewisian, on the eastern part of which is a thin patch of Cambrian quartzite truncated by the Moine thrust. Heddle (1880–1) writing of Whiten Head says, 'Regarded from this headland, the sombre lines of precipices, which, with strongly-contrasted blotchings of green and red, extend for miles across the frontlet of Ben Thutaig, seem to possess outlines of both dignity and grace, which must surpass those of Whitten (*sic*). They are uniformly of surpassing altitude, and attain, under Ben Thutaig, a height of 878 feet.'

Loch Eriboll is deeper and penetrates farther inland than does the Kyle of Tongue. The Cambrian rocks on its eastern side are in a synclinal fold and have been pushed forward by the Arnaboll thrust; between Heilam and the Hope river they display imbricate structure very well. On the western shore a thin layer of basal Cambrians rests on Lewisian Gneiss. The loch shores display many traces of raised beaches and fossil cliffs. Ard Neackie is tied to the mainland by a raised tombolo. The stream in Strath Beag drains into the loch, and there is an extent of fresh-water marsh a little above the junction.

The coast from the entrance of the loch to the Kyle of Durness is especially interesting. The Lewisian Gneiss around Rispond and Ceannabeinne is characteristic. It does not form ordinary cliffs but rounded slopes which plunge down into the sea. At Rispond there are several geos, and to the west is a larger inlet, Tràigh Allt Chailgeag, which corresponds with a fault running to the north-east. Three small streams reach the bay; only the most easterly is graded to sea-level. The beach is good, but much exposed to the north-east. Eilean Hoan, built of two groups of Cambrian rocks separated by a thrust, protects the bay from the north to some extent. Ritchie and Mather (1969) note that 'the age of the cliff line [in the bay] is evidenced by the plug of glacial till in the geos to the west of the beach, indicating that the cliff-line must have been cut in pre-glacial [? inter-glacial] times'. The present cliffs are eroding but slowly. At the east of the beach there is a raised shingle bar (cf. Sangobeg). There are small dunes and a little machair in their rear.

About a mile farther west there is a good beach at Sangobeg. The rock hereabouts is the Pipe-rock of the Cambrian, but on the eastern headland there is Cambrian quartzite and that on the west is limestone. A small mass of gneiss outcrops on the east part of the beach. The western headland suffers some erosion, but contains caves, notches, and benches cut at a higher sea-level than that of today. In the limestone to the west is Smoo Cave, one of the best-known features on the north coast. The cave is at the head of a narrow gorge nearly half a mile long. The gorge is very like a geo, but is of somewhat different origin since it is in limestone. It was partly cut by the Allt Smoo and partly by marine erosion, both acting together with solution. There are two main chambers in the cave, and a visitor's impression will depend much on whether he enters the cave in wet or dry weather. To the west of the small limestone protuberance in which the gorge is cut is Sango Bay (Sangomore). There is a beach divided by small headlands into three parts, and the structure is complex. The upper part of the eastern side is limestone; this straight shore coincides with a fault; the bay itself is in gneiss and schists, and the west (or northern) headland, Creag Thairbhe, is limestone. This headland partly coincides with a fault, and is subject to some erosion. The beach sand is probably derived from offshore deposits; a little may come from the streams. Since, however, till is in the Sangomore valley and can also be followed on the cliff edge, it is more than probable that it continued farther seawards.

Between Craig Thairbhe and Balnakiel is the great projection of Faraid Head. This line corresponds closely with a fault which separates the limestones, which form all the ground on the east side of the Kyle of Durness, from the Moine Schists which underlie the dunes of An Fharaid up to the line where they are cut off by a thrust. This thrust follows fairly closely a direct line joining A'Chléit and the south-east end of the bay south of Faraid Head and separates the schists from a narrow band of Lewisian Gneiss. The north-western headland is in the mylonized rocks above the Moine thrust. The northern cliffs and stacks are imposing. On the east coast there is a wide erosion platform, and the cliffs, if not completely fossil, are certainly not subject to any severe erosion. The middle and southern parts of the headland are sand covered; the sand is arranged in several ridges trending north-east, and reaching the 100-foot contour. They are usually stabilized by vegetation. Marram is dominant on the seaward edge; more mature communities follow inland. There are some slacks, and near Faraid Head are some typical machair plants. There are fine beaches on the west of the peninsula.

The south side of Balnakiel Bay is fronted by a conspicuous rock platform and low cliffs in the Cambrian beds. These reach as far as the eastern entrance of the Kyle of Durness. The roughly oval area between the bay, the Kyle and the road from Keoldale to Balnakiel is a low plateau (100–50 feet: 31–16 m) in which there are three small lochs. The remainder is covered by blown sand some feet thick. This is held by a close cover of vegetation which includes an interesting flora. The sand does not appear to increase, but at low water wide sand flats are exposed in the Kyle of Durness. They are not exposed long enough to allow of much wind action. The sward is grazed; the lochs, and occasional pits or scratches shew its foundation. The surrounding area is deserving of further study. Recent investigations by Reid et al. (1966–7), have shown evidence of two or even three soil surfaces in the sands. Near Loch Lanlish a circular hut had been built on the lowest of the three surfaces. At other places on the higher soils, there are traces of dry-stone dykes, cairns and a hut circle. The evidence suggests one or more periods of sand movement, and a long period of stability. The remains are probably of the Iron or Bronze Ages. The sketch map (Fig. 7) shows that a fault runs from the Kyle of Durness north-westward, and the coast itself is partly in Lewisian Gneiss, with a thin streak of Cambrian rocks around A'Ghoil, and then Torridonian rocks as far as Kearvaig. The Torridonian cliffs in this part of the coast are fine and nearly vertical. The bedding is more or less horizontal, there is marked vertical jointing and the waves have cut deep caves and chasms, locally called gloups or gyoes. Macculloch (1819) regarded the twin stacks at Clò Kearvaig as the finest in Scot-

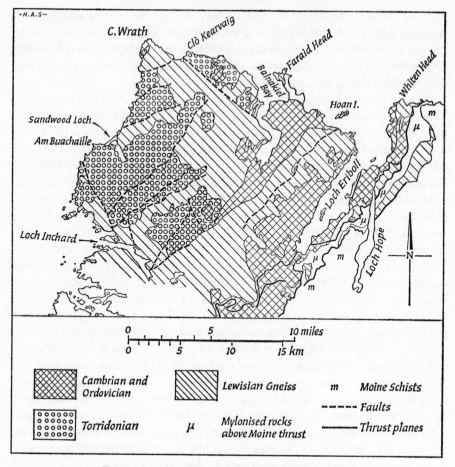

Fig. 7. Cape Wrath area, based on Geological Survey

land. At Clo Mór the cliffs are almost vertical and 600 or more feet high (183 + m). They form two distinct headlands. At their summit false bedding is conspicuous. The inlet of Geodha na Seamraig is at the junction of two faults, the one just referred to and the second which runs to the south-west.

CAPE WRATH TO LOCH HOURN

Cape Wrath is a steep cliff of Lewisian Gneiss and shows alternating bands of granite gneiss and veins of pegmatite (see Plate 29). A prominent cliff is unusual in the gneiss areas of the mainland; nearly everywhere this rock makes small cliffs and knobbly headlands and islands. There are, however, more prominent gneiss cliffs in the island of Lewis–Harris. From the cape to Geodha na Seamraig, where there is a good beach, the coast is parallel to the strike of the granitoid gneiss, but at Clais Chàrnach there is another inlet cut to sea-level, the western side of which is a mass of red granite which weathers grey. The gneiss contains many dark bands. The cliff scenery is bold, and the cliffs themselves are usually inaccessible, except in the small inlet just west of the lighthouse. A little south from the lighthouse there is a fault which brings in a small patch of Torridonian conglomerate. A short stretch of gneiss cliffs follows to the south, but a fault which runs to the north-east from a small inlet a little more than half a mile north of Geodha Ruadh na Fola cuts off the triangle of which Cape Wrath forms the apex (Fig. 7). Torridonian cliffs, interrupted by inlets including the Bay of Keisgaig, where another north-east fault reaches the coast, extend to within about half a mile of Sandwood Loch. The loch is dammed by a beach of coarse sand, and the stream makes an exit at the northern end, possibly indicating a northerly drift of beach material (Plate 31). The loch and the depression in which it lies trend south-east to north-west, parallel with the junction of the Torridonian rocks on the south and the Lewisian to the north. At the exit of the loch about 50 or 60 yards (46–55 m) of rounded cobbles are exposed, and perhaps the original closing of the lake was the formation of a spit of shingle. Today, however, it is the abundance of sand which is so conspicuous. The beach extends northwards of the depression for at least a mile, and blown sand has covered the coastal slopes of the gneiss to the north of the loch. The main dunes between the loch and the sea are marram covered, but the dunes are not fully stabilized. The whole area is very remote; there was once a clachan at Sandwood. Today, with its variety of landforms, plants and wild life, it is one of the most attractive parts of the coast of Scotland. The sandstones make the cliffs to the south of the loch, and there is a conspicuous and fine stack, Am Buachaille. Two rocks now awash, indicate the position of two former stacks close to Am Buachaille and also that erosion is severe. The off-lying islets, Am Balg, are in Lewisian Gneiss. Small streams which reach the coast near Rubha nan Cùl Gheodhachan and Rubh an Fhir Leith follow lines of fault and leave minor headlands to the west of their mouths. In general the dip of the rocks is slightly seaward, and the cliffs are formed by the faces of strike joints and tend to overhang. At Cnoc an Staca, a little to the north-west of Sheigra, there is a small but prominent cliff fall or minor landslip in the Torridon Sandstone. Southwards the cliffs are less high. At Sheigra is the first of four bays. Sheigra Bay, like those which follow, lies in a south-westward trending depression. In this bay which is cut along the margin of gneiss and sandstone, the northern wall shows typical bare glaciated rock, and the southern a steep ridge of sandstone. The gneiss continues seawards as skerries, and the depression itself extends between Dubh Sgeir and Eilean an Roin Mór. The coast hereabouts is exposed, and the headlands, especially the one in the sandstone, is being eroded. There is a small beach in the bay head; because of its exposed position, it consists of relatively coarse material. The cobbles at the inner margin of the beach are large, still mobile, and well-rounded. There is also a good deal of shell sand on the beach, and behind the upper part of the beach is a development of machair. The small inlet of Port Chaligaig, with a cobble beach to the south, is shut in by the Torridonian spur and cliffs on the north and the prominent headland of gneiss which separates it from Bàgh Phollain (Old Shore Beg). This is wholly in the gneiss; the south-east margin is steeper than that to the north-west. The fine beach is fairly well sheltered, but since the gneiss does not easily form cliffs the enclosing spurs are rounded. There is some local cliffing, presumably where the rock has been weakened. The beach is all fine sand, mostly shell sand. There are dunes behind the beach, and sand and machair spread up the slope of the depression. Bàgh Phollain is separated from Am Meallan (Old Shore More) by another ridge of gneiss which is mainly grassed over and bears a rich vegetation. It is shut in by Eilean na h-Aiteig, an off-shore rock of Torridon Sandstone subject to active erosion. It is possible that its eastern edge corresponds with a fault. Most of the beach consists of shell sand, apart from some shingle and cobbles in the north-west. There are dunes behind the beach and sand continues some distance up the slope of the depression.

The coast along Eilear na Molo is fairly simple; the back of Loch Clash appears to be fault-controlled, and the fault continues through the inner loch at Kinloch-bervie and then along the straight southern shore of Loch Inchard.

The Ceathramh Garbh, or rough country, between Loch Inchard and Loch Laxford is, in its western parts, between 300 and 450 feet (91 and 137 m) in altitude and slopes gently to the west. It is cut up by innumerable valleys and craggy features running in diverse directions. 'many . . . are almost straight and mark lines of fault or crush. The majority . . . run north-east and south-west; but a good many have a north-west and south-east trend, while a few run nearly due north and south' (Peach and Horne, 1907). It may be presumed that some of the coastal features are, at least in part, controlled by these fractures, but glaciation and submergence of this particularly rugged district are mainly responsible for giving the coast its picturesque outline.

Loch Laxford to Loch Cairnbawn (Loch A'Chàirn Bhàin)

This part of the mainland coast is wholly in the Lewisian Gneiss, but Handa island is Torridonian. It is a district in which there are numerous west-north-west to east-south-east dykes mainly of epidiorite. These give a marked grain to the country. Many can be traced for miles. They may split or give off branches. Some are remarkably broad, often reaching or even exceeding 100 yards (91 m). In addition to the pattern given directly by the dykes, there is also another system of lineation produced by pre-Torridonian movements. To the south of Scourie two main directions can be traced, one trending roughly north-west to south-east, the other nearly east and west. The first series is almost parallel with that of the dykes, but the east–west lines are dominant, and near the coast may be bent as much as 20° to 30° south of west. These movements have had an effect upon the dykes, so that several are shifted to the westward of their normal course. Similar movements occur throughout the district. There are also variations in the nature of the gneiss itself, but these need not concern us since they probably have no appreciable effect on the nature of the coast.

The whole area shows a great number of minor features, crags and valleys, which follows lines of dyke and fault. If they reach the coast they produce minor details, such as are well displayed on Creag a' Mhàil, north-west of Scourie, where a dyke makes two small notches, separating two minor headlands from the coast. But the main features of the coast are the result of glaciation and submergence of a mass of Lewisian Gneiss. The intricacies of the coast are often related to the dykes and lines of movement, but the submergence undoubtedly gives it its characteristic appearance.

From Tarbet to Creag a' Mhàil the coast is simple in outline, and follows a line of fault which separates Handa island from the mainland along the Sound of Handa. The island is entirely formed of Torridon Sandstone. On the western side of the island there are bold cliffs reaching 400 feet (122 m). The strata in these cliffs are nearly horizontal, but the dip increases to the east to about 20° or 25°. The line of fault along the Sound of Handa is prolonged through the low ground which runs from Tarbet to Fanagmore on Loch Laxford.

Scourie Bay is picturesque and sheltered. On its eastern side is a cobble beach which faces due west and is therefore subject to some considerable wave action. The bar has blocked the outlet of Loch a'Bhadaidh Daraich and has also made a small lagoon immediately behind it. The inner bay on the south and west is enclosed by a knoll of gneiss and is sandy and sheltered. Since there is no active cliff erosion, and no material brought into the bay by streams it is thought that both the shingle and sand are derived from some former glacial outwash. The surrounding landscape shows the hummocky rounded and smoothed domes so typical of the gneiss districts. A little farther south is Badcall Bay, one of the most beautiful features on the west coast. There are many off-lying islands of gneiss which extend into Edrachillis Bay. Near the entrance to Loch Cairnbawn are two larger gneiss islands. The stretch of coast from Badcall to Loch Cairnbawn shows the effect of submergence of the gneiss to perfection. There are, however, no inlets which can be called true fiords in this piece of coast, although there is no doubt that ice has modified the inlets just as it has the surrounding country.

Loch Cairnbawn to Enard Bay

In many respects this part of the coastal area resembles that just described. The Lewisian Gneiss is almost continuous except for the Stoer peninsula which is formed of Torridon Sandstone and corresponds with Handa island. Once again there are numerous bare domes and ridges enclosing small lochans. Moreover, in the coastal part there is little drift. The height is moderate, up to about 500 feet (152 m), but increases considerably farther east. Like the coastal area farther north, the western part is characterized by many straight and nearly parallel features. These may be dykes, or may be produced by normal fault and lines of thrust. The dykes trend west-north-west, the shear planes rather more westerly. Where these features reach the coast there is usually a small headland or re-entrant, because it by no means follows that a dyke forms a promontory; it is often the reverse. Many small faults reach the coast, but do not control its outline; most perhaps run at a high angle to it. Loch Cairnbawn (see also Chapter II, p. 30) is a major inlet. Lochs Glendhu and Glencoul unite to form the major loch which, like Lochs Laxford and Inchard, runs south-east and north-west, parallel to the dykes which are so numerous in this part of Scotland. The general trend of Lochs Cairnbawn and Glencoul is continued to the south-east by Loch Shin and represents an important structural line. Loch Cairnbawn debouches into Edrachillis Bay; some features of its northern side have already been described. The southern shore is broken by several small inlets, and many small faults with a south-west to north-east trend cut the coast. Oldany island may be separated from the mainland along one of them. In Lochs Ardbhair and Nedd there is a fair development of salt marsh. The islands which form such a characteristic feature on the opposite shore of Edrachillis Bay find numerous representatives on the south, and the coast around Drumbeg and Culkein Drumbeg is delightful. To the west of Oldany island the coast is somewhat simpler in outline, and in Clashnessie Bay there is a small beach at its head. The depression of the bay is continued to the south-west to Stoer Bay. The present beach consists of fine material, but Ritchie and Mather (1969) point out that at the back of it there is a shingle beach covered by later deposits. This beach indicates a higher sea-level. Winds funnel both from the north-east and south-west through the trough between the Stoer peninsula and the mainland; there is an absence of a dune and machair belt, but a sharp transition from bare sand to mature machair which is being eroded rapidly.

The Stoer peninsula is wholly formed of Torridonian rocks. The general dip is to the west. This means that the older beds are to the east, at the base of the peninsula. The division between the Applecross group to the west and the Diabaig group to the east follows fairly closely a line joining Culkein and Balchladich Bays. Here, as elsewhere, the sandstones and conglomerates rest on an irregular surface of gneiss. The finest cliff scenery is on the north coast where the rocks consist mainly of massive bands of arkoses coloured purple and red. Near Rubh' an Dunain the dip is 5° to 15° west of north, and reaches 50° where the coast turns to the north. There it decreases and at the Point of Stoer is reversed so as to make a small syncline, which favours the formation of cliffs which hereabouts are 200 to 400 feet (61–122 m) high. The Old Man of Stoer is a prominent stack cut out along two main systems of vertical joints. The south-west-facing coast is far less impressive.

Stoer Bay is large, and contains a fair beach about 500 yards (457 m) long between two headlands of Torridon rocks. The northern headland, Stac Fada, runs north and south, and the dip of the rocks in it is at 20° to the west. This dip and strike prevails in both Stoer and Clachtoll Bays. In Clachtoll there is only a slightly smaller beach, but it is divided by a knob of sandstone. The sand is reddish in colour in both bays, and inland gives way to machair which reaches back to the gneiss which lies behind the Torridonian fringe. The machair in Stoer Bay is more stable largely because there is a wide and high cobble beach on its seaward margin. There are no dunes. On the headland between the two bays there are remains of a broch and some cairns. A mile or two farther south is Achmelvich Bay. It contains a small and popular beach at its southern end. The bay is broken into smaller indentations which have a north-west trend. The country rock is the Lewisian Gneiss, and because of this and especially because of the jointing pattern in the gneiss, cliffs are poor or absent. A low pass separates the southern part of the bay from Loch Roe and focuses wind action on the dunes and machair at the back of the bay. There is serious erosion of the machair as a result of caravans, and especially cars, when turning on the machair itself.

Inlets like Loch Roe, Loch Inver and Loch Kirkaig, into all of which streams flow, have been modified by ice, but they cannot be regarded as fiords. The two river valleys entering Loch Inver are not typically glacial in form, and the head of the loch, on which the village stands, is a nearly straight north–south shore at the foot of a moderate slope. Inverkirkaig stands in a somewhat similar position. There have been interesting changes in the River Kirkaig. The history of its basin has been discussed by Boyd (1956). He maintains that before glaciation occurred, tectonic features were mainly responsible for the drainage of the area. The present Kirkaig river is regarded as a Post-glacial watercourse, and the adjustment of the drainage to this line had widespread consequences. The original stream, the Muirichinn, reached the sea in Badnaban Bay. Its old lower course is now obscured by crofts and fields. At a later stage the river was diverted to Strathan Bay, along the present course of the Allt a' Mhuillin. Today the Kirkaig is receding headward rapidly, and in a relatively short time the Falls of Kirkaig will be cut back to the loch (Fig. 8).

Moreover, certain topographical features between Fionn Loch and Kirkaig Head (Point) suggest that the whole Kirkaig basin is part of a large Tertiary valley in which a river flowed to the east.

If . . . this great through valley, stretching uninterruptedly from the heads of [the Muirichinn and the present day Kirkaig] . . . to the Dornoch Firth . . . was originally occupied by a pre-glacial eastward-flowing river; that as a result of glacial erosion, the watershed of this river was shifted miles to the east, thereby causing a local reversal of drainage, the facts no longer conflict and the paradoxes disappear. We realize that the greater Kirkaig is a composite river and that in this catchment area we are dealing with two river systems now amalgamated into one (Boyd, 1956).

South of Loch Kirkaig the gneiss coast is broken by a number of picturesque bays and islands. Fossil cliffs border most of the coast. The interior country is wild

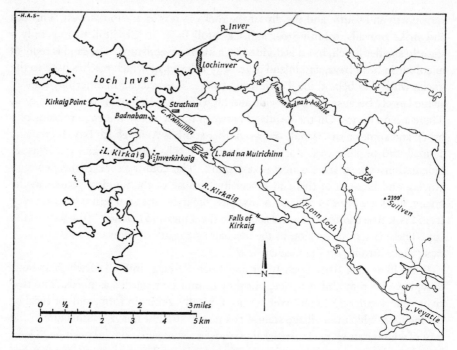

Fig. 8. Rivers near Lochinver (based on Ordnance Survey; Crown Copyright Reserved)

and beautiful, and the background of the coast both north and south of Loch
Inver is dominated by the series of isolated Torridon peaks resting on the hummocky
and heavily glaciated gneiss. Suilven, Canisp, Cul Mór, Cul Beag and Stac
Pollaidh (Stack Polly) make a landscape unique in these islands, and it is impossible
to think of the coast without considering them. At Loch Inver and Inverkirkaig
there are good beaches, but in the smaller inlets farther south beaches are inconspicuous. Rubh' a'Bhrocaire is tied to the mainland by a short tombolo submerged at high water. Behind it is the remarkably straight ? fault valley named Allt
Gleann an Strathain. The small Lochan Sal is almost enclosed by a spit. The Polly
river reaches the sea in Polly Bay. The separation of gneiss and sandstone is abrupt
in Lag na Saille, the straight eastern side of which coincides with a fault. In this and
in Garvie Bay there are small beaches, but the rock platform and fossil cliffs become
more noticeable along this part of the coast.

Enard Bay to Loch Hourn

Achnahaird Bay penetrates far south, and the line of the bay is continued by two
lochs and low ground to Badentarbat Bay. This would have been a strait in higher
raised beach times, and what is now the peninsula of Rubha Mór would have been
an island. At the south end of this low ground there is an outcrop of Cambrian

rocks. The peninsula is formed of Torridonian rocks and introduces the almost continuous belt of those rocks which fringe the coast as far as Loch Alsh. Rubha Mór is largely drift covered. Many small faults reach the coast and give rise to minor features. There are cliffs, often behind a bench, almost all round the peninsula, and the crenulate pattern of the coast depends much on the jointing and other lines of weakness in the sandstone. On the west coast the Loch of Reiff is enclosed by sand now traversed by a small road. Ristol is connected with a low-water tombolo to the mainland (see Plate 32).

The Summer Isles represent a continuation of the sandstone rocks of Rubha Mór and Coigach, but in the northern parts of Horse island and its north and south neighbours the gneiss foundation shows. The major features of the Summer Isles, like those on the mainland to the north, trend with the strike. They are rugged and rocky, and their coasts are irregular, with many caves and headlands. Macculloch (1819) thought these features were only too common. On p. 83 of vol. 2 of his *Western Isles* he writes:

The variety and frequency of this class of objects throughout the western coast, almost destroys the interest first excited by their novelty and effect. The voyager who has passed through the arches of Cape Wrath, and has visited the innumerable specimens of gloomy and grand scenery presented by the caves of Whitten Head and Loch Eriboll, will scarcely turn his boat again, even to view the towering pinnacles of Rockill and Ru Storr!

Macculloch was perhaps a little bored when he visited the Summer Isles; there are indeed many features of interest hereabouts.

To return to the Summer Isles: these, with the exceptions already mentioned, consist of Torridonian rocks, but not all of the same age or type. The rocks of the Applecross group appear in the peninsula of Rubha Mór and in the outer Summer Islands which, except Priest island, are formed of coarse felspathic sandstones. The dips are usually to east-south-east at 30° to 40°. The island of Glasleac Mór is crossed by many faults and joints, and the dip is here west of south. Eilean Mullagrach is also much faulted. Farther north, on the coast of Rubha Mór, the effects of faults may be examined in Faochag Bay, and the coast west and east of Rubha Còigeach is cut by several small faults which give rise to many of its details. The highest Torridonian strata are named after the village of Aultbea where they are characteristically developed. The Aultbea sediments are paler, and sometimes much brighter, red than the arkoses. They are finer in texture and there are not infrequent bands of shale. Many of the Summer Islands are in the Aultbea group, including Tanera More, most of Tanera Beag, nearly all Priest island. Gruinard island and much of the Isle of Ewe are also in the Aultbea group. In all the dip is usually to east-south-east or south-east at 20° to 30°, and sometimes as much as 50°. It is these high dips which are responsible for much of the appearance of Coigach, the Summer Isles and the southern part of Rubha Mór – a succession of ridges and hollows trending north-north-east to south-south west. The scarps of cliffs face west, features well seen in Priest island, the coast of which is clearly related to this structure. Similar structures occur in the south of Gruinard Bay.

Into the unnamed bay in which the Summer Isles are located, two major sea-lochs debouch, Loch Broom and Little Loch Broom. At the mouth of the former Isle Martin is almost tied to the mainland by two spits, the one jutting out from its south-eastern corner, the other and longer one running out from the raised spit of Ard na h-Eighe. To the north is the fine scarp of Torridonian rocks forming the straight northern coast of Camas Mór, and to the north-east is the silted mouth of the Kanaird-Runie river situated within the low peninsula of North Keanchulish. This low ground, as well as that on the east side of the river, is mostly below 50 feet (15 m) and would have been covered by the sea in higher raised beach times. In Loch Broom there are many traces of raised beaches, and raised deltas. The town of Ullapool stands on one, and there are smaller ones at Leckmelm and Ard-charnich. The lower raised beach is conspicuous in Camas A'Chonnaidh. The River Broom has filled the upper part of the loch with alluvium, a process much aided by the smaller stream which enters the loch near Inverlael House. The loch shows the effect of ice-moulding particularly above Ullapool and in Strath More. The Cambrian beds and the Moine Thrust cross the loch at and immediately south of Ullapool. Little Loch Broom lies wholly within the Torridonian rocks and is less impressive than its bigger neighbour. The slopes are locally steep, but rather monotonous. The upper part of the loch is filled with detritus brought down by the Dundonnell river.

In the south and east of Gruinard Bay there is a fine outcrop of the Lewisian Gneiss. The high rounded hills south of Mungasdale, the small bays and beaches, the wooded country behind Gruinard House, and the magnificent raised beaches cut in superficial deposits near Little Gruinard offer one of the most interesting and instructive scenes on the west coast. The gneiss gives way to sandstone near First Coast, and within another mile there is a faulted band of Trias and Lias which forms the low ground between Laide and Aultbea, and also makes a clear-cut boundary to the peninsula, also named Rubha Mór, which terminates in Greenstone Point (Rubha na Lice Uaine). The peninsula closely resembles that other Rubha Mór to the north of the Summer Isles. It is entirely in the Torridonian area and is surrounded by a low rock platform. The cliffs are often fossil, but in exposed places are also being worked upon by the present sea. In the north there are numerous geos. A little north of Laide there is a good example of an unconformity in an arch eroded by the waves. The sides of the arch are in steeply inclined Torridon Sandstone, and the top is formed of gently dipping Mesozoic strata. On the west side of the peninsula there is an extensive foreland, below 50 feet (15 m), at Mellon Charles and Culconich, near Aultbea.

Loch Ewe to some considerable extent owes its formation to the great fault which follows the north coast of Loch Maree and then follows the shore of Loch Ewe as far as Midtown Brae and then runs, in the same line, direct to Camas Mór. The triangular area north of Midtown Brae is downfaulted and forms lower country than the rest of the peninsula. Most of the northern part of the coast resembles that of Rubha Mór to the north; it is somewhat less intricate, but fossil cliffs and geos, especially in the north-east, are present. At the north-western extremity, Rubha Réidh, and also in Caolas Beag, just opposite to Longa island, there are two small inliers of Trias. Along the west coast, south of Melvaig, there is a marked change in the nature of the coast. There is a good deal of boulder clay overlying sandstone rocks dipping seawards. In the north there are coarse pebble beaches, and a low sandy plateau farther south. Good sand beaches front fossil cliffs, and raised beaches – three are recognizable near Melvaig – are conspicuous. Similar features, on a smaller scale, are present in Port Erradale.

Longa island, sandstone, marks the turn of the coast into Loch Gairloch, the north and south shores of which are in the Torridonian. The north shore is simple in outline, and the old cliff can easily be followed along much of it. The south shore is more broken, partly because the gneiss, which also forms the eastern shore, extends as far as Badachro on the south. Gairloch is, in shape, unlike the other lochs; it has certainly been glaciated, but its form seems to depend little on ice erosion. The east side, behind a narrow platform, is steep and shows the usual rounded features of the gneiss.

Loch Gairloch and Loch Torridon are the two major inlets in the long line of Torridon coast from Rudh Ré to Loch Kishorn. The type of coast varies somewhat from place to place as we have already seen north of Loch Gairloch, but in general the outer coast is not impressive. This is partly because it is locally a low coast, often overspread with boulder clay and covered with bracken and other vegetation, partly because it is less broken by small inlets, but even more because much of it lies behind a rock bench and the cliffs do not often show the fresh outlines of active erosion. To the south of Gairloch there are good beaches at Port Henderson, Opinan and Redpoint farm. The peninsula of Red Point was an island in high raised beach times, and today there is sandy heath leading down to the entrance of Outer Loch Torridon.

Loch Torridon is one of the most beautiful of the west coast lochs. The outer part on its northern side slopes fairly steeply to a beach; the southern side is much more broken almost as far as Arinacrinachd; there are geos and other features typical of a sandstone coast. Then the scenery changes abruptly. Between, approximately, Arinacrinachd and Diabaig a broad ridge of gneiss crosses the loch. This is one of the many ridges of higher ground over which the Torridon Sandstones were deposited. On the north shore there is only one outcrop which extends from Loch Diabaig to Ob a Bràighe; on the south there are four or five outcrops, and it is to this fact that we can ascribe the much greater variety of coastal scenery on that side. Loch Shieldaig lies between two masses of gneiss, and the irregular and picturesque bays east of Shieldaig owe their forms largely to the alternation of sandstone and gneiss. The opposite shore in the sandstone is relatively straight. At the head of the loch there is a broad strand on which there are also many cobbles and boulders. The setting of the loch, especially the upper loch, is magnificent. Beinn Eighe and Liathach are a little inland from its north-eastern corner, and to the south there are Beinn Shieldaig and Ben Damh and several peaks up to or exceeding 3,000 feet (915 m).

From Loch Torridon to Applecross the coast is but slightly indented. It is also sheltered by Rona and Raasay. A rock platform lines most of it and there are fossil cliffs. There are several miles along which the 'roo-ft' beach is developed, and the old cliffs are boulder-clay covered in places. The sandy inlet at An Cruinn-leum lies on a line of fault. The main fault in the area reaches the sea in Applecross Bay. The coast makes a major bend into the bay on its northern side. There is an extensive sand flat rimmed with cobble ridges in the bay which lies in Triassic and Liassic strata; these beds continue southwards beyond Camusteel, and make lower ground against the steep face of the Torridon Sandstone on their landward side (Lee, 1920). On their seaward side they are faced with boulder beaches. The coast at the southern end of the Applecross peninsula is irregular. Several inlets run roughly north and south parallel to the general trend of the coast. They are all in Torridon Sandstone, but in the northern part of Poll Creadha moraine reaches the coast. In the upper part of Loch Toscaig there is an extensive spread of the lowest raised beach. The south coast of the peninsula is a steep slope, and a narrow rock platform is more or less continuous at its foot and extends into Loch Kishorn. This part of the coast may owe its general appearance to the Kishorn fault. The head of the loch is, on the Kishorn river side, largely filled with sands and marsh. Here, and particularly on the beach south of Ardarroch, there are great spreads of boulders. No comment on Loch Kishorn, however, can omit the views to the north; the great walls of sandstone leading up to Bheinn Bhàn, and seen to advantage from the road up the Bealach-na Bà, are truly impressive.

The Crowlin Islands form a detached part of the Applecross peninsula. The two main islands are separated by a long and narrow channel. They are built of hard sandstones and shales of the Aultbea group, and traversed by basalt dykes. Jointing, faults and dykes give rise to gullies of which the harbour is the best example.

The straight northern shore of Loch Kishorn follows closely a line of fault parallel to that of the Applecross river. Loch Kishorn in its upper parts gives way to an extensive development of salt marsh, and is separated from Loch Carron by a headland of Torridonian. The narrows in Loch Carron at Strome Ferry are in some ways conparable to those in Loch Torridon; a ridge of gneiss crosses the loch near that place and, although no distinction of name is made, it effectively creates an inner and outer Loch Carron. On the south bank, and to the west of this ridge, the Torridonian coast is much indented and picturesque. Plockton, facing up the loch, is well known, and the numerous small islands, nearly all belonging to the National Trust for Scotland, which stretch from Plockton to Kyle of Lochalsh are a true skerry-guard. There are numerous traces of raised beaches hereabouts.

The south side of the peninsula, facing Loch Alsh, is simple in outline, but around Dornie and Eileann Donnan another ridge of gneiss causes a contraction and makes Loch Long and Loch Duich distinct from the outer loch, i.e. Loch Alsh. The Torridonian extends on both sides of Kyle Rhea (see p. 32), but on the mainland side it is but a narrow strip which ends at the entrance to Glenelg Bay.

We must return to Loch Kishorn and Loch Carron. At the head of the former are extensive depositional terraces at 85 to 95 feet (26 to 29 m) above sea-level. They are fronted by a lower terrace, cut into the older and higher one, at 13–17 feet (4–7 m). Wright (1937) discussed at some length the beach-like features in Loch Carron near Strome Ferry. At the narrows he said that 'there are immense gravel embankments which have been regarded by former investigators [i.e. the Geological Survey] as deposited in the 100-foot [30 m] sea . . . These are not beaches at all . . . ' McCann (1963) has re-examined this area in detail. He is convinced that the terraces at the Narrows are marine and related, just as those in Loch Kishorn and in Skye (near Kyleakin), to the Late-glacial sea. He maintains that in origin the gravels are glacial outwash; there is no river which could have deposited them. Later, they were re-worked by the sea when it stood at its maximum level of 85–95 feet (26–9 m) above the present. McCann, therefore, discards Wright's view that the gravels represent 'a sub-aerial outwash face following a fall in sea-level'. Moreover, McCann was quite unable to trace any gravel embankments near Strome Ferry at the 130-foot (40 m) level noted by Wright. At the head of Loch Carron there is a wide spread of gravels at 13–17 feet (7–8 m), backed by a greater mass at 50 feet (15 m) which are largely peat-covered. The well-known sketch map by Wright showing his views on the exclusion of the '100-ft' beach from Loch Carron, and the very similar map of the Carron area by Charlesworth must, in view of the later observations of McCann, both be regarded as open to correction.

The valleys separating Skye from the mainland near Kyle of Lochalsh appear, according to Godard (1965), to have been formed by a glacier following the line of the Allt Gleann Udulain and presumably joining a larger one travelling westwards from Lochs Duich and Long. This combined glacier then passed through Kyle Rhea. An overflow from this glacier extended westwards and in Godard's opinion reached the sea through a col of diffluence which is now Kyle Akin. He does not appear to contemplate the possibility that the main glacier followed Loch Alsh and Kyle Akin, and that a southern tributary overflowed at a col, where now is Kyle Rhea. He notes that in all the northern part of the Sound of Sleat including presumably Kyle Rhea, the cliffs are old landslopes, and there is no doubt that Kyle Rhea was once a river course. The final separation of Skye from the mainland was the result of submergence and the flooding of glacial troughs (Peach, Horne, *et al.*, 1910) (see Plate 43).

LOCH HOURN TO LOCH LINNHE

On the mainland coast the Lewisian Gneiss is mostly north of Loch Hourn, apart
from a small area to the south of the mouth of the loch. Moinian rocks form a
narrow fringe opposite Port Aslaig, and there is also a narrow fringe of Torridonian along most of the east side of Kyle Rhea. Behind both Moinian and Torridonian there is a broader belt of Lewisian Gneiss. South of Loch Nevis Moinian
rocks form the outer coast as far as the inner end of the Ardnamurchan peninsula.
Locally, along the Sound of Sleat, the Lewisian rocks make greener and less rocky
features than those associated with the Moines which, between Loch Alsh and Loch
Hourn, are mainly psammitic. To the south of Loch Hourn psammitic and pelitic
varieties are more or less equally abundant. Throughout the whole of the area, on
both sides of the Sound, there are numerous dykes. In the Lewisian part of Sleat,
especially, they locally give considerable detail to the coast. All this coast is
relatively sheltered, and it is almost impossible to say precisely how much of its
form is the work of marine erosion. That wave action is far from negligible is
clearly demonstrated at Kirkton of Glenelg, where there are considerable expanses
of raised beach shingle, much of which is frequently re-arranged by the present
waves. There are many examples of rock platforms at Femaig, Plockton, Duirinish,
Portnacloich, Erbusaig, Kyle of Loch Alsh, Balmacara, and other places. These
vary in elevation, but traces of the '100-ft' and lower beaches occur. On the south
side of Kyle Akin there are prominent raised beach phenomena extending as far as
Broadford Bay. In most of these examples marine cliffs may be seen behind the
beaches, but the arms of water separating Skye from the mainland are primarily
parts of former river courses which have been greatly modified by ice action and
later by submergence so that marine cliffs are absent. The slopes, often steep, of
these waterways probably owe more to ice erosion than to any other form of
denudation, but fluvial and to some extent marine agencies have played significant
parts. The beaches in Glenelg Bay, the flats between the Sandaig Islands and the
mainland, those joining Isle Ornsay to the mainland, and the partial filling up of
Loch na Dal and Camas Croise are examples of marine action.

Two of the most beautiful sea-lochs enter the Sound of Sleat, Loch Hourn and
Loch Nevis. In Chapter II attention is drawn to their bottom topography and also
to the several basins into which each is divided. Differential erosion by ice can be
invoked to explain anomalous features, but such explanations are often inadequate
or difficult to believe. To some extent this is true in Loch Hourn. Differential
erosion can certainly produce basins on the floor of a loch, but it is less easy to
explain the considerable number of islands, most of them very small, in the loch
and also the marked narrowing at Caolas Mór. The steep west-facing cliff at
Barrisdale may to some extent be connected with a small glacier in Glen Barrisdale,
but undoubtedly a more comprehensive reason must be found. The silting up of
Lower Glen Barrisdale and the extensive sand flats at the foot of the cliff are note-

worthy. The setting of Loch Hourn is extremely fine. High mountains enclose it; Beinn Sgritheall on the north and Ladhar Bheinn on the south both exceed 3,000 feet (915 m). Loch Nevis is in rather less high country and is somewhat shorter than Loch Hourn. There is a marked narrowing at Kylesknoydart, and at Inverie Bay the river of the same name makes a deep valley, and in the bay are extensive sands.

The structure of the mainland coast between Loch Hourn and Ardnamurchan is complicated, and is interpreted differently by, for example, Kennedy (1955) and Lambert (1958, 1959). From the present point of view it is sufficient to note that the Moinian rocks of which it is formed show increasing deformation inland. The coast itself is broken and picturesque. Between the Morar river and Loch nan Ceall it is relatively low, and many of the bays have fine beaches, mainly of calcareous sand. Even more striking in this section are the numerous low-lying islands and reefs which make a true skerry-guard (see Plate 46). They are usually flat and are parts of raised platforms. North of Morar the coast as far as Mallaig and Loch Nevis is cliffed, but is not high. The coast of the Sound of Arisaig, together with Loch nan Uamh and Loch Ailort, is much indented; there are numerous minor headlands, enclosing small bays, and many islands. Loch Moidart and its northern outlet (North Channel) is straight and follows the line of a presumed fault. The whole complex of channels and islands and small peninsulas between North Channel and Kentra Bay is but another fine example of a submerged coast. The inner part of Loch Moidart, the mouth of the River Shiel (see below) and Kentra Bay are shallow and filled with sands. There are, however, two places of particular interest, the outlets of Loch Morar and Loch Shiel. McCann (1966) has reinterpreted the deposits at the lower ends of these two lakes. Fig. 9. shows the western edge of the ice at Charlesworth's (1955) stage M, together with emendations by McCann. The spreads of gravel are regarded as glacial outwash deposits (cf. those in Loch Linnhe, inset) and not as raised beach gravels. Loch Morar is the deepest loch in Scotland, and its waters now reach the sea, through a rock trench cut by the River Morar. Originally the outlet must have been much wider, but still shallow, and followed the line of the gravels and the low ground between Camas Ruadh and Keppoch. 'The main mass of gravel . . . has a surface slope southward and south-westwards, from a maximum height of 61 ft O.D. [19 m] in Blàr na Caillich Buidhe, to 26 ft [8 m] at the top of the low bluff at the rear of the Post-Glacial raised beach flat at the Back of Keppoch.' The material is poorly sorted; north of Camas Ruadh it is moundy (Fig. 10).

A rather similar deposit is found at the mouth of the Shiel. The Shiel gravels are more extensive than those of Morar and show greater variety of form. Gravels also extend along about five miles (8 km) of the southern shore of the loch. The River Shiel is entrenched partly in the older gravels and has cut a wide alluvial plain about 10–12 feet (3–4 m) above present water level. This same level is cut in the older gravels farther up the loch, and may indicate a Post-glacial incursion of the sea. McCann adds that 'it is interesting to record that the height of this raised

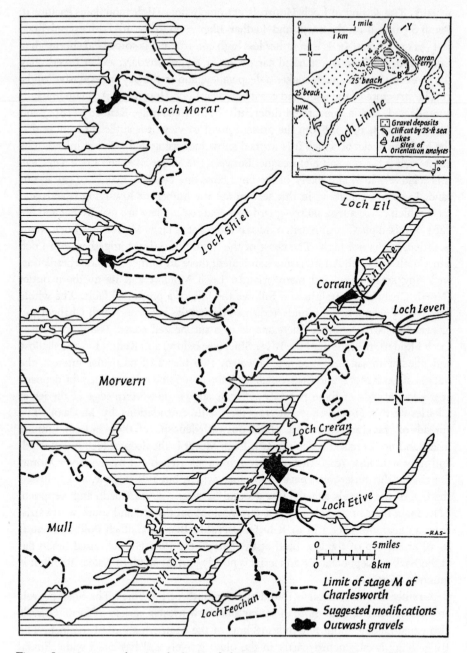

Fig. 9. Location map, showing the limits of glaciation at stage M according to Charlesworth
(after S. B. McCann); Inset: the Corran gravels, Loch Linnhe (after S. B. McCann)

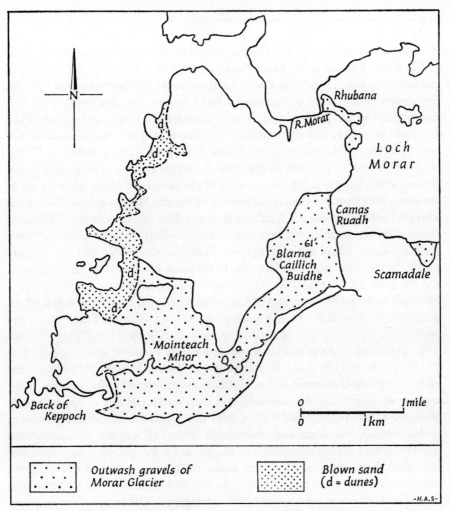

Fig. 10. The Loch Morar gravel deposits (after S. B. McCann)

beach along Loch Sunart to the south is about 26–28 ft O.D. (8–9 m)'. Although the Shiel has destroyed some of the original outwash fan, the inner margin and associated eskers and kettle holes are preserved north of Acharacle church. Kentra Moss, between the Shiel and Kentra Bay, is below 50 feet (15 m) and was covered by the sea at the period of the highest raised beach. The coast of Moidart deserves much greater study; much remains to be done on the relations between tectonics and glacial and marine erosion in its evolution. Along it there are remains of old cliffs and rock benches. Whilst it is true to say that the coasts of Moidart and Arisaig are drowned coasts, it is very far from accounting for the great variety of features present on and off the shore.

Ardnamurchan

This peninsula is one of the Central Intrusion Complexes on the west coast (Richey and Thomas *et al.*, 1930; Richey, 1961). The highest point, Ben Hiant, reaches 1,729 feet (527 m); there are several summits exceeding 1,000 feet (305 m), and somewhat to the east of the complex Ben Laga reaches 1,679 feet (512 m). It is not relevant to the present discussion to consider the origin and nature of the complex (see p. 123), but to call attention to some coastal features. These, as can be seen from Fig. 11, depend to a large extent on the form of the complex itself. The ring-like outcrops of almost all the rocks is obvious, and it is undoubtedly that pattern which has given the main outline of the peninsula. Along several parts of the coast, for example, in the north near Meall Buidhe Mór, in the south-west and south between Port Mìn and Kilchoan there is a discontinuous fringe of Mesozoic rocks; there are occasionally outcrops of Moine Schists including that in Port a Chimais. Cone sheets and ring dykes enclose the main vents and run roughly parallel to the coast, or cut it obliquely. In the south coast and also north of Sanna Bay the great gabbro ring dyke is a dominant feature apart from patches of Mesozoic beds as at the northern end of Garbhlach Mhór and on the north side of An Acairseid. Between this inlet and Kilchoan there is a profusion of cone sheets and basic dykes which cut through Mesozoic rocks and Tertiary lavas. Fig. 12 shows their appearance south of Kilchoan and on the little peninsula that runs eastwards to Mingary pier. They also follow the coast westwards to Sròn Bheag where the cliffs, in Jurassic limestones and sandstones, are traversed by many sheets. They are less abundant near An Acairseid, and their strike changes to approximately north-west. Apart from a single cone sheet which cuts the gabbro at the Point of Ardnamurchan, the sheets disappear under the sea to reappear at Sanna Point. But in this area the metamorphism of the sheets by the gabbro is so intense that individual sheets are determined with difficulty. There is no doubt that along the coast north of the great eucrite ring dyke the cone-sheet complex is continuous, even if its component parts are not easy to recognize. The complex can be traced inland around the main intrusion centre to join that on the south coast (see Plate 47).

The Mesozoic rocks around the plutonic complex are all tilted outwards at about 30°; farther east they are horizontal or nearly so. We have seen that the Mesozoic outcrops are discontinuous, but wherever they occur the outward dip is noticeable. This dip is associated with the injection of the great number of cone sheets. It has been estimated (Richey and Thomas *et al.*, 1930, Memoir) that an upward movement of 4,000 feet (1,219 m) is probable. This figure is based on the accurately mapped shore-section at Kilchoan (Fig. 12). 'The combined thicknesses of the cone sheets in the half mile section (in this figure) are approximately 950 ft [290 m]. The total for the whole belt may then be estimated at about 3,300 feet [1,006 m]. Taking the average inclination of the cone-sheets to be as low as

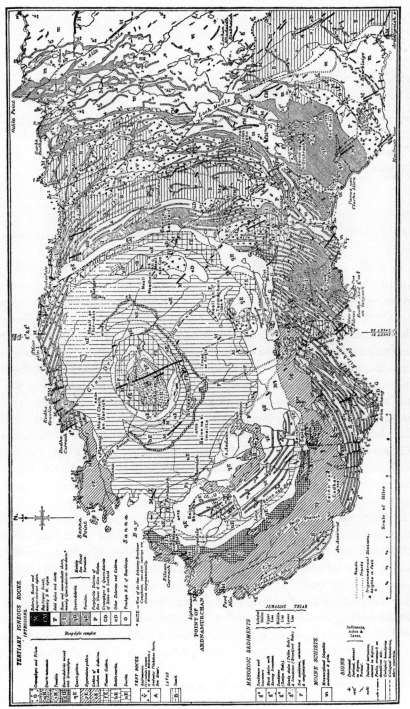

Fig. 11. Map of Tertiary igneous complex of Ardnamurchan, rep. from Pl. VIII, *The Geology of Ardnamurchan* (Mem. Geol. Surv., 1930)

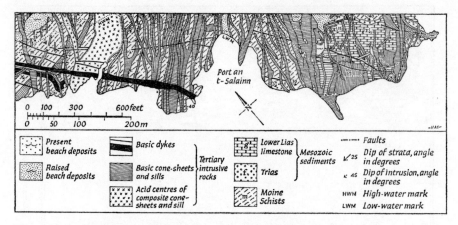

Fig. 12. Part of the south coast of the Ardnamurchan peninsula (based on Geological Survey)

35°, we obtain 4,000 ft. [1,219 m] as an estimate of the total uplift' (Richey, *ibid.*). It will be noticed on Fig. 12 that cone sheets, even in detail, do not always make a pronounced minor shore feature. The group of sheets just west of Port an t'Salainn forms a small projection; many others do not. It is difficult to assess the effect of marine denudation on the outline of the peninsula. At most it seems to be little more than a mild trimming of an outline produced by uplift caused by the intrusive phenomena, and by submergence at a subsequent time. But there are also examples of raised beaches and benches, clearly implying that the submergence was not a simple movement. On the west side of Kilchoan Bay there is a marked notch at about 140 feet (43 m). This seems to equate (see p. 47) with the 'Pre-glacial' beach of the islands. The same notch is also found on the north coast near Achateny. In Kilchoan Bay there are other beaches, the lowest being 10 to 12 feet (3 to 4 m) above high-water mark. There are traces of higher beaches up to about 85 feet OD (26 m) about half-a-mile inland from the bay head, near the Free Church manse. On other parts of the coast raised beaches are uncommon.

The effects of faulting both inland and on the coast of Ardnamurchan are not great. Lines of faulting, or crush, often trend north-west or north-north-west, and their influence is seen in the direction of a number of small streams. Possibly Faskadale Bay is influenced in this way. A study of the One Inch map will suggest, but not prove, the origin of certain other coastal features. Ockle Point, Rubha a'Choit, Garbh Rubha and Ardtoe Island all correspond with quartz dolerite dykes. In the intervening bays and westwards of Faskadale and eastwards to Rubha Aird Druimnich there are fossil cliffs, benches, caves and scars. The straight coast from Rubha Aird Druimnich to Camas an Lighe is almost parallel with Upper Loch Moidart and North Channel.

On the south coast Mesozoic rocks fringe the coast eastwards of Kilchoan as far as the Allt Choire Mhuillin. A little farther east the main feature in the landscape of the northern coast of Loch Sunart is Ben Hiant, a great intrusion of quartz dolerite which extends to the coast to form a steep slope (Plate 48). To the east it is succeeded by agglomerates and tuffs which also make the prominent cliffs of Maclean's Nose. These cliffs 'may be pictured as the cast of the crater, of which the wall has been eroded away' (Richey and Thomas, et al., 1930, Memoir, p. 124). These agglomerates are separated by a vertical wall from the Moine Schists which follow and form both sides of Loch Sunart up to the granite mass around Strontian. The southern side of Loch Sunart west of Portabhata, and the Sound of Mull are almost entirely in basaltic rocks. The Sound of Mull, like the Sound of Sleat, is probably part of a former river system. It is narrow, seldom exceeding two miles (3 km) in width, and usually less. The mainland coast of the Sound of Mull is almost entirely built of Tertiary lavas which show trap (step) features very well. The Sound is more than 20 miles (32 km) long; there are three larger and three smaller basins on its floor, each exceeding 50 fathoms (91 m) in depth. Along both sides of Loch Aline and for two or three miles east of Ardtornish Point, Mesozoic sediments outcrop. Scott (1928–31) points out that these rocks are important because they are sandwiched between the underlying gneiss and overlying basalts and so help in the production of the steep cliffs in Loch Aline. These cliffs are behind the '25-ft' beach. West and north of Loch Aline there are no sediments, and the mainland slopes show no cliffs apart from a small raised beach scarp. Just east of Auliston Point, at the junction of the Sound of Mull and Loch Sunart, there are abrupt cliffs which are associated with a line of fault (Plate 48).

The line of Lochs Teacuis, Doire nam Mart, Arienas and Aline, formed a long and narrow strait in higher raised beach times. There are traces of the '100-ft' beach found along it (see pp. 144 ff).

LOCH LINNHE TO WEST LOCH TARBERT INCLUDING THE ADJACENT ISLANDS OF LISMORE, KERRERA, SEIL, LUING AND SHUNA

The Firth of Lorne and Loch Linnhe owe their origin primarily to the great fault, the Great Glen Fault, which cuts Scotland from the Atlantic to the Moray Firth. This was appreciated many years ago when it was also assumed that the fault was a major displacement in a vertical sense. The work of Kennedy (1946) revolutionized our views. He suggested that the fault represented a great lateral shift of rock masses. His reason for this was based in his view on the similarity between the granite masses at Strontian and Foyers. He suggested that originally they were one mass, and that as a result of lateral movement they are now 83 miles (134 km) apart. He thought that this displacement took place in Lower or early Middle Old Red Sandstone times. This view has been challenged by other writers, Sir E. B. Bailey (Mem. Geol. Surv. U.K., Ben Nevis and Glen Coe (1960)); M. Munro (Scott. Journ. Geol., 1 (1965), 152); P. A. Sabine (Bull. Geol. Surv. Gt. Britain, 20 (1963), 6);

and M. U. Ahmed (*Nature*, **213** (1967), 275). The detailed reasons for or against the displacement as Kennedy described it need not concern us here. In 1969 Holgate (1969) brought forward evidence not only supporting the Kennedy displacement but also strongly in favour of a later shift of 18 miles (29 km) in the opposite direction. Both these movements, but particularly the second one, bear directly on the coastal features of Loch Linnhe, the Firth of Lorne and, at the other end of the Great Glen, in the Moray and Dornoch Firths.

There are several tributary lochs in Loch Linnhe. These run at a high angle to the trend of the main loch, and many years ago it was suggested (Bailey and Maufe, 1916) that they were parts of an ancient river system. It was assumed, and this view is unchanged, that a major south-west flowing stream along the line of the Great Glen cut back to the north-east and disrupted this earlier system of drainage which flowed approximately north-west to south-east. That there have been later modifications by ice and submergence does not contradict the main thesis. A study of a physical map of Scotland suggests that Loch Eil may be part of a stream that passed to the north of Fort William in to the Spean valley and so to Loch Laggan and the Spey. The Cona and Scaddle, which both enter Loch Linnhe through Inverscaddle Bay, may have run into the lower part of the valley now occupied by the Kiachnish river and then through a col into Upper Glen Nevis, the Amhainn Rath, and then in a general sense following the line of the present West Highland railway via Corrour to the Gaur and Loch Rannoch. The Loch Sunart–Tarbert line found its continuation in Loch Leven, Glen Coe, and Loch Laidon. From the eastern end of Loch Leven a stream may have followed the line of the River Leven through what is now the Blackwater reservoir and thence to the Gaur and Loch Rannoch. Holgate (1969) enlarges upon this view. In many parts of the Glen it is possible to recognize comparable valley lines with similar orientations on either side, nevertheless 'any approach to alignment of these across the Great Glen is exceptional at this present time'. The valleys reaching the Glen from the north and west have been modified by ice, and also hang into it. More significant is the fact that the rivers in these valleys are all deflected to the south-west or south about a mile or two above their junction with the Great Glen. Holgate argues that 'the change . . . most obviously competent to produce river deviations of such uniform sense is the reactivation of the Great Glen fault at a time when the major river valleys across the Northern Highlands were already in being'.

This same or a similar movement helps greatly to explain very different features lower down the loch. In Mull, Morvern and Ardgour on the one side and in Lorne, Lismore island, and Benderloch on the other side of the *main* line of the Great Glen Fault there are the great dyke-swarms, the dykes being generally orientated south-east and north-west. At the present time there is a lack of agreement in the geological structure between the two sides of Loch Linnhe. The dykes in the north-eastern half of Morvern and those in Ardgour find no corresponding features immediately opposite in Appin and Lochaber, and the great number of dykes in

Lismore island stand in marked contrast to the very few in Benderloch. If, however, Holgate's thesis of a later movement is invoked, there is then a complete agreement between the two sides of Loch Linnhe. The vast number of dykes in Mull and Morvern are then brought directly opposite to those in Jura and Lorne, and Lismore island lies exactly between Morvern and Lorne, and its dykes fit in closely with those on either side. Kennedy and earlier writers all visualized the main Great Glen Fault running between Morvern and Lismore. Holgate finds local evidence for another lateral fault between Lismore and Appin-Benderloch-Lorne (Figs. 13 and 14).

Some of this local evidence is of direct interest to a study of the coast. At the end of the old sea-cliffs south-west of Port Appin there is an arch on the '25-ft' beach eroded in Appin quartzites. Holgate notes that the arch itself has been determined by dislocation. The absence of brecciation, the disappearance of three small quartz veins above the plane of slip, and comparable phenomena a little to the north-west are all indicative of significant dislocation. In Lismore island dykes in the old sea-cliffs on the south-east side of the island are frequent, but cannot be traced in the steep foreshore and raised beach platform in front of the cliffs. There is no doubt that the strait between Lismore and Appin-Benderloch coincides with a fault, and its magnitude is suggested by bringing the Lismore dykes opposite to those near Oban (Plate 50).

Better agreement is possible between those of Lismore and Morvern; this implies a later movement of five miles (8 km). 'The amount of dextral shift along the Great Glen fault proper, along the line of Loch Linnhe north-west of Lismore, is therefore taken as five miles; the remainder of the overall 18 miles (29 km) shift, i.e. 13 miles (21 km), must then be represented by movement on the associated, divergent, Firth of Lorne dislocation' (Holgate, 1969). In short, Holgate is of the opinion that movement has taken place along both faults, but that the amount of movement has not been the same on either side of Lismore island.

It will be convenient at this stage to look at certain other features in Loch Linnhe and in Loch Leven which are of more recent origin and are not in any way connected with the major movement already discussed. The features are often glacial in origin but they also include a number of raised beach features some of which may be Pre- or Inter-glacial. The gravels at the narrows at Corran and Ballachulish modify the shape of the lochs. McCann (1966) interprets these features as glacial outwash fans. The Corran fan is now nearly 80 feet (24 m) above water level, and was modified by waves at a later stage. The fan was deposited by the Loch Leven glacier. The smaller fans at Rudha Charnuis and North Ballachulish mark, in McCann's view, the limit at stage M of the Loch Leven glacier. In an earlier paper (1961) McCann showed by careful analysis that the Corran gravels were definitely outwash gravels and in no sense a raised beach (see Fig. 9). There is another interesting feature near Corran. At Onich there is a short but marked dry valley, named Dubh Ghlac, cut in the Appin quartzites. At present the Amhainn

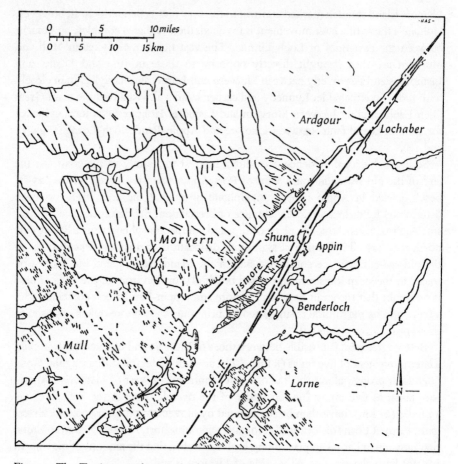

Fig. 13. The Tertiary north-westerly dyke-swarms astride the Great Glen Fault system at Loch Linnhe and the Firth of Lorne (after Norman Holgate)

Righ reaches Loch Linnhe over a waterfall east of the main road. The Amhainn Righ is separated from Dubh Ghlac by a ridge formed by a landslip. There is no doubt that the dry valley is the old course of the Amhainn Righ, and that it was diverted to its present course by the landslip. The fertile part of Onich is probably the alluvial fan of the old Amhainn Righ (see Walter 1924–31).

Farther south in Loch Linnhe there are extensive gravel deposits at Benderloch, between Loch Creran and Ardmucknish Bay. The solid ground of the Benderloch peninsula consists of one major and several minor ridges of slates and schists trending south-west and north-east. These are now joined to one another and to the mainland by what are marked on the geological map as raised beaches, largely covered by peat. The highest are indicated as the '100-ft' beach near South Shian and Balure; most of the remainder are indicated as the Post-glacial raised beach.

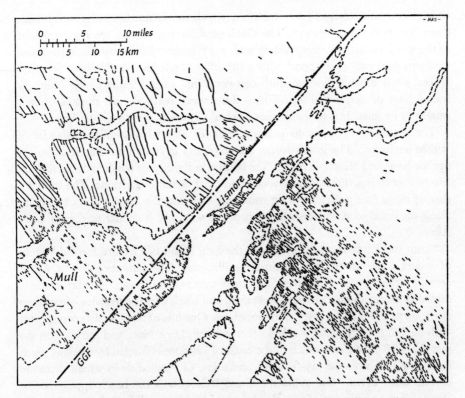

Fig. 14. The Tertiary north-westerly dykes astride the Great Glen Fault prior to the initiation of an 18-mile dextral wrench (after Norman Holgate)

McCann (1966) gives good reason to suppose that here again we are dealing with glacial outwash gravels. They slope gradually to the south-west,

As at Corran and the entrance to Loch Etive, the development of the Post-glacial raised beach along the inner part of the outwash fan differs from the development of the beach along the lower outer part, thereby emphasising the original depositional form of the older gravels. Thus, on the northern or Loch Creran side, the Post-glacial shore-feature is a fairly distinct eroded terrace, backed by a low cliff and cut into the highest part of the older gravel deposit. On the southern Ardmucknish side, however, it is depositional, consisting of a series of shingle ridges, the highest of which is a pronounced feature superimposed on the gentle south-westward slope of the older gravels.

The Morvern coast of Loch Linnhe south of Cilmalieu is a mass of granite (the Strontian granite). There is a smaller mass, about a mile to the north and east, which forms the peninsula of Rubha na h-Earba. Between the two granites is Camas Chil-Mhalieu with its boulder-strewn sandflat. Steep coastal slopes extend south-westwards from the bay as far as the double inlet of Camas na Croise and Loch a'Choire. The steep-sided Glen Galmadale runs southwards into the former, and into the much larger Loch a'Choire, which trends north-west to south-east, there flow two minor streams. The Geological Survey regards the northern part of the granite coast as corresponding with the Highland Boundary Fault; the more southern part may correspond with a subsidiary fault line. In this part there are several small streams flowing south-eastwards into Loch Linnhe which definitely follow lines of minor faulting. Where the streams reach the sea there is often a small bay or inlet. It is probable that Loch a' Choire follows a similar line.

The granite gives place to psammitic gneiss about half-a-mile north-east of Rubha an Ridire. The gneiss forms the point and in Inninmore Bay it is faulted against lavas and Mesozoic rocks. The section in the bay is: basalt lavas resting on Liassic limestones, Triassic sandstones, and Coal Measure sandstone. The lavas are part of those (see p. 81) covering much of Morvern; the Mesozoic rocks extend along the coast to Ardtornish Bay where the sequence is similar to that in Loch Aline.

South-eastwards from the granite is the long island of Lismore formed largely of a dark blue limestone, sometimes argillaceous. The limestone (see p. 83) is crossed by numerous dykes trending north-west and south-east. The limestone is folded along north-east to south-west axes, and black schists and slates often come to the surface along the crests of the anticlines. One line of black schist corresponds with the long valley which holds Kilcheran Loch, Loch Fiart, and the inlet on the east side of Rubha Fiart. Much of the coast of Lismore is fringed by a raised beach corresponding with the '25-ft' level, and a line of fossil cliffs in which there are several small caves. Near Acnacroish and Balnagowan the beach appears to be absent. In the north-east of the island the bedding in the cliffs is shown up by the difference in weathering of the impure argillaceous limestone and quartzose layers which alternate with bands of purer limestone. Close folding is often visible. Black schists, especially in the north-east, are mainly found on the west side; their outcrop runs along the base of the cliffs and may extend upwards into cliffs and folds. The islets, Eilean Ramsay and its neighbours, and also Eilean Dubh, are similar. Fossil cliffs are found in Eilean Dubh. The group of islands, including Eilean nan Gamhna and Eilean na Cloiche which is joined to Eilean Dubh (not the island mentioned above) by a narrow tombolo, is rather more complex. In Eilean nan Gamhna grey phyllites represent a transition from Ballachulish slates to Appin quartzites, and in Eilean na Cloiche the descending sequence is black limestone, black schist, crushed quartzite, white limestone and grey phyllites. Each island in the group is partly surrounded by a rock platform and a tombolo, covered at high water, joins Creag island and Pladda island.

On the other side of Loch Linnhe, south of Oban, in Loch Feochan, there is at the head of the loch a spread of the '100-ft' raised beach and alluvium. But this does not reach Loch Nell or the part of Glen Feochan near Craigentaggart. In both these localities McCann finds outwash gravels associated with fragments of terminal moraine (see Fig. 9). To the north of Oban there is another great spread of gravels and peat, the Moss of Achnacree, at the mouth of Loch Etive. These are also interpreted by McCann (1961) in the same way. If all these gravels were removed Loch Linnhe would have a much more open appearance and the entrances to certain tributary lochs would be much wider.

Apart from the glacial outwash features described by McCann, there are also numerous traces of raised beaches, rock notches and benches. They are well displayed around Oban (Kynaston and Hill, 1908). Several of the older roads and houses of the town are built on a rock platform, behind which is a high cliff (see Plate 53). The platform varies from about 30 to 100 yards (27 to 91 m) in breadth, and implies a long period for its formation. There is a similar platform on Kerrera island, as well as on all the other nearby islands, two of which are named the Dutchman's Hat (Bach island) and Shepherd's Hat (Eilean nan Gamhna); the rim of each 'hat' is a bench around the crown of the island. North of Oban there is a fine stack on the bench near Dunollie Castle, and to the south there is marked undercutting seen on the roadside, to the south-west of Dungallan House. Kerrera is separated from the mainland by the narrow Sound of Kerrera. The mainland side of this channel is very irregular, but its main pattern is the result of faults running either approximately parallel to the sound or almost at right angles to it. The lavas and occasional vents, the raised beach deposits in Port na Tràigh-linne and other inlets which are in large part related to faults, make this an interesting coast. The structure of Kerrera island is also largely determined by faults trending to the north-east or to the north-west. The inclination of the Old Red Sandstone together with the faulting largely governs the pattern of the rock outcrops. There is a remarkable band, about 300 yards wide (274 m), of lavas let down in a rift valley between schists. The valley is close to the east coast of the island, and in its northern part is open to the sea. The island is irregular both in outline and in contour. The highest point, Carn Breugach, is 617 feet (188 m). It is difficult to say in detail how much the coastal outline is the result of faulting; Port a'Chroinn is the submerged southern end of the rift valley mentioned above. There are abundant remains of raised beaches on all sides of the island.

South of Loch Feochan black slates (Dalradian) reach the coast for three or four miles (5–6 km). They are cut through by the very narrow channel which separates Seil from the mainland. At Clachan Bridge the nature of the channel is easily appreciated, and the most striking feature is the narrow, but well-marked, rock terrace on either side of the water. The terrace is backed by steep cliffs, and it is not easy to understand how the terrace was cut. At and beyond Rubha Garbh Airde the lavas return to the coast as far as Easdale Sound. Throughout this stretch, both in the slates and the lavas, there are numerous dykes of Tertiary age which trend south-east to north-west, and form many minor features in the coast. This is also true of Insh island, a detached fragment of the lavas.

The coastal area from Loch Feochan to the Crinan Canal and Loch Sween is one of much interest. North of Loch Melfort the rocks are mainly the Lorne lavas

(Peach *et al.,* 1911). The ground is not high, and is of the nature of a denuded plateau averaging about 450 feet (137 m). The seaward slopes are often steep. The grain of the country is distinctly north-north-east and south-south-west, and shows as a series of escarpments, corresponding with lava flows. Individual scarps can be traced for several miles and may locally exceed 900 feet (274 m) in height. A large number of dykes, and some faults, cut these lavas at approximately right angles. There is a considerable amount of landslipping along the steep coast of Seil Sound between Ardmaddy Bay and Port na Morachd. One or two of the northern peninsulas and also Eilean Coltair (in Loch Melfort) correspond with outcrops of epidiorite. The Oude enters the head of Loch Melfort in Fearnach Bay, around which there is a considerable development of raised beaches. South of Loch Melfort the country is higher and bolder, and reaches 1,199 feet (366 m) in Tom Soilleir. The ridge running from this mountain to An Cnap is a watershed. An Cnap itself is an outcrop of granite. South of this headland the coast is fringed by raised beaches which, at Tràigh nam Musgan, extend far inland along the Staing Mhór. In '100-ft' times sea-water penetrating up this inlet must also have met that running up the Barbreck river from the head of Loch Craignish.

The Craignish peninsula, Loch Craignish, the islands in it and those in continuation of the peninsula, as well as the eastern shore of the loch all follow the same north-north-east trend. The same is true of several of the bays and inlets in the peninsula. This pronounced pattern is produced by the sharp folding, and subsequent erosion has led to the formation of a number of ridges which are nearly always in the more resistant igneous rocks, whereas the lower ground usually corresponds with slates, phyllites and limestones. Tertiary dykes, running almost at right angles to the trend of the rocks, often weather into gaps and lines of low ground. Some of these correspond with lines of crush.

In the islands of Shuna, Luing, Torsa and Seil there is a slight modification of this type of scenery. The folding of the sedimentary rocks in these islands was just as intense, but there is much less igneous rock and hard quartzite. This means that, in general, the ground is lower and more uniform in height. At the same time, the Tertiary dykes in these islands stand out, and on the raised beaches often form walls or rock notches along the coast. The trend of these islands is rather more north–south and they are separated from the mainland by narrow straits along which there are numerous traces of raised beach platforms. It is difficult to understand how these could have been cut in such restricted waters; even in raised beach times the channels were not appreciably wider. Cuan Sound separates Seil and Luing, but in '100-ft' times Luing was divided into two by an east–west strait between Black Mill Bay and Toberonochy (see Plate 52). This east–west sound of raised beach times was joined on its north side by a large gulf which reached Achafolla, and thence a narrow arm of the sea, following the line of the present road, led to another small bay at Bardrishaig. The group of islands in the bay and at Poll Gorm are low and flat and elongated north and south parallel to the coast of Luing. The fossil cliffs and rock bench are conspicuous between Poll Gorm and Cobblers of Lorn. In Seil, to the north of Cuan Sound, the highest raised beach is also prominent. At that time a narrow strait connected Balvicar Bay to Cuan Sound in such a way that a small island to the east of the strait was separated from the main island. Raised beaches and fossil cliffs are conspicuous on the west coast of Seil, especially south of Easdale. On the mainland side of Seil Sound, Ardmaddy Bay is backed by a wide extent of elevated beaches, and its lower part filled with a wide modern beach. On the northern end of Seil a large number of basalt dykes cut the coast and the numerous islands, and also Insh island a mile to the west. These dykes, the fossil cliffs and rock bench are the most significant features on this broken coast. On the eastern side of Luing, the smaller islands of Torsa and Shuna are both cut by north-west to south-east dykes, and in both raised beaches, especially the highest, are conspicuous around almost all the periphery of both islands. In all these islands the general strike of the rocks is north-east to south-west, and the direction is reflected in the trend of most of the sounds and straits.

The characteristics of the mainland coast extend to Loch Crinan where they are interrupted for more than a mile. Loch Crinan was formerly much larger in area, and occupied all the low ground called Mòine Mhór (Peach et al., 1911). Today the mounds of epidiorite that rise above its surface clearly indicate the islands of former days. Before, however, discussing coastal features we must retrace our steps to explain the drainage of Loch Awe and its relation to the Kilmartin burn and to the Pass of Brander. Today Loch Awe drains to the sea by the River Awe which follows the marked tectonic line of the Pass of Brander and reaches Loch Etive near Bonawe. This great line of fracture then follows Gleann Salach to Loch Creran, and possibly via the Strath of Appin to Loch Linnhe at Portnacroish. However, the natural outlet for Loch Awe is to Loch Crinan. Raised beaches extend several miles up the burn which, a little above Kilmartin, flows through the Creagantairbh Pass. But a little earlier than this the overflow water from what was then the Loch Awe glacier passed over the watershed into the River Add, which to-day enters Loch Crinan jointly with the Kilmartin burn. Earlier, however, it would have had a separate entry into the loch a little farther east. It was undoubtedly this precursor of the Add which helped so effectively to silt up Loch Crinan. But the main entry of Loch Awe waters was via the Pass of Creagantairbh which is narrow, with sheer rocky walls some 300 feet high (91 m) (Fig. 15).

The southern side of Loch Crinan is marked by an important fault which is now followed by the general line of the Crinan Canal (see Plate 59). To the south of the canal the country is similar to that farther north. Long lines and ridges of epidiorite and quartzite give it a characteristic appearance. The parallelism of ridge and furrow is the result of the isoclinal folding of the quartzite and sills. The pitch of the folds is to the south-south-west. The deeper furrows contain sea-lochs. Loch Sween and its many tributary lochs, of which Linnhe Mhuirich and Caol Scotnish are the largest, emphasize the pattern. At Tayvallich the peninsula is

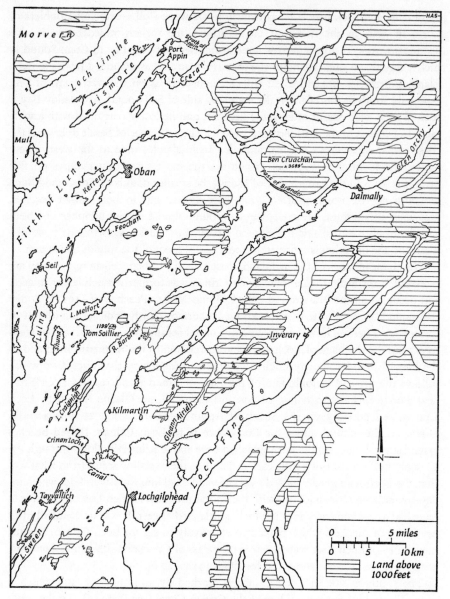

Fig. 15. Loch Awe drainage

nearly severed, and would have been completely severed in higher raised beach times. The narrows probably lie along a fault, parallel to the Crinan Fault. As a result of submergence, the outer folds are now islands, the long axes of which run with the coast, and are separated from it by narrow and straight channels. The island of Danna, joined by a causeway to the mainland, forms the southern end of the Tayvallich peninsula. It is low-lying; the highest point reaches 178 feet (54 m). The same structure prevails, and is continued seawards into Corr Eilean, Eilean Mór and smaller islands. The small rock of Dubh Sgeir may be regarded as the termination of the Tayvallich complex. The larger peninsula between Lochs Sween and Caolisport extends a little farther south. Its western half structurally resembles the Tayvallich peninsula, but in its eastern half phyllites and slates replace the quartzite. The bands of epidiorite are still present. Much of the coast is fringed by raised benches or beaches so that, here as elsewhere on this part of the coast, old cliffs usually stand well back from the sea. The prevailing south-south-westerly trend leads to the formation of capes where more resistant rocks run out to sea, and also to the formation of islands on both sides of the peninsula, the southern end of which is a bold knob of epidiorite partly surrounded by raised beach.

Loch Caolisport is parallel to Loch Sween and West Loch Tarbert and is in the phyllites (Peach *et al.*, 1911). There are extensive sand flats at its head and also some salt marsh. The coast on both sides of the loch, but particularly on the northern side, shows by the south-westerly trending headlands, for example Rubh' an Tubhaidh, the outcrops of epidiorite or resistant ridges of phyllite. The Point of Knap is a mass of epidiorite. Where the loch gives place to more open water, the southern coast is fringed by fossil cliffs, beaches and benches. The headland on the north side of Miller's Bay and Kilberry Point correspond with epidiorite dykes. North and south of Kilberry Head the lowest raised benches are conspicuous, and near Gourach the old cliffs are prominent and shingle of the '100-ft' beach occurs on their tops. The lowest beach is wide on the north and east of Loch Stornoway which, at its maximum, extended about a mile-and-a-half (2.4 km) inland, parallel to West Loch Tarbert. The remainder of the Knapdale coast is more irregular; at Ardpatrick Point there is an epidiorite dyke, but the headland is quartzitic. In West Loch Tarbert the coastal slopes are gentle, and there is a considerable extent of sandy beach on both shores. The old cliff begins to become prominent in Dunskeig Bay.

THE KINTYRE PENINSULA, COWAL AND BUTE

The Kintyre peninsula, south of East and West Lochs Tarbert, is virtually an island, and was one in higher raised beach times; the highest point of the isthmus is just more than 50 feet (15 m). The word Kintyre means head or end of land, or land's end (see Plate 55).

Around Tarbert the inland scenery is pleasant but not striking. McCallien (1919–26) suggests that the hills are a continuation of the plain of marine denudation seen so well in Arran. The details of the topography depend almost entirely on the strike of the rocks which runs from north-east to south-west. The general trend of West Loch Tarbert agrees with the strike, and in part at any rate it follows a line of weakness, a strike fault which can be traced along the valley between Glenralloch and Stonefield. In the harbour of Tarbert (East Loch) joints are responsible for most of the coastal features. There are two sets of joints, the one trending north-east and south-west, and the other north-west and south-east. The interior of the peninsula has been much moulded by ice. In the north the land reaches 500 to 600 feet (152–183 m), and is somewhat dome-shaped. In the northern part the metamorphic rocks are responsible for a series of ridges and valleys trending north-east and south-west. Farther south it is cut by a few rivers which run in deep valleys, e.g. Carradale, Saddell, Lussa and Glen Barr. South of the Laggan the appearance of the country is very different. It is higher – Cnoc Moy is 1,462 feet (446 m) – and cut by broad valleys. The watershed and the rougher ground of the metamorphic and igneous rocks, stand in contrast to the much smoother topography to the south-east, composed largely of Old Red Sandstone and Carboniferous lavas.

The coast of Kintyre is particularly interesting, especially because of the fine development of raised beaches, usually the lower or '25-ft' beach, between the mouth of West Loch Tarbert and Machrihanish Bay. The main road follows the beach almost all the way, only once or twice does it deviate from it. Although there are many similar roads in Ayrshire and Galloway, few allow the traveller to appreciate as well as this does the details of the old platform with its stacks and other features, and the old cliffs, often interrupted by what were small inlets. The road is close to the sea from Ronachan Bay to Auchinadrain, and then for about six miles (10 km) there is a wide stretch of flat land on its seaward side. This low ground reaches a maximum width at Rhunahaorine Point. This is a nearly symmetrical sandy foreland. The outer seaward part is modern; nearer the road there are raised beach traces probably at two levels. The position of the point is interesting. It is exactly opposite to the most easterly point of Gigha island. No detailed field work seems to have been done on this part of the coast. Rhunahaorine Point on a big scale resembles the small triangular accumulations that grow seawards from a straight coast behind a small island or even a wreck (Zenkovich, 1967). South of the triangle of the point there is a broad expanse of flat land of recent origin which narrows southwards until at A'Chléit the coast and the road coincide again. From this place to Westport, where the nature of the coast changes abruptly, old cliffs, stacks, geos, ravines and raised benches are magnificently developed. The road runs inland near Glencardoch Point which shows the same features.

Gigha island is about six miles long, and has a maximum breadth of about two miles. It lies about three miles to the west of the Kintyre peninsula. It is surrounded by many islets and rocks; Cara island, about a mile long and a third of a mile wide, lies to the south of it. The highest point in the main island, Creag Bhàn, is 331 feet (101 m). For the most part it is low and made up of one or two well-defined ridges which are separated by valleys filled with raised beach deposits. There are many rocky points round the coast all of which trend in much the same direction as the island itself. The west coast is abrupt and rocky; the cliffs, often fossil, are parallel to the strike of the schists which dip steeply. This is one reason for the many small landslips. The west coast is cut by several gullies, 20 to 30 feet wide (6 to 10 m), and running at right angles to the cliff which reaches 100 feet (30 m) or more in height. Caves often continue the line of the gullies inland.

Much of the shoreline is difficult to traverse on account of differential weathering. This has removed soft micaceous schists, and left the more quartzose parts and the epidiorite projecting, often as knife-edges. Since these, in their turn, are cut by joints along which marine erosion has worked, the resulting surface is jagged, especially in the quartzites. The epidiorites often show a honeycombed surface, and they form Eilean Garbh (a peninsula in the north-west) and Ardminish Point, where smooth glaciated surfaces are retained, although they are often broken into cuboidal blocks. There are numerous dykes all round the shore; they project from the cliffs as high walls, and often large gullies are worn along their edges (McCallien, 1926–8).

The west coast of Kintyre, south of the Sound of Gigha, shows the raised beach and old cliffs to perfection; only for a mile or two inside Glencardoch Point does the road leave the coast. If the traveller takes the trouble to climb, at some vantage point, the old cliff, he will be able to appreciate the contact between the former and the present coast, and note the ravines made by the small streams which cut through the cliffs, and the detailed effects of former marine erosion upon them. He will also be surprised at the little work done by present waves.

Between Machrihanish Bay and Campbeltown is the Laggan, a down-faulted area. It is a small rift valley which preserves coal-bearing sediments of Carboniferous age. There is a fine beach in Machrihanish Bay which is backed by dunes and then extends inland as a series of terraces. Near Campbeltown these give way to somewhat higher ground rising to about 200 feet (61 m). The knolls hereabouts are connected by the highest terrace or raised beach. Lower beaches can be followed to north and south of the Laggan, and are locally separated by distinct slopes. At Campbeltown pre-Neolithic, probably Mesolithic, artifacts have been found in the upper parts of the raised beach, and may be the equivalent of the Larnian in Ireland (McCallien and Lacaille, 1940–1).

South of the Laggan the coast is of a different type. The coastal slopes are in metamorphic and igneous rocks and often fall sharply to the sea from heights of 1,000 feet (305 m) or more. West of Largybaan the coast is spectacular. There are caves near Largybaan in schists and igneous rocks, but the most southerly one is cut along a belt of Loch Tay limestone. The caves farther north are almost impossible to enter. The cliff at Aignish, in limestone, green schists, and intrusives, reaches 1,185 feet (361 m). Along many parts of this coast there are large masses of slipped material consisting of great blocks of rock. At the Mull and for some distance along the south coast, the same type prevails. Strone Glen and Breakerie water mark a distinct change in the coast; they reach it across a small level plain.

Breakerie water has been diverted in Carskey Bay from an outlet rather more to the east. A short line of higher ground, followed at its foot by a road, separates Carskey from Dunaverty Bay. Interesting changes have taken place here. The Conieglen Water originally reached the sea in Dunaverty Bay. As a result of coastal drift and the piling up of beach material this exit was blocked, so that the water now escapes in Brunerican Bay. The headland of Dunaverty is a small mass of Lower Old Red conglomerate. Keil Cave, near Southend, is of some interest. It is the largest of several in this headland and contains a strange assemblage of materials. Ritchie (1966–7) thinks that occupation began in the third or fourth centuries and continued intermittently: 'in view of the abundance of iron slag recorded and the unusual range of finds it could even be suggested that the cave was used by an itinerant tinker or group of smiths . . .'

Sanda island is wild and rugged and is built of Lower Old Red Sandstone and is more than 400 feet high (122 m). It is by far the largest of a group of islands. The eastern part of Sanda is separated from the west by a fault. At the eastern end there are big landslips. Around Fliuchach on the south coast three raised beaches can be seen, but in other parts only the '25-ft' level is noteworthy. Usually the coast is steep but around Fliuchach and the Black Rocks shales and sandstones form low-lying reefs, the harder layers of which project upwards. Caves are common, usually along faults, although in the east they are cut along the dip. Just east of the lighthouse is a fine arch, the Elephant Rock. It is so called because a mass of breccia shaped like an elephant's trunk passes into the main body of the rock (McCallien, 1928–31).

The south-east coast from Macharioch Bay to Davaar island, at the entrance to Campbeltown loch, is generally steep. The coastal slopes and cliffs are largely in conglomerates of Lower Old Red Sandstone age, and at The Bastard the coast is steepest. A few small streams break the line of high ground. Davaar island is joined to the mainland by a spit named the Dhorlin; it is covered at high water, but dries at about half tide. The suggestion that it owes its formation to the effect of tidal eddies along the outer coast meeting the main tidal streams flowing into and out of Campbeltown Loch seems inadequate. It ignores wave action, and the spit is not formed only of fine material. A careful analysis of its features and a study of wave incidence would be repaying. The cliffs of the island are conglomerates of Lower Old Red age, and in the south-west there are several caves, all associated with vertical joints.

North of Campbeltown the east coast is less steep than farther south, but is varied and picturesque. It is the more attractive because of the views of Arran. There are several bays and small inlets; between the larger bays there is frequently a raised beach platform devoid of marine alluvium, but often marshy and unsuitable for cultivation. The Lussa, Saddel Water, Carradale Water, Crossaig Glen, Claonaig Water and Skipness Water make deep glens, and in some, especially Saddell Bay, there are good storm beaches. The upper ones formed at a higher sea-level are now grass-covered. In Skipness Bay the stream has been diverted. In 1927 a heavy storm caused the stream to break the beach, and McCallien (1928) states that material was encroaching on the present mouth from the east and deflecting the river to the west. In this same bay there are remnants of beaches at '25-', '50-' and '100-ft'. Between Claonaig and Skipness there is a fine raised beach and rock platform, with some stacks and cliffs behind it. This is one of the many examples in western Scotland where a platform in resistant rock has been cut in relatively sheltered water.

Just north of Skipness Point is the entrance to Loch Fyne. To what extent this loch is of fault origin is debatable; the Tyndrum fault runs into its northern end, and may well continue farther south (see Fig. 1). The loch has been glaciated, and much of its present form is the work of ice. Loch Gilp, the chief irregularity on its western shore, is associated with the fault that runs between Crinan and Lochgilphead. It is also possible that the basin of Loch Gilp and Lower Loch Fyne, which have no clear relation to the strike of the schists, may owe their nature to a shatter-belt following a line of fault. In Pre-glacial times Loch Fyne and other sealochs in Cowal may have been part of a river system, a matter which is discussed on p. 40. Rock structure and the nature of the rocks play an important role in this area. Between Ardno and Otter Ferry, Loch Fyne trends very closely with the strike of the rocks and is in easily eroded Ardrishaig phyllites.

Loch Riddon, and its northern continuation in Glen Daruel, Loch Striven, the Kyles of Bute, Loch Long and Loch Goil, and Gare Loch owe their present appearance largely to the work of ice and subsequent submergence. Nevertheless in Cowal there is often a close connection between valley direction and geological structure. The north-west to south-east valleys, e.g., Glen Finart, Glen Massan and Holy Loch, run approximately with the great number of basalt dykes. The valleys trending north-east and south-west, e.g., Glen Fyne, Inverchaolin Glen, Invervegain Glen, coincide more or less with the strike of the rocks. On the other hand Lochs Riddon and Striven run nearly north and south; as do also Loch Eck and Loch Goil. Loch Long, however, trends more nearly north-east to south-west, and is a direct continuation of the Firth of Clyde.

The Highland Boundary Fault reaches the Clyde near Helensburgh and cuts the southern end of the Dunoon peninsula and then divides Bute and passing through the north of Arran follows the east coast of the Kintyre peninsula to disappear to the east of Sanda island. This great structural line brings in Lower Old Red Sandstone conglomerates around Rosneath Point and Toward Point. On the other hand the north-east to south-west strike of the metamorphic rocks immediately to the north cuts almost at right angles across the lower parts of Lochs Lomond, Gareloch and Loch Striven.

It is convenient to consider Bute in this context. The northern part consists of resistant metamorphic rocks which make fine scenery in the Kyles. Kames and Ettrick bays are the submerged parts of the stretch of low ground which corresponds fairly closely with easily weathered phyllites. There follows a higher part, formed of schists, west of Loch Fad. The Highland Boundary Fault follows the low line of Rothesay Bay, Loch Fad and Scalpsie Bay, to the east of which is an undulating plateau of sandstones and conglomerates of Old Red Sandstone age. The southern end, with a rocky coastline, is built of volcanic rocks and lavas. There are many traces of raised beaches; the highest is well developed in the low ground near Kilchattan and Ettrick. 'Before the deposit of these sands and gravels south Bute was separated by sea from the tract to the north, and this in turn from the north point of the island. With the rise of the gravels above the " 100" ft sea Bute became built of a series of islands connected by tombolos' (McCallien, 1939). The lower beach can easily be followed around much of the coast, and, as elsewhere, is backed by cliffs in which a number of caves are cut. Dunagoil cave is well known and illustrates clearly the relation between marine erosion and jointing. The Old Red Sandstone presents some interesting coastal features. Just east of Kilchattan it is columnar; the columns are up to seven inches (18 cm) in diameter and nearly at right angles to the bedding. The Haystack is a piece of Old Red Sandstone, about 20 feet high (6 m) and 20 feet in diameter (6 m) a little south of Ardscalpsie farm. It stands on the raised beach platform, but is on the wrong side of the Highland Boundary Fault. McCallien thinks it may be the remnant of the fill of a great fissure which ran approximately along the fault. Perhaps, too, the present raised beach cliff here represents the north-west wall of this fissure. At the southern tip of Bute is a composite sheet which gives some striking scenery. The outcrop runs north from Roinn Clùmhach to the summit of Torr Mór (485 ft; 148 m). It then turn to the south to form a great scarp to Barr Buidhe and then runs round the summit of St Blane's Hill and reaches the sea at Uamh Capuill. Dykes of igneous rock are not uncommon round the coast, and are conspicuous at Ascog and Bogany Points near Rothesay.

The Kyles of Bute and Loch Striven are among the few places in Scotland where observations have been made on the movements of beach material. Lamont (1945) rightly assumes that waves are the main agent causing movement, and he analyses the accumulation of beach in relation to the fetch of the waves acting upon them. Reference to Fig. 16 will make it easier to follow his views. Waves approaching from the direction of Wemyss Bay have a considerable effect around Ascog; Toward Point shelters Rothesay to some extent, but waves from the north-east must have been largely responsible for the raised beach cliffs near Craigmore. The cliffs a little farther west come, and came, under the influence of waves from Loch Striven, and these and the waves from the north-east carry material into Rothesay Bay, a factor that presumably accounts for a large part of the town being built on raised beach. Ardbeg Point is not sheltered by Bogany Point from waves coming from the east-south-east. These waves, together with the material which has been carried down by the Ardbeg Burn, have led to an accumulation here, which does not move to the north since Loch Striven waves prevent this. Nevertheless in Kames

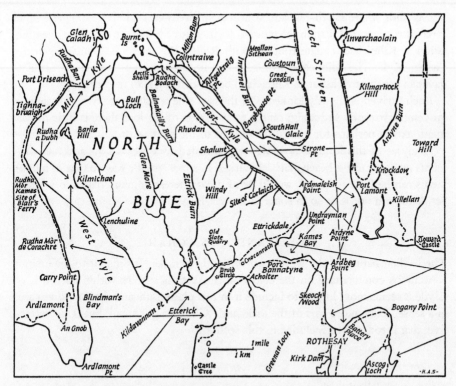

Fig. 16. Coastline of North Bute. Heavy line shows high-water mark; broken line, landward boundary of raised-beach deposits. Lines of fetch of waves at right angles to which accumulation is taking place also are indicated (after A. Lamont)

Bay there is a marked movement of beach material to the west, and in raised beach times there was a good deal of erosion of the slaty cliffs at Port Bannatyne – and this helped considerably in building the low ground joining north and south Bute. The drift of material on the north side of Kames Bay has also helped in this way. North of Ardmaleish waves from the Largs area affect the coast, but Lamont thinks that the coast between Undraynian and Ardmaleish Points is suffering erosion as a result of oblique waves both from Loch Striven and Bogany Point directions.

On the east side of the eastern Kyle Bargehouse Point is symmetrical. There is no stream, and the point seems to result from the balance of material travelling to this place from north-west and south-east. Farther north, at Colintraive, in '25-ft' times, when the Burnt Islands were covered, the waves travelling down Loch Riddon diverted the Milton burn to the south-east. Rudha Bodach (in Bute) has also been built up, but at the present time small waves from the Shalunt direction drive a little material toward Colintraive. In the mid-Kyle (Buttock Point to Kames) there is little detritus; there was more in raised beach times, and Rudha

4

Ban is built where north-east and south-west waves exerted approximately equal force. It is suggested that, in the western Kyle, Rudha Mór Kames and the accumulation at Kilmichael equate roughly with Rudha Bodach and Colintraive in the Eastern Kyle. The Kilmichael feature began to form in '100-ft' times. From Kildavanan northwards as far as Kilmichael beach-drifting is to the north.

Loch Striven is slightly kinked. On the east side south-westerly winds cause a movement towards Ardtaraig; on the western side of the loch there is little movement, except perhaps for a local movement to the south-east towards Ardbeg.

Analyses of this kind, if possible supplemented by observations and measurements made by using beach material marked in some specific way so that its movements can be fully recorded, would be of great interest in many places in western Scotland, and not least so in some of the sea-lochs.

ARRAN AND AILSA CRAIG

The broken nature of the west coast of Scotland makes it difficult to maintain a continuous commentary on the coast. Since Arran is wholly within the Firth of Clyde it seems more logical to include it in this place. Bute and the Cumbraes are, in a sense, much more part of the mainland. Ailsa Craig is small, but is so distinctive, that it too will be included in this section.

Arran

Arran is an island with an area of about 165 square miles (427 sq. km), and a relatively simple coast about 60 miles long (97 km). The geology and interior physiography of Arran have been studied in detail, but somewhat less attention has been paid to the coast. Attention need only be drawn to the general contrast between the high granitic area in the north, reaching 2,866 feet (874 m) in Goat Fell and the much less mountainous southern half. Between the two there is, from the geological point of view, the extremely interesting central ring complex. A feature particularly associated with the northern part of the island is the marked development of the 1,000-foot (301 m) platform around Goat Fell. The granite, the central ring complex, and the 1,000-foot platform do not directly affect the coast (Fig. 17).

The coast is of great interest. A marked raised beach platform is nearly continuous around the island; it may be absent at some headlands or covered with debris as at The Fallen Rocks near North Sannox. A glance at Fig. 17 shows that a variety of rocks reach the coast, and that the granite areas are wholly inland. The coastal rocks according to their resistance and disposition present cliffs of many forms but except at some headlands they are wholly fossil cliffs. The rock platform in front of them is primarily associated with what is referred to as the '25-ft' beach, but a clear distinction, in the platform, can seldom be made between that level and traces of a higher beach at about 40 feet (12 m).

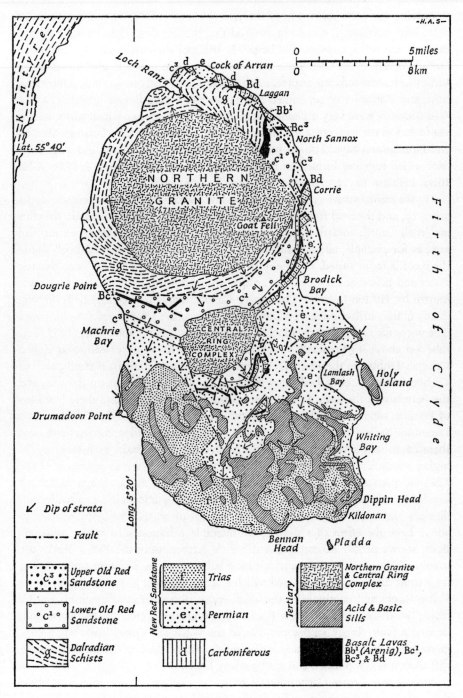

Fig. 17. General map of Arran (based on Geological Survey)

Leaving aside the formation of the 1,000-foot (301 m) level and any traces of other high surfaces, it should be realized that the existing inlets were, in higher raised beach times, considerably larger. In some of these there are remains of the '100-ft' beach. Then Glen Rosa and Glen Croy were sea-lochs and Brodick Bay had a much more intricate outline. Loch Ranza was then about one mile longer than now, and Catacol Bay penetrated more than half-a-mile farther inland. The two Glens Sannox were very different, especially the northern one which was a narrow sea-loch. On the west coast the greatest change would have been between Machrie and Drumadoon bays. The sea penetrated behind Torr Righ Beag and Torr Righ Mór which together formed an island. Deeper inlets were present in lower Glen Iorsa, Lamlash Bay, and on a smaller scale in other places.

On the north-western coast, that formed of Dalradian schists, the coastal slopes are steep, and the road follows the narrow coastal platform. The rocks in the cliffs are locally much folded; in places coarse pebbly bands are exposed in the cliff face, as for example, about a mile north of the Iorsa mouth. Beyond Loch Ranza the coast is more varied. On the north-eastern side of Loch Ranza, near Newton Point and below north Newton Farm, is an unconformity which was first made known by Hutton in 1787. The Lower Carboniferous (cornstone) beds dipping gently to the north-west rest on Dalradian schists with a steep dip to the south-east. The unconformity does not coincide with a marked gash in the rocks, but is some four feet above it (Tomkieff, 1953). A little farther to the north-east, near Rubha Crèagan Dubha, is the Fairy Dell, a small bay, which contains a small patch of Carboniferous rocks faulted down between the Dalradian rocks on the west and the Permian on the east. At the most northerly part of the island there is a mass of Permian sandstone enclosed within two narrow outcrops of the Carboniferous Limestone Series. The sandstones, bright red and false-bedded, dip seawards, and there have been several landslips which have left their marks as fissures in the higher ground. To the east of this sandstone outcrop various members of the Carboniferous are faulted down alongside the Dalradians, and are themselves much divided by smaller faults trending roughly north and south. In this Laggan section dips are high – 40° to 60° – and generally to east of north. The faults referred to above have the effect of repeating considerable thicknesses of strata. Millstone Point shows pebbly sandstones, and a little farther south the Fallen Rocks are masses of Old Red Sandstone which have fallen on to the platform and hidden it for a short distance. The Old Red which reaches the coast just to the north of the Fallen Rocks and continues to the south end of Sannox Bay is much broken by faults. South-east of the Fallen Rocks conglomerates appear on the shore and, beyond a fault, flaggy sandstones overlie the volcanics. Other faults and a conglomerate follow, but for nearly half-a-mile north of the North Sannox burn the cliff section is complicated and cut by many faults.

At and near Corrie Carboniferous rocks are exposed on the foreshore; these rocks are both sedimentary and volcanic. Several steep streams cut the cliffs hereabouts, and here, too, the rocks are much faulted. The cliffs are behind the main road and the village, and the Carboniferous rocks together with the Old Red Sandstone to the north and the New Red to the south offer a striking piece of coastal scenery.

The Permian reaches the coast along most of the remainder of eastern Arran. These rocks are unconformable to the older ones beneath. Between Corrie and Brodick the lowest division of these rocks is exposed on the foreshore, and consists of bright red false-bedded sandstones which, largely on account of harder ridges and veins, weather into honeycomb structures. There are also several dykes crossing the foreshore. Similar features show in the cliffs in which caves were cut. On the north side of Maol Donn the old cliff is noteworthy; the scarp above is largely the result of an extensive slip. Near Brodick Old Quay there are caves in the cliffs behind the '25-ft' beach. The former extensions of Brodick Bay were mentioned above. On the south side of the bay breccias, conglomerates and sandstone rest on false-bedded sandstones, and many faults and dykes cut the platform. Similar features are apparent in the cliffs. A major fault south of the pier raises the conglomerate so that it does not reach the shore again until the dip brings it down at Corrygills Point. To the north of this point for about half-a-mile in the old cliffs and in the raised and modern shores false-bedded sandstones are seen, and also curious ridges and spires like those found on the north side of Brodick Bay. Conglomerates and sandstones persist south of Corrygills Point, and the dip of the beds increases to as much as 60° near Clauchlands Point. Before that point is reached there is a sill of felsite, part of which is beautifully spherulitic. Clauchlands Point and Hamilton Rock are parts of a great crinanite sill. The north face of the Clauchland Hills, which end in the point, is the escarpment of the sill. The prehistoric fort, Dun Fionn, stands on a prominent scarp of the sill.

Lamlash Bay between Clauchlands Point and Kingscross Point is the seaward end of another crinanite sill, which reappears in the southern end of Holy island. Although the raised beach is present in the bay, it has been much cut up by the streams that drain into the bay, and it follows that in several places marine and fresh-water alluvia are inextricably mingled. The intermediate beach is well developed to the west of the present bay, the part that was open water in those times. Holy island is mainly a great sill of riebeckite-trachyte. The sill rests on sandstone apparent round much of the coast, except on the east and north-east. These sandstones, in the south, have been intruded by the crinanite sill. The island reaches 1,030 feet (314 m) and the total thickness of the sill is about 800 feet (244 m). In the sandstones several caves have been cut, e.g. St Molio's and Sheep Cave, and in places screes have more or less obliterated the platform. In Whiting Bay the general features resemble those in Lamlash Bay; the main road is forced close to the shore and the platform and old cliffs behind are present. Creagh Dhubh is a dolerite sill intruded in the sandstones, and from this point to the Dippin is a good section of the Upper Sandstones, but there is a considerable amount of shingle in this stretch. At Largybeg Point there is an interesting arch in the raised beach.

Beyond Largybeg Point red marls are the country rock, but for some miles igneous intrusions are of peculiar interest. Dippin Head is another crinanite sill. There are good cliffs between the two points, and at Dippin they are about 200 feet high (61 m). The sill is rudely columnar and rests on shales dipping to the north-east. Kildonan Castle stands on a 20-feet thick (6 m) sill of craignurite (see Plate 60). It is near here that the great dyke-swarm of south Arran begins. It is part of a much greater series connected with Skye, Mull and Rhum. In general the dykes trend north-west to south-east. The Arran dykes are basaltic and belong to a later phase in the volcanic activity. Along the whole of the south coast of the island they are close together and numerous, and they cross the shore platform and beaches like a series of groynes. They also cut the raised beach platform. They make the stretch of coast from Dippin Head to Sliddery unique in Scotland. Many other coasts, for example Muck and parts of the Ardnamurchan peninsula and Mull, show numerous dykes, but both in members and regularity, as well as in the way they rise above the beach in low and almost parallel black ridges, they stand apart. The view from Bennan Head is remarkable. The Head is the seaward end of a large intrusion of felsite, and shows good cliffs. West of Sliddery the dykes cease, and the coast is locally low and boulder clay reaches it. At Brown Head is another felsite intrusion which forms cliffs. To the north the cliffs fall in level and at Kilpatrick Point the igneous mass is buried beneath shingle. There are caves, including the Preaching Cave, in the felsite of the raised beach cliffs and there are well-marked bedding joints in the rock. Drumadoon Bay enclosed between Kilpatrick and Drumadoon Point is in the marls, and was the entrance to the former strait behind Drumadoon Point (see Plate 57). There is now a beach and some scars on it. Drumadoon Point is another quartz-porphyry intrusion; just behind it is a fine sill of columnar rock. The igneous rock is surrounded by a narrow belt of marls, and both south and north of the point several dykes cut the foreshore including a massive quartz-porphyry one at Cleiteadh nan Sgarbh. A little further north, at King's Cave, or rather caves, the sandstones are red or yellow and are intruded by pitchstones which show both on the foreshore and in the cliffs behind the raised beaches. From An Cumhann to King's Cave the '25-' and '50-ft' beaches are present, and make themselves apparent chiefly by the caves cut in the sandstones now above high-water mark. To the immediate south of King's Cave a gully followed by a stream marks a fault separating the Permian (north) from the Trias (south). About 500 yards south of Leacan Ruadha a pitchstone dyke makes a prominent feature on the raised beach platform. Machrie Water now reaches the sea near the middle of the flat area representing the old northern inlet of the strait behind Torr Righ. It is here that there are extensive remains of the '100-ft' beach as well as of the intermediate and lowest ones. The high ground reaches the coast again at the north end of Machrie Bay; a number of basalt dykes cut the shore and the break between the New and the Old Red Sandstones occurs a few hundred yards north of Auchagallon. There is no very noticeable difference in the appearance of the cliffs in the rocks, nor is there at the junction of the Old Red and the Dalradian rocks a little north of the mouth of Glen Iorsa, where there is a considerable extent of the '100-ft' raised beach behind the lower platform.

Ailsa Craig

Ailsa Craig (see Plate 61) is a conspicuous and isolated peak in the Firth of Clyde; it is nine and a half miles (15 km) west-north-west of Girvan. It has an area of 225 acres (10 ha) and its summit is 1,114 feet (340 m) above sea-level. It is made of a fine grained micro-granite; the hornblende in it takes the form of riebeckite. The plug was formed in the Carboniferous. On the north, south, and west the rock rises steeply from the sea; the east is somewhat less steep, and the lighthouse is built on a triangle of relatively flat land where there is a considerable amount of coarse shingle. There are several dykes of dolerite which weathers more easily than the granite, and so the dykes usually mark caves and small ravines. They are mostly on the south and west. The Swine Cave, on the north, is cut in a dyke fifty-seven feet (17 m) thick; most of the dykes are about six feet thick. In the ice age it was enveloped, and a thin spread of glacial detritus remains. Apart from the low ground on the east, the lower slopes of the plug are cliffed but, as is shown so well in Vevers's (1936) map (Fig. 18), the cliffs, except at Stranny Point on the south-west, are above high-water mark. This presumably means that they are fossil cliffs and that the caves, except perhaps the water cave at Stranny Point, are also fossil or partly re-worked at the present day. This is borne out by the vegetation on the sides of the caves. The following plants are found: *Asplenium marinum*, *Umbilicus rupestris*, *Marchantia hemisphaerica*, *Mnium hornum* and *Porotrichum alopecurum*. On the beach above high tide, a distinctly rupestral habitat, the plants include *Armeria maritima*, *Cochlearia officinalis*, *Senecio jacobaea*, and *Bryum alpinum* and also a number of lichens; and one moss, *Rhacomitrium lanuginosum*. At a lower level the upper limit of the intertidal zone is marked by a dark band of *Pelvetia canaliculata*. There is often a talus slope at the foot of the cliffs, and on the south-west the rock known as Little Ailsa is a stack which originated when the sea-level was higher; it is, however, still washed by the waves (Plate 61).

THE FIRTH OF CLYDE, GALLOWAY AND THE SOLWAY FIRTH

On both sides of the upper part of the Firth of Clyde, between Dunoon and Toward Point on the west and Cloch Point to Skelmorlie on the east, the raised beach platform corresponding with the '25-ft' beach is more or less continuous, but often very narrow. Cloch Point is a basaltic porphyritic mass interbedded in Carboniferous sediments. The raised beach widens at Inverkip. Wemyss Point, a projection in the upper Old Red Sandstone, is surrounded by a narrow fringe of

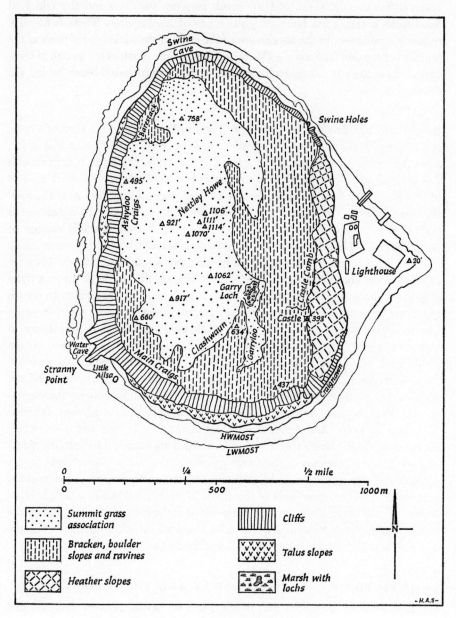

Fig. 18. Ailsa Craig (after H. G. Vevers)

raised beach, and just south of Wemyss pier there are several north-west to south-east dykes cutting the coast. At Largs the raised beach deposits widen to a mile or even more, and two levels are found.

The country rock adjacent to the coast from Lunderston Bay to Ardrossan is, with the exception of certain dykes and plugs, the Old Red Sandstone. It is usually composed of red and red-grey sandstones, marls and conglomerates. At Farland Head, the main prominence on this part of the coast, the strata are on end, and strike nearly north and south. This may indicate an anticlinal axis or a fault line. Raised beach phenomena are almost continuous, and locally impressive, along this coast. As in so many other places, the beach platform is accentuated by the main road, except where it runs behind the Farland Head peninsula. Beaches or platforms at different levels can be recognized. One at about 15 feet (5 m) is well developed between Largs and Fence Bay, and again from Sea Mill to Ardrossan. A slight cliff, usually in the form of a grassy slope in boulder clay, and somewhat steeper in solid rock, backs this terrace. The next higher one at about 25 feet (8 m) is clearer, and is nearly everywhere a rock platform, and the old cliffs behind it are steeper and higher. This beach is conspicuous north of Farland Head and Portencross. Along this almost north–south coast it is less than 70 yards (64 m) wide, and well-wooded cliffs which may reach 300 feet (91 m) stand behind it. Beyond Hawking Craig the cliffs fall in level, and the low ground around Hunterston House was a bay in which there are traces of a higher beach at about 50 feet (15 m) or a little less. It will be easily appreciated that there is not always a clear distinction between the 15- and 25-feet levels.

There are extensive sand flats south of Fairlie, and some blown sand at the southern end of Fence Bay. The sands are often boulder strewn. Black Rock is a small volcanic vent in the Calciferous Sandstones, and Horse island is an isolated fragment of Upper Old Red Sandstone. The other detached rocks, and also the rocky outcrops on the foreshore north of Ardrossan, are similar.

Great and Little Cumbrae Islands (Tyrrell, 1926) present several points of interest. They are situated between southern Bute and Farland Head, and the deeper water passage is on their western sides. The main part of Great Cumbrae is built of Upper Old Red Sandstone. On the west side the dip is to the south-east, on the east it is similar, but less constant. Almost all round the island there is a marked beach platform with old cliffs behind it (see Plate 56). It is followed, and accentuated by the road which encircles the island. At Millport and on the larger islands in its bay there are outcrops of Calciferous Sandstone. There are also intrusive Carboniferous rocks in the form of dykes on the coast, especially in the north-east, and as two small bosses, The Clach and The Miller's Thumb, in Millport Bay. The straight eastern side of Millport Bay coincides with an important fault which is continuous in a south–north direction through the island. The highest point on the island is 415 feet (126 m). The numerous dykes, sills, and bosses of igneous rock give the island a ribbed appearance and also a rough and diversified surface. Aird Hill, at the north-eastern end, was a separate island in high raised beach times. Little Cumbrae is almost wholly composed of Lower Carboniferous trap rocks, locally columnar. The successive flows make good terraces and are arranged in a shallow syncline, the axis of which trends approximately north-north-west. Older and separate flows form Castle island and Trail island; the latter is awash at high water. Dykes of basalt and dolerite also form the Broad Islands, most of which are awash at high water. A good raised beach platform encircles the island. There are six or more caves in Little Cumbrae; the largest is the Monk's Cave which is about 100 feet long (30 m), but only three in width. The others are small, but in Waterloo Cave three distinct layers of shell and bone deposit were found (Marshall, 1939). In Great Cumbrae there is only one cave of any size in the cliff of Farland Hill. The fine development of platforms cut in solid rock in these confined waters deserves much more consideration than has been so far given to the problem – it is only part of the puzzle that is present along so much of the west coast of Scotland.

From Ardrossan–Saltcoats to Ayr the coast is generally low, and there is a remarkable development of raised beaches and, locally, of sand dunes. The projections on either side of South Bay on which the towns of Ardrossan and Saltcoats are built, are both low and largely covered by raised beach but owe their existence in part to outcrops of Carboniferous rocks into which some dykes and sills of post-Old Red Sandstone age are intruded. There is a wide sandy foreshore in South Bay. South and east of Saltcoats there is a long sweep of sandy foreshore extending as far as Troon (Richey, Thomas et al., 1930). Industrial development has greatly altered the appearance of the northern part of Irvine Bay (see Fig. 19 (b)). The area covered by raised beach deposits which, at Ardrossan, is a mile or more wide, extends inland to Kilwinning, runs behind the town of Irvine, extends almost to Kilmarnock up the Irvine river and narrows to about a mile near Troon, and expands again behind Prestwick and Ayr (Fig. 19(a)).

In this extensive area several levels of beaches are recognized. In this paragraph the heights given *refer only to this district*. At Saltcoats a level of about 110 feet (34 m) is formed. Between Saltcoats and Kilwinning levels above 40 feet (12 m) abound, and extend perhaps two miles beyond Kilwinning. Hereabouts the old coast is cut in hummocky boulder clay, which was eroded relatively easily. Up the Irvine valley the old beaches reach to 100 + feet (30 + m), although 80–90 feet (24–7 m) is more common. Brick-clays of marine origin occur, e.g. at Ban Rhead, at about 80 feet (24 m) OD. There are beach-like features at 95 and 75 feet (29 and 33 m) between Drybridge Station and Dundonald. Near Warrix a 'forest bed' is known in the Irvine river, and 'overlies sandy clay that may belong to one of the [higher] beaches . . . and is covered by sand that is probably part of the 40-ft. [12 m] beach'. However, the great part of the raised beach deposits in this area are less than 40 feet OD. There is often a small cliff marking their inner edge. This is not continuous, but can be traced at frequent intervals from Ardrossan to west of

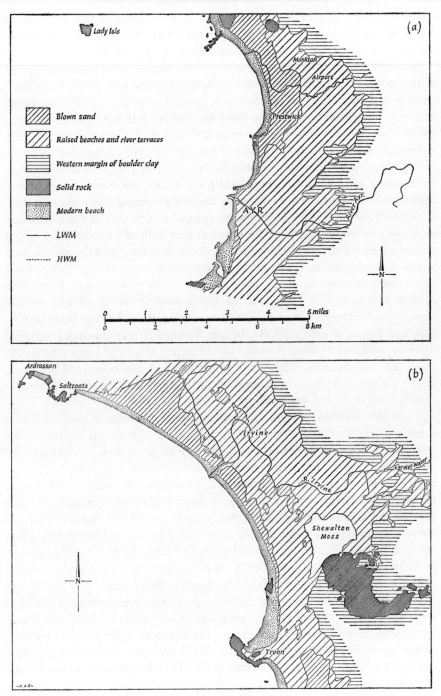

Fig. 19. Sketch map of the coast at (a) Ayr and (b) Irvine (based on Geological Survey)

Dundonald. This 40-foot level was regarded by the Geological Survey as part of the '25-ft' beach. Sheets 22 and 14 of the One Inch (Drift) Geological Survey show the extent and distribution of these features very clearly.

John Smith (1896) made some interesting comments on these beaches in a paper of the Irvine Whale Bed. He noted that between Irvine and Troon there was a series of low ridges running approximately parallel to the curve of the bay. South of Meadowhead and along a line from Shewalton Moss towards the shore he counted 26 ridges. They were round-backed, usually about two feet above the intervening hollows. Some of the larger ones reached seven or eight feet above the swales. Occasionally they were double, in the sense that a slight hollow ran along the crest of a ridge. They do not reach the present coast, and there is a considerable width of level ground between them and the modern dunes. Smith regarded them as wave-formed, and thought they were situated in that part of the former bay most open to storms. They are for the most part built of fine material, much of which he thought might have been brought down from the glacial deposits through which the Irvine flows. There are no similar ridges being formed today, and a broad plain, as noted above, separates them from the present coast. At the time of their formation it is likely that the Irvine reached the sea somewhere near Shewalton; the Garnock, which now enters the sea about halfway between Salt-coats and Troon, then debouched at or near Kilwinning. The Garnock is deflected south for about two miles. It would be interesting if we had more information about the changes in the joint mouth of the Garnock and Irvine; unfortunately, from a physiographical point of view, the whole area has been much altered by development in the last few decades.

The southern end of Irvine Bay is formed by the small peninsula on which Troon is built. This, like that at Ardrossan, is an outcrop of post-Old Red Sandstone dolerite sills largely covered by raised beach deposits. A wide sandy beach has accumulated to the north suggesting, as does the deflection of the Garnock, a southerly drift of material at the present time. Between the Garnock mouth and Barassie (the northern part of Troon) the coast is in a semi-natural state, since between the railway and the beach there is a golf course. The Stinking Rocks at Barassie are basaltic and awash at high water. Lady Isle, two to three miles (3–5 km) west-south-west of Troon, is low and formed of post-Old Red Sandstone intrusives. Meikle Craigs and Little Craigs are of similar origin, but like the Stinking Rocks are in or close to the beach and are awash at high water.

In the bay between Troon and Ayr the raised beach phenomena are similar to those north of Troon (Eyles et al., 1949). Beaches at various levels are recognized. A well-marked beach exists at about 80 feet (24 m) (an average figure); it may reach 90 feet (27 m) or fall to 70 feet (21 m). It is most conspicuous between Monkton Hill and the Heads of Ayr. In this stretch the old cliff is cut in boulder clay, and because of the softness of the clay the cliff form is poorly developed. South-west of the Heads of Ayr, where hard and resistant rocks reach the coast, the level is seen in occasional rock platforms as at Dunure, Culzean Bay (traces) and especially around Maidenhead Bay. A beach at about 70 feet (21 m) is present only to the south of Maidens. The 40-foot (12 m) level is the best developed; Prestwick and Ayr are built on it. As in the higher levels, its inner margin is poorly developed in glacial deposits, but in the resistant rocks farther south a marked cliff is present.

In raised beach times, especially when the higher beaches were being formed, the coastline of central Ayrshire was deeply indented, and the rivers Garnock, Irvine and Ayr were tidal for some distance. The cliffs in boulder clay were somewhat more prominent since they were being actively eroded. As the level of the sea fell extensive plains were exposed, and at still-stands, or re-advances of the sea, new cliffs were cut. But as the sea retreated to its present level, the old cliffs were worn down, and today are only conspicuous in areas of resistant rock. It follows from this that the extensive outcrop of lavas forming the high ground to the south of Bracken Bay was then even more prominent than it is today.

The present coast south of Ayr is diversified and locally attractive. Before the Heads of Ayr are reached the coast, apart from the small outcrop of agglomerates on which Greenan Castle stands, is low, and dykes cut the fringe of the Calciferous Sandstones which lines the foreshore. Greenan Castle stands on the cliff edge. It is known that, about the beginning of the nineteenth century or a little earlier, a carriage could be driven round the castle. This gives some idea of the rate of erosion (W. Burns, *Trans. Geol. Soc. Glasgow*, 8, 1884–88, 287). The Heads of Ayr represent a volcanic vent in sediments of Calciferous Sandstone age. The vent makes a sea-cliff about 200 feet (61 m) high and half-a-mile long. The foreshore is rocky, and the platform is cut in ashes intruded by dykes, alongside the vent, and in the calciferous beds on either side of the vent. The double aspect of the vent is produced by the weathering of the cement stones which are exposed on the shore and break the continuity of the agglomerates. In Bracken Bay there are traces of a beach at about 40 feet (12 m), and at about 10 feet (3 m) there are old sea-caves which are probably associated with a beach at about 20 feet (6 m) which can be traced on either side of Dunure. At this place higher level beaches up to 80 feet (24 m) are present. These may be followed in Culzean Bay when the coastal slopes are somewhat gentler and cut in boulder clay. The same level is also found in Maidenhead Bay, where with the lower beach there is a width of about half-a-mile of flat ground which extends into the disused airfield and the golf course at Turnberry which are on the lower beach. The offshore rocks in Maidenhead Bay are lavas.

The coastal tract from Turnberry to Girvan averages about 30 feet (9 m) OD, the main road follows its inner edge which is often marked by an old cliff which is cut in rocks of Lower Old Red Sandstone age. The rocky outcrops on, or just off, the beach are also in these rocks. There is much blown sand on Turnberry warren, now Turnberry golf course. The Silurian rocks first reach the shore a little to the south of Girvan, but they are for the most part covered with raised beach deposits which extend as far as Ardwell Bay. However, at various places, e.g. Horse Rock and

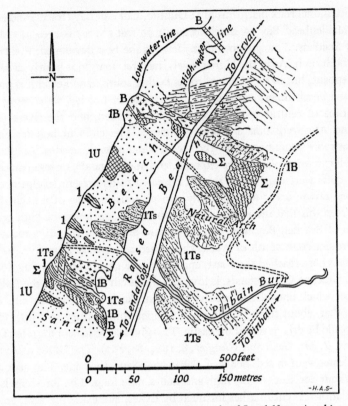

Fig. 20. Strata on the shore at Pinbain, one mile north of Lendalfoot, Ayrshire. 1, Arenig black shale; 1B, Diabase lava; 1Ts, Agglomerate; Σ, Serpentine; 1U, Intrusive basic rock (Silurian); BB, Dolerite dykes (Tertiary) (based on Geological Survey)

Craigs Kelly, there are outcrops of Llandovery rocks which are thrown down by a strike fault parallel with the shore. Between Shalloch Mill and Kennedy's Pass the graptolite flagstones show a series of rapid folds in which the beds are often vertical or nearly so, and locally much contorted. Kennedy's Pass affords an excellent viewpoint. From it the relatively narrow rock platform at the point and to the south of it can be seen to widen out into the extensive beach plain toward Girvan. The nature of the folding is exposed in the rock platform, and also on the walls of numerous gullies eroded by the waves along joints, the general trend of which is to the north-west. Although there is an important fault in Ardwell Bay it seems to have little effect on the physiography. On the other hand, about a quarter-of-a-mile south of Kennedy's Pass wrench faulting plays an important part on the foreshore. About 600 yards (540 m) north of the Pinbain burn, there are prominent cliffs, and somewhat nearer Kennedy's Pass the conglomerate forms stacks, but where a fault brings conglomerate and volcanics together there is a small north–south hollow. Fig. 20 shows the details at Pinbain Shore.

Nearer Lendalfoot* the old cliff recedes inland a little way, and the Water of Lendal reaches the sea in what was a small bay, now filled with raised beach deposits (see Plate 58). The main road follows the cliff foot as far as Burnfoot; the coastline is for the most part cut in hard and brecciated lava; on the seaward side is a hollow running parallel with the cliff, and beyond the hollow there are some stacks in serpentine. As far as Balcreuchan Port the coast cuts the strike obliquely, and the Port itself is a cove enclosed by steep cliffs. From this place to Bennane Head the shore platform shows great and irregular folding in lavas, pyroclastics, shales and cherts. There are several conspicuous stacks, including one of folded radiolarian chert at Bennane Cove, and on a stack in Port Vad pillow structure can be examined. About half-a-mile south of Bennane Head the shore platform widens and is followed by the main road as far as Ballantrae. The old cliff along this stretch is nearly a quarter-of-a-mile inland. The seaward part of the platform is similar to that farther north. The River Stinchar is deflected a little way to the south at its mouth. The map suggests that there have been several changes in this outlet, but there seems to be no account of them.

To the south the coast is steeper, and but a very narrow platform separates the old cliff from the sea. Between Downan Point and Currarie cliff there is a fine section of Silurian lavas; at Dove Cove a synclinal fold shows sedimentaries bounded by lavas, and about 300 yards (274 m) farther north a deep gulley has been cut along a dyke. At Wilson's burn beaches at approximately 100 and 25 feet (30 and 8 m) are present. Currarie Port is a deep-cut cove where the Currarie Glen and the Shallochwreck burn reach the sea. As far as Breakness Hole the cliffs are steep. At the Hole red cherts embedded in sandstones and greywackes, much folded, run out to sea. In a small cove a little farther south there are three anticlines of cherts in mudstones, and there is a constant repetition of mudstones and greywackes as far as Glendrishaig. Portandea is a small bay on the north side of which is a sloping beach which merges upward into raised beach which separates a ridge of rock on its seaward side from the old sea cliff. This hollow can be followed for about 350 yards (320 m). This section of coast ends at Glen App which reaches the coast at Finnarts Bay. At Finnarts Point the coast is steep and broken by a narrow cleft. Glen App itself is on the line of the main Southern Uplands Fault.

This fault traverses Loch Ryan and cuts through the northern part of the peninsula to the west, along a line running from Lady Bay to Dounan Bay. It will, therefore, be more convenient to consider the northern and western coast of Wigtownshire before dealing with Loch Ryan and Luce Bay. The coast between Milleur and Corsewall points is rocky, and for the most part consists of a low sloping surface without marked cliffs. Nearer Corsewall point the raised beach platform is well developed. Coarse conglomerates are often prominent, as in the

* Between approximately Girvan and Luce Bay frequent reference is made to Peach and Horne (1899).

lighthouse cliff. Massive grits continue for about a mile south of the lighthouse; they are much folded, often nearly vertical, and traversed by many faults which usually trend north-west to south-east. These are succeeded by mudstones in which there are hard ribs. Conglomerates come in again opposite Genoch rocks, to be followed by more mudstones and hard bands as far as Portnaughan Bay where the smashed rocks indicate a fault. For a little beyond the fault the south-west dip prevails, but at the southern headland of the bay the dip is to the north at 5°–17°. The same general sequence may be followed to Dounan Bay, the hard ribs making characteristic features. Because the Glen App fault occurs in Dounan Bay, the beds are much shattered. Greywackes and shales run on to Salt Pan Bay; sometimes there is overfolding in the cliffs as at March Pot and Slocknamorrow (about 300 yards (274 m) north of Portobello). Again, the alternation of greywackes and pebbly grits, often with shaley portions, can be followed to within about a quarter-of-a-mile south of Juniper Point. Little change, apart from local overfolding, occurs in Slouchnawen and Weeport bays. Near Saltpans Bay much folded shales extend along the shore; most of the folds are inverted. Broadsea Bay is rather similar. A prominent narrow ridge of igneous rock runs out from the cliffs and forms skerries. At Dove Cove the decomposition of calcareous nodules takes place more readily than that of the rock in which they occur and produces a honeycomb appearance. The shore from Knock Bay to the south of Killantringan Bay shows the usual greywackes and shales. In all this stretch the cliffs are fossil and stand a little back from the sea. In places, for example, Saltpans Bay and Killantringan Bay, there is a more marked re-entrant, and the cliffs run behind a raised beach which rests on the wave-cut platform. Although the whole of this stretch of coast is simple in general outline, there is much detailed irregularity; the harder ribs of greywacke or other rock make minor headlands, and the softer rocks are worn slightly back. But much of this applies to the fossil cliff; the modern sea is modifying slightly the wave-cut bench.

Black Head is somewhat more prominent and is followed southwards by two minor bays, Portamaggie and Portavaddie, which face south-west and carry small beaches. In Port Mora there are two small caves in a faulted zone of steeply dipping greywackes. One cave corresponds with a waterfall. Ouchtriemakain Cave is 21 feet (6 m) above present sea-level. Portpatrick harbour is picturesque; the Portpatrick rocks are mainly greywackes with seams of siltstone and mudstone, and the wrench fault which runs north-west and south-east, has caused much shattering so that for about half-a-mile north of the harbour sedimentary structures are often obscured. In Morroch Bay, enclosed by dykes, there is intense minor folding and faulting; the cherts at the south end of the bay are thought to be the oldest beds in the Rhinns. A headland of greywackes separates Morroch Bay from Port of Spittal Bay in which there is a shingle beach. The bay itself is cut in shales. Its southern headland is again in cherts, shales, greywackes and siltstones, and these, repeated by many minor folds, run onto Portayew. In this bay is the Ordovician–

Silurian boundary. Cairngarroch Bay is similar, but is wide open. It is bounded on its south by Cairnmon Fell, Scarty Head and Money Head, where greywackes and grits occur in the cliffs (see Plate 63).

Southwards from Money Head the coast is more irregular, and is broken by several large bays in which there are some good beaches. The raised beach platform and the fossil cliff occur all along the coast. Ardwell Bay is a good feature and Float Bay, a mile or two farther north, is cut more deeply into the land. Drumbreddan Bay is double in the sense that it is divided by a rocky promontory; the bay itself is in the Moffat Shales. Port Nessock (Port Lagan) Bay is wide open, and is backed by a good beach and the fossil cliff. Clanyard Bay is generally similar; Moffat shales outcrop in the bay and its midpart is sandy. Between these two bays the coast is relatively straight and resembles several other stretches farther north. In Portencorkrie the cliff is absent at the head and rather sloping elsewhere. Crammag Head projects a little from the general line of the coast, but from there southwards to West Tarbet the coast is fairly steep and between Laggantalloch Head and Portdown Bay, nearly a mile south of Crammag Head, a small mass of granite reaches the coast and gives a different type of scenery. Nevertheless, the fossil cliff can be traced in the granite as elsewhere. Between West and East Tarbet the ground is below 70 feet (21 m), and the Mull of Galloway was an island in high raised beach times. A narrow band of felsite follows this line of low ground; the Mull, which reaches 278 feet (85 m), is made of rocks similar to those immediately north of the isthmus.

The low belt between West and East Loch Tarbet is one of three such features in the Rhinns. The other two run between Port Nessock (Port Lagan) Bay and Terally Bay, and between Clanyard Bay and Kilstay Bay. The west coast is usually steep, and at Dunman the old cliffs are about 400 feet (122 m) high. The general slope of the ground is to the east and all the coast between East Tarbet and Sandhead is much lower. The fossil cliff can be traced in many places but is not as conspicuous nor as high as on the west coast. For long stretches on the east coast the beach is fairly wide, but often coarse and covered with boulders. There is some sand in Maryport Bay, but from there until Sandhead Bay, the south-western end of Luce Sands, there is relatively little sand on the foreshore. Occasionally there are rocky platforms, as between Chapel Rossan and Portacree. All this coast is sheltered from the prevailing winds.

This part of Scotland enjoys a mild climate; along the coast of Wigtownshire the average minimum night temperature in January is 35 °F or more, and maximum day temperatures in July about 64 °F. Vegetation is luxuriant on and near the coast. On the cliffs the halophytes include *Crithmum maritimum*, *Silene maritima*, *Plantago* spp., *Cochlearia officinalis*, *Armeria maritima*, *Tripleurospermum maritimum*, *Spergularia maritima*, *Sedum acre*, *S. anglicum*, *Lotus corniculatus* and in rocky pools at the foot of cliffs, *Glaux*. The non-halophytes are prolific and numerous; shore and inland types mingle. 'On the summits of the cliffs and on ledges a little below the summit the flora depends on the nature of the vegetation type that comes down to and over the cliffs. In places grassland reaches the cliff edge, and here the soil is thin and dries quickly, the flora being that of a dry pasture' (Sutherland, 1925).

In high raised beach times the Rhinns peninsula was an island. The flat ground between Loch Ryan and Luce Bay consists largely of raised beach deposits. The southern part of this area is made up of the extensive dunes and beach of Luce Bay (see below). Most of the town of Stranraer is built on Triassic rocks which also form a band a mile or more wide along the western side of Loch Ryan which is sheltered from all but northerly winds. Along the eastern side a ribbon of raised beach material fringes the loch to a little beyond Carn Point, but a narrow platform carries the main road up to the Water of App. The cliffs are steep, and the grey-wackes and shales dip steeply, and are often nearly vertical. They are cut by many small strike faults which run at right angles to the shore, and are often the site of caves which vary much in size. On the western side the small peninsula at Kirkcolm is a raised beach, and is prolonged into the spit called the Scar. The western slopes of the loch are gentle. Along this western shore the Permian (? Triassic) breccias rest conformably on Ordovician and Silurian rocks, and are cut by many faults along which small caves have been excavated. The fact that on both sides of Loch Ryan so many caves have been cut in the fossil cliffs is an indication of the greater power of the waves in former times. It is not easy to understand how caves, even along lines of weakness, could be so readily cut in such sheltered water.

In Luce Bay a belt of sand, often exceeding half a mile in width, is exposed at low water all the way between Sandhead and Glenluce (Smith, 1889–93). (The Royal Commission on Coast Erosion (1911) commented on the erosion between Sandhead and Drummore; the main road was partially cut.) Behind the beach is an extensive area of dunes. On the dune-beach border strand plants such as *Salsola kali* and (*Honkenya*) *Arenaria peploides* are abundant. Sea rocket and sea holly grow well in front of *Ammophila* dunes. *Calystegia soldanella* is significant, and various thistles, including *Carlina vulgaris* and *Cirsium arvense*, are common. Torr's Warren covers more than four square miles (10 sq. km). There is clear evidence of old shorelines. Dunes reach the 50-foot (15 m) contour; the sand is sub-angular, and is arranged in a series of ridges more or less parallel to the shore. The inner dunes are fixed; the outer are mobile to some extent. Farther in there is a strip of marshland which remains wet even in summer. Behind it are more ridges and minor strips of marsh. This succession does not hold everywhere since it is subject to modification by wind, by rabbits (perhaps more in the past than now) and it is in part a Ministry of Defence area. Ling and heather are abundant on fixed dunes: *Juncus communis* and *Scirpus repens* occur in many marshes. Other plants include *Galium palustre*, *Hydrocotyle vulgaris*, *Littorella lacustris* and *Poly-trichum*. Pearlwort (*Sagina procumbens*) is found in the rather drier marshes. The vegetation is locally more luxuriant where a small burn flows in the dunes – Marsh pennywort, *Juncus communis*, even *Potamogeton natans*. Bracken is usually abundant, and so too is bog myrtle.

There is salt marsh at the mouth of Piltanton burn. The plants are those common to new marshes, *Salicornia europaea*, *Glaux maritima*, *Spergularia*. The algae include *Rhizoclonium*, *Vaucheria* and *Phormidium* (Sutherland, 1925).

In Luce dunes and in several other places around Luce Bay, Wigtown Bay and farther east there are Mesolithic sites. Coles (1964) suggests that these sites cannot be 'directly dated in relation to the raised beach, but on the basis of probable midden material, and proximity to the upper margin of the high sea-level, they are believed to date from a time of maximum or near-maximum sea . . . and in absolute terms within the brackets 5000–3500 B.C.' He prefers to describe the artefacts as 'south-west Scottish coastal Mesolithic'; this does not make any direct link with other groups. In a detailed study of Low Clone, a locality a little north of Port William, Cormack and Coles (1968) show that the site is at the top of the Heugh (the old cliff) at about 50 feet (15 m) OD and 500 feet (152 m) from present high-water mark. It is near a stream which cuts through the Heugh. It was concluded after a careful study of the artefacts that this and similar sites 'may represent strand-looping activities of groups who also occupied inland areas, and perhaps islands, at appropriate seasons'.

The eastern coast of Luce Bay affords another example of a main road built along a raised beach platform backed by a cliff. It is not a particularly attractive shore, and many parts of the beach are covered with cobbles and boulders. The slopes above the cliffs are usually gentle except near Corwall Port, and the rocks composing the cliffs closely resemble those in the Rhinns of western Wigtown-shire and, like them, show numerous folds. The fossil cliff, again like those farther west, is cut by small gullies and caves, and shows differential weathering. Between Burrow Head and Isle of Whithorn the shore shows a good section of Hawick beds, mainly grey and purple shales. Burrow Head is a rocky headland. Thieves' Hole and the Devil's Bridge are cut in graptolitic shales. The coast north of Isle of Whithorn is comparatively low. There is frequently a rocky bench, very well developed in Portyerrock Bay and Port Allen. Along much of the shore there is a fossil cliff, and the major inlets, Cruggleton (Rigg) Bay and Garlieston Bay, mark the outcrop of bands of softer shales and flags which are more easily worn away. This was the view expressed in the survey Memoir of 1878; it would be interesting to know the relative parts played in their formation by marine and subaerial denudation. In both bays there are boulder-strewn beaches, and at Eggerness Point and northwards to Innerwell Port a rocky beach. Beyond this place a great flat of sands fills the upper part of Wigtown Bay. Along the western side, bordering the disused airfield, and near the Moss of Cree there is a good deal of salt marsh. The Moss of Cree and the inner part of the airfield is raised beach, although there is also a large spread of peat in the middle part of the Moss of Cree. On the east side of the bay, south of Creetown, there is a narrow fringe of raised beach, once again followed by a main road. Two small outcrops of granite make minor features on the coast (see Plate 62).

Beyond Ravenshall Point the coast is far more broken and picturesque. Fleet Bay, filled with sands exposed at low water, is followed by a spread of raised beach extending up to Gatehouse of Fleet. The indented and rocky coast from Fleet Bay to Kirkudbright Bay deserves far more study than it seems to have had. The Islands of Fleet are good features and Ardwall and Barlocco are approachable on foot at low water. Kirkcudbright Bay, like Fleet Bay, is largely sand filled. The east side is in well-bedded greywackes, grits and flags. These beds also outcrop on the Meikle Ross peninsula, and Little Ross Island at the western entrance of the bay. Gipsy Point, on the east side of the mouth of the bay, is in coarse grits and conglomerates; in Howwell Bay there is a good deal of faulting in the shales and there are also some dykes. In Mullock Bay there is some beach, but the skerries are perhaps more noteworthy. Conglomerates and shales outcrop at Netherlaw Point. A little farther east the Calciferous Sandstone Series forms a strip along the shore, and the very broken coast containing Auchencairn Bay, Orchardton Bay, and Rough Firth is cut back into the western parts of the Criffel granite. The bays are all sand filled and at their upper ends there are small expanses of raised beach. The craggy and broken coast between Rough Firth and Sandyhills Bay is another part of the Calciferous Sandstone Series. There are traces of fossil cliffs. The whole setting of the coast between Roscarrel Bay and Sandyhills Bay is good, and the recent development at Rockcliffe and the holdings of the Scottish National Trust emphasize its attraction.

Farther east the coast is entirely different. It is a low coast; much of it is marshland, either in its natural state or reclaimed. Most of Preston Merse is reclaimed; it rests largely on marine sands some of which may be part of a low raised beach. There is some higher ground from Southerness Point to Barron Point and again near South and North Corbelly. But strips of marsh fringe the coast which is the sand-filled mouth of the Nith. On the east side of the estuary is Caerlaverock Marsh which has been studied in some detail, and may perhaps be taken as typical of the Upper Solway Marshes, except that it is rather more exposed than others (Fig. 21).

Caerlaverock Merse (Marsh) (Plate 64) is about six miles (10 km) long, measured along its outer curve between the Nith and Lochar Water (Marshall, 1962b). It fringes the raised beach which forms the lowland of Caerlaverock. The beach shows two levels, an upper one at about 23 feet (8 m) OD separated by a four- to five-foot (1.2 to 1.5 m) bank from a lower one. Marshall thinks that both beaches were at one time salt marsh since old creeks on them can be seen on aerial photographs. These marshes are nearly all composed of fine sand (grains 0.2 to 0.02 mm); the remainder is fine silt and clay. Today only exceptional tides cover the higher beach; the lower is inundated somewhat more frequently. Erosion is mainly concentrated on the south-west part of the marsh. In the thirteenth and fourteenth

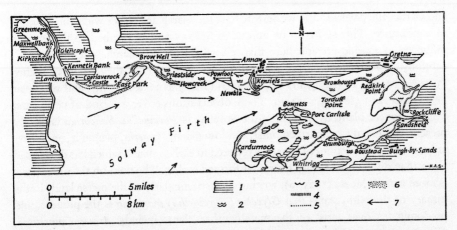

Fig. 21. The coastlands of the Upper Solway. 1, Inner edge of raised beach; 2, 'second' warp terrace; 3, 'first' warp terrace; 4, bank separating two warp terraces (where present); 5, inner edge of modern alluvium, i.e. salt marshes; 6, areas of accretion; 7, direction of drift (after J. R. Marshall)

centuries the moat of Caerlaverock Castle was filled by the sea. The castle, however, is defended from erosion by scars of solid rock, and in early centuries there may have been but little marsh around it. The modern marsh is terraced; usually three steps are present 12 to 24 inches high (30 to 61 cm). This has given rise to two points of view. Similar features on the Cumberland coast have been interpreted by the Geological Survey as indicative of changes of level. Marshall thinks this is improbable, and regards the steps as having been produced by meandering of the marsh channels; nevertheless she agrees that small changes of sea-level must be taken into account.

Marshes develop quickly in the Solway, and sand banks in favourable localities are soon colonized by plants. It seems probable that the whole marsh seawards of the lower raised beach at Caerlaverock has developed in a maximum of 140 years (i.e. before 1962). Three main phases can be traced in the growth of the marsh. Before 1856 there was accretion under the lee of Saltcot Hill; then, until 1898, accretion took place mainly at Bowhouse Scar, and subsequently up to 1927 to the west of that Scar. The oldest part of the marsh is at the eastern end. This is higher and carries a somewhat different vegetation. In 1960 there was severe erosion between Midtown Creek and Kenneth Bank, accretion east of Midtown Creek, and changes in the Lochar Water. Since there is a cover of plants in summer, it follows that accretion is mainly at that time of the year, whereas erosion is more likely to occur in winter.

Marshall points out that the form of the marsh indicates a drift of material to the east. The first diversion of the Lochar Water was the result of a shingle spit, in raised beach times, which formed near Castle corner; the process continues. The Lochar accounts for some erosion near its mouth. A glance at a One Inch map shows that the creeks in the western half of the marsh are rather wide and stumpy in appearance. This seems to be unique in Solway and is ascribed to the greater influence of marine erosion here than elsewhere on the marsh. Accretion rates, carefully measured, vary from place to place. At East Park they reach 1¼ inches (3 cm) a year at 15 feet (5 m) OD, 0.5 inch (1 cm) at 16 feet (5 m), and nil, or at best a trace, at 17 feet (5.2 m) OD. These observations cover a period of two years.

The pioneer plant, as is usual in the Solway, is *Puccinellia*. *Salicornia* is locally present, but is not a primary colonist. On low marsh *Armeria*, plantains, and *Aster* are common. On account of the steps referred to earlier, marsh which has developed since 1940 differs floristically from that formed earlier. The flora on the older parts is more varied. Morss (1925–6), writing more particularly of the merse lands in the estuary of the Nith, states that *Glyceria* (*Puccinellia*) *maritima* is the pioneer, and predominates over most of the marsh; other plants include *Armeria maritima, Glaux maritima, Plantago maritima, Cochlearia officinalis, Spergularia, Aster tripolium, Triglochin maritima, Suaeda maritima* and *Salicornia herbacea*.

Caerlaverock Marsh, at any rate since 1846, has been used mainly for pasturage, and this is the probable reason for the absence of such species as *Halimione portulacoides* and *Limonium vulgare*, and in heavily grazed parts *Cochlearia* and *Aster* and some other species grow only on the slumped sides of creeks. One factor that must be fully appreciated in the relatively recent development of Caerlaverock Marsh is the activity of the Nith Navigation Company in strengthening and confining the meanders of the Nith in the 1850s and 1860s.

In a paper on the Upper Solway Marshes Marshall (1962 b) considers other marshes, some of which are on the English side and do not concern us here. Again, she emphasizes the stepped nature of most marshes, and in general maintains her view that the steps are primarily erosion features. She does, however, state that accretion on a low-level step, where it may be expected to be relatively rapid, seldom catches up with the higher level above it so as to eliminate the step. The occurrence of a small step in such a situation is, therefore, suggestive of minor uplift. Kirkconnell Marsh, on the western side of the Nith, is regarded as a one-level marsh because the rubble wall built along the river kept it in course and so prevented lateral erosion. However, by 1961, some of the stones have led to local erosion. Some other causes of erosion include floods, heavy rainfall, sudden thaws in the hills, and easterly winds.

Since the rivers draining into the Solway reach it over gentle gradients they carry into it little material in suspension, at any rate during most of the year. It is suggested that the rather higher percentage of clay than silt in the marshes means that the rivers can carry some of this finer debris which is deposited when flocculation takes place when it reaches salt water. Since the marshes, in this respect, are all much alike Marshall maintains that almost all of the fine sand of which they are built is of marine origin.

There is some development of marsh to the east of the mouth of the Lochar

Water. Above Cummertrees slightly higher ground reaches the coast, probably largely associated with raised beach times. The coast is fringed by sands, and in the Upper Solway the low-water channels are narrow. There is little or no true marsh beyond the Annan, until the peninsula of Rockliffe Marsh divides the Esk and the Eden. On the other hand, the low ground on the Scottish side is all part of former raised beaches.

An interesting point on the Solway marshes is made by Perkins (1968). The lack of silt in these marshes not only renders the sand more mobile, but its absence also deprives the substratum of a vital nutrient material for plant growth.

Chapter V

The Inner Hebrides

SKYE AND ADJACENT ISLANDS

Before discussing the coasts of Skye it will be convenient to give a brief account of
the structure of the whole island (Fig. 22). The south-eastern portion is structurally
part of the North-west Highlands. The Moine thrust separates the Lewisian Gneiss
from the Torridonian south of Loch na Dal. There are fragments of Mesozoic
rocks near the Aird, and the Cambrian (quartzite and grits) occupies a small area
on the south side of Loch Eishort. A narrow belt of Lias running from Broadford
Bay to the peninsula between Lochs Eishort and Slapin is succeeded by a smaller
outcrop of Cambrian rocks. On the west side of Loch Slapin Middle Jurassic rocks
occur, but the belt is not continuous with the corresponding outcrop on the north
coast near Strollamus because it is interrupted by the granite mass of the Red Hills.
This mass, and that of the gabbro forming the Cuillin Hills make the highest and
most spectacular ground in Skye. All the rest of the island is covered by great
sheets of lava and sills, which, on the east and north-east coasts cover an interesting
sequence of Mesozoic rocks. (The sills of dolerite which were emphasized so much
by A. Harker (1904), are now regarded as the more resistant inner parts of lava
flows. This view does not invalidate Harker's conclusions, except that his sills
were not intruded.) Occasional smaller outcrops of these rocks are seen in
Vaternish and near Waterstein Head. Simplification can only too easily be over-
emphasized, but broadly we may divide Skye into three distinct parts – the northern
basalts and underlying Mesozoics, the central and extremely complicated area of
the Cuillin and Red Hills together with the limited outcrops of Mesozoic rocks
on their eastern sides, and, thirdly, the pre-Cambrian area of the east and south-east.

The whole island is so irregular and indented that no place is more than five
miles (8 km) from salt water. Hence it is difficult to know precisely how to define
its coast. It is perhaps better not do do so, but to call attention, for example, to the
magnificent views of the Cuillins as seen from the higher ground of the Sleat
peninsula across Lochs Slapin and Scavaig, a view which shows the difficulty as
well as the futility of rigid definition of coast in this context. The gabbro barely
reaches the sea, but it is in every sense a real part of the coastal scenery. Less
spectacularly the Red Hills, as seen from parts of the north coast, can rightly be
claimed as belonging to the coastal scenery. The great difference between the
rounded forms of the granite Red Hills and the numerous sharp peaks of the Cuillins make not only one of the most striking but also one of the most beautiful

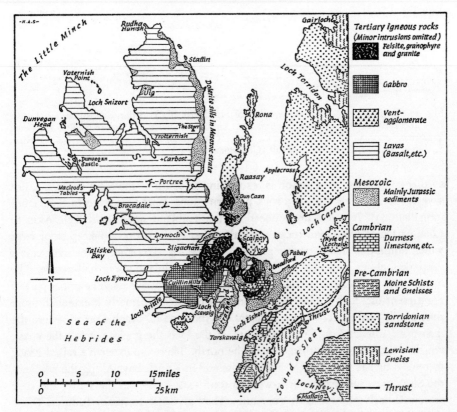

Fig. 22. Geological map of Skye (Geological Survey)

contrasts on the whole coast of Britain. The much indented coastline of Skye is associated with the fact that the surface relief was inaugurated when the land stood higher than at present. There is no doubt that the lavas once covered greater areas than they now do. In Soay the underlying rocks have been fully exposed. It is assumed that Torridonian and perhaps Mesozoic rocks underlie the lavas of most of Skye.

A major tectonic feature in Skye is the great Camasunary fault. This runs between Loch Scavaig and Loch Ainort. Part of its course is buried under the Blaven (Blath bheinn). As suggested on p. 132, its northern continuation may be along the east coast of Raasay; to the south it is suggested (George, 1966), that it continues between Rhum and Eigg. Its relation to the structure of the Strathaird peninsula is shown in Fig. 23. This figure also emphasizes the width of the raised beach in the upper part of Loch Scavaig.

Those parts of the coast of Skye along Loch Alsh, Kyle Rhea and the Sound of Sleat resemble parts of the mainland formed of Torridon Sandstone and Lewisian Gneiss (Peach, Horne, et al., 1910). They face, however, relatively sheltered water

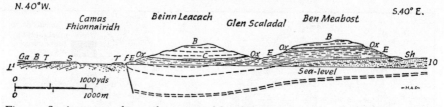

Fig. 23. Section across the southern part of Strathaird. T, Torridon Sandstone; L¹, Lower Lias under Basalt; IO, Inferior Oolite; Sh., Ammonite-shales in Inferior Oolite; E., Great Estuarine Series; Ox., Oxford Clay; C, Upper Cretaceous; B, Tertiary Basalt; Ga., Gabbro; S, Raised beach; f, Camasunary Fault (Geological Survey)

so that erosion by wave action is, except in the south of the Sleat peninsula, not severe. Nevertheless there is a rock platform along most of it, even in Kyle Rhea (see Plate 43). It is less pronounced on the mainland side. From Kyle Akin to Loch na Dal Torridonian rocks reach the coast in the form of a steep slope over which numerous small streams plunge into the sea. The Kyle Rhea river cuts a deep glen, along which raised beach features reach inland about half-a-mile. The major break is at Loch na Dal which lies on a line of weakness. The upper part of the present-day loch is filled with sands and boulders; formerly it extended more than a mile farther. There are remnants of the '100-ft' beach near to where the main road crosses Lòn Creadha. South of the loch the gneiss reaches the coast, which is more irregular than that to the north. There are extensive raised beach features near Isle Ornsay in Camas Croise and in Knock Bay. Along the whole of the Sleat coast numerous north-west to south-east dykes cut the shore, and they are particularly numerous in the south. They give rise to minor features. For a short distance south of Armadale, at the Point of Sleat and along the coast almost to Tarskavaig Bay, Moine Schists reach the coast. The rock bench is prominent on the western side, and in higher raised beach times the sea must have extended two to three miles (3–5 km) up Gleann Meadhonach. Tarskavaig Bay, with its fine inner beach, was also a bigger inlet; raised beach extends upwards beyond Achnacloich. Conditions in Òb Gauscavaig are somewhat similar. Along all this coast dykes are even more numerous than on the east.

The outer parts of the peninsulas between Lochs Eishort and Slapin and Lochs Slapin and Scavaig are formed of Mesozoic rocks, mainly Liassic. The Suishnish peninsula is faced by a broad rock platform, and there are also several waterfalls. The east and south of the Strathaird peninsula consists of low country, largely peat-covered, near the coast. The main road to Elgol divides this sharply from the high ground, largely basaltic, to the west. The coast also changes in appearance. From Elgol westwards basalts fringe the coast, but the coastal scenery is dominated by the Cuillins which rise up just behind. Fossil cliffs line much of the coast, and locally the rock platform is present, but is probably covered in places with the modern beach. Loch Brittle contains a wide beach of dark sand.

The island of Soay is formed entirely of Torridonian rocks apart from a few

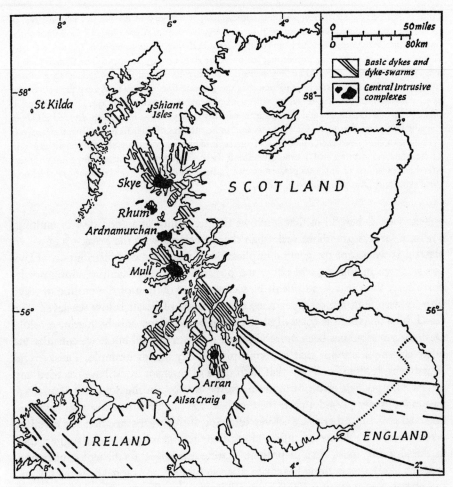

Fig. 24. Tertiary north-west dykes in relation to Tertiary plutonic districts of the British Isles (rep. from *The Geology of Ardnamurchan* (Mem. Geol. Surv., 1930), Fig. 4)

minor intrusions of Tertiary age. The coast is steep and the cliffs bold and rugged. Their tops are locally peat-covered. The island is nearly split in two by the deep narrow gash of Soay Harbour, and the wider Camas nan Gall. These and the low ground between them lie along a major fault with a downthrow to the east. A smaller fault runs from An Dubh chamas to An Dubh-sgeire; it has much less effect on the topography. Along the west coast there is a long string of rocks awash at high water; they, together with those on the south-east, are presumably separated parts of the coastal bench.

Recent work (Anderson and Dunham, 1966) has modified considerably the old view of a vast North Atlantic lava plateau. The lavas in this great area are basaltic, but there are often big differences in detail (Fig. 24).

Their present day distribution also militates against the theory of a common origin since in every case they are grouped about or near an intrusive centre. Moreover, the isolation of the various plateau remnants would require large-scale faults, whereas so far as is known, the late-Tertiary faults, though very numerous, are nowhere of great displacement. The distribution of lava types in Skye suggests that they were extruded from several fissures related to a central volcano ... There is abundant evidence that the Jurassic floor is nowhere very far below sea-level, and the introduction of faults with a displacement of more than 1,000 feet [305 m] is difficult. Moreover, the several groups into which the Skye lavas have been divided all show thinning away from centres which are assumed to be the sites of their feeders. Thus the lava-pile is thought to have been built up by flows from several fissures operating at different times and in different places, meeting and overlapping along their margins and eventually forming an inter-leaved series which in any one place may not have exceeded 4000 feet [1,120 m] in thickness (Anderson and Dunham, Memoir, 1966).

This view is helpful in that it allows the possibility of a good deal of faulting, but on a far less grandiose scale than that associated with the break-up of a vast plateau. If we accept the general implication of the scheme of distribution of lava flows shown in Fig. 24 and allow the possibility of some faulting along north-north-west lines it is less difficult to picture the evolution of the outline of Skye than by assuming the disappearance of vast areas of basalt below sea-level. The sea-cliffs of much of the coast of Skye cannot be explained only by marine erosion. Doubtless erosion has been more vigorous at certain times, but if we consider the relatively small amount that is taking place today at, for example, Talisker and Waterstein, it seems unlikely that high cliffs of resistant rock have receded any distance simply as a result of modern marine erosion. In the Sound of Raasay ice has undoubtedly played a part; so, too, has land-slipping. Ice may also have had no small effect in other places. Even allowing for submergence that has certainly taken place, the relation of cliffs and sea-floor topography does not suggest that, in the past any more than at present, marine erosion is a sufficient explanation. It seems likely that faulting has taken place along or near to some lines of cliff, and that the present form of the cliff is a modification, locally considerable, of the fault scarp. The occurrence of former sea-levels (see p. 146) bears this out to some extent. The Memoir already quoted remarks that 'Since Tertiary times ... the coast of Skye must have been, as it now is, for the most part rocky and precipitous.' (These arguments also apply to many other parts of the Hebridean area. The steep cliffs of Canna, some of those on Mull and Eigg, and those on many small islands seem likely to be of this nature.) The irregular outline of the coast of northern Skye dates back to a time when the land stood higher than it now does. It may also be con-nected with the formation of numerous faults, and also with a tilting of the strips between pairs of faults to the west. Harker (1904) also thought that there was a tilting of the whole country in this direction. Whether it is possible to connect the general trends of Loch Harport, Loch Pooltiel, Loch Dunvegan and Loch Snizort with these movements is, at the best, problematical. But these lochs, as well as the general trend of the western and eastern coasts of the island, demand explanation.

The admirable One Inch Geological Map of northern Skye marks a number of faults having an approximately north-westerly trend. The detailed cross-section between Mointeach nan Tarbh (just north of Moonen Bay) to the east coast at a point about one-third of a mile north of Prince Charles's Cave indicates the nature of the faulting clearly, and suggests, for example, Loch Dunvegan is in a down-faulted strip (Fig. 25). Faulting roughly parallel with parts of Loch Snizort and Loch Bracadale is indicated. This does not prove that the major outlines of these lochs are fault-controlled in any direct sense, but it is difficult not to draw the conclusion that faulting has been a controlling factor in shaping the main outlines of Skye north of the Cuillins. It can also be assumed that locally small faults have given detail to the coast. It is likely that Gesto Bay, Loch Beag (?), Loch Varkasaig and Brandarsaig Bay, all in Loch Bracadale, Lorgill Bay and, on the east coast, Tianavaig Bay, Invertote, Staffin Bay, and farther north, Lùb Score and Camas Mór and Camas Beag are to some extent fault-controlled. That is not the same as saying that they owe their origin entirely to faulting.

There are also numerous small dykes which cut the coast. In many cases they may have a minor effect. On the other hand the importance of intrusive sills is often paramount in the coastal scenery. The Memoir (Anderson and Dunham, 1966), makes clear that it is wrong to use sills in the plural; the sills are almost always leaves of a single major sill. This great sill is magnificently displayed in the eastern sea-cliff between Loch Sligachan and Staffin Bay. Some of it is submerged in north Trotternish, but it reappears in many off-shore islands. In the west it is almost always submerged. The leaves of the sill frequently transgress the Mesozoic rocks. The sill is in general an olivine-dolerite, although certain petrological variations enter in farther north. The leaves of the sill, especially where they are thicker, usually display good columnar jointing.

The sill can be traced along the coast from Loch Sligachan northwards. The highest leaf is at sea-level between Loch Sligachan and Camastianavaig. Lower leaves show in the sea-cliff north of Port a'Bhata, but farther north fall below sea-level except in Holm Island which is part of the sill. The sill is well exposed in Bearreraig Bay where columnar dolerite overlies Inferior Oolite. The manner in which the sill divides into leaves is seen near Upper Tote. Half-a-mile north of Rubha Sùghar this uppermost leaf divides into a lower portion which transgresses downwards in the Inferior Oolite and appears to die out at Rigg Burn, and an upper portion which continues at more or less the same stratigraphical level as far as Upper Tote. Just north of Armishader, however, this upper leaf has divided into three and it is the lowest of these which is seen in the cliff from Rigg to Upper Tote. The middle limb, at a higher level, is only seen in the sea-cliff just north of Rigg, whilst the uppermost sheet, occupying a still higher position in the Jurassic sediments, forms the great inland outcrop at Creag Langall (Memoir). The lowest leaf is faulted down north of Dùn Grianan and makes the waterfall in the Lealt river. The middle leaf makes the cap of Dùn Dearg, rejoins the lower at Loch

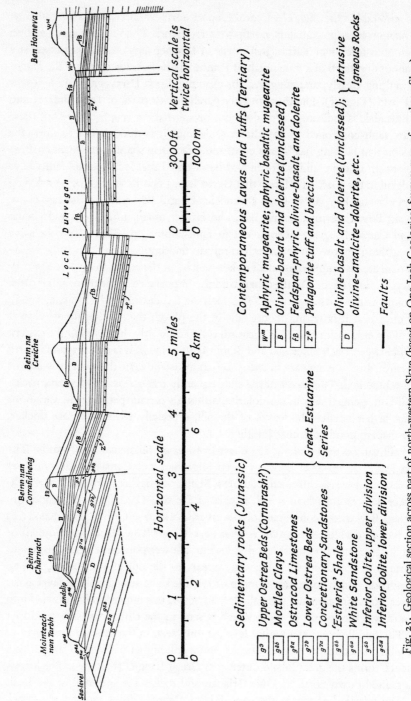

Fig. 25. Geological section across part of north-western Skye (based on One Inch Geological Survey map of northern Skye)

Mealt and, north of the fall, the leaves split once again. The lower one forms the lower part of the cliff at Kilt Rock and the platform at Sgeir Bhàn and, farther north, Sgeir nam Faoileann and Staffin island. Rubha Garbhaig and An Corran are in the down-faulted upper leaf. In this locality the sill shows columns up to 150 feet (46 m) in length. At the Kilt Rock there are columns of 100 feet (31 m). These places all show vertical columns; at Creag na h'Eiginn (north of Flodigarry) and on Trodday and Fladda-Chùain islands the columns are inclined as much as 15°–20°. In the cliffs east of Rubha Hunish the jointing is irregular and at Lùb a'Sgiathain it is fan-shaped. The sill continues along the coast almost to Uig Bay. In this northern part the rock is more teschenitic. Eileann Trodday, Sgeir nan Maol, the Ascrib Islands and Mingay island are of this nature.

The three main northern peninsulas, Trotternish, Vaternish and Duirinish, are unlike one another. Trotternish is followed almost throughout its length by a fine lava scarp, which reaches 2,360 feet (720 m) at The Storr. The ice (see p. 32) which followed the east coast exerted great pressure on this peninsula, and the main scarp is cut by several passes (bealachs) which separate peaks such as The Storr and Beinn Edra. The eastern slopes of Trotternish are steep, the western more gentle and follow the dip of the lavas. In Vaternish there is also a central lava ridge, but lower, usually below 100 feet (31 m). In Duirinish the lavas are almost flat, or slightly inclined to the east. Macleod's Tables, Healaval Mhór and Healaval Bheag are flat-topped hills which show up the nature of the lavas.

Partly because the lavas, especially on the east coast, rest on Mesozoic rocks, partly because of the steep eastern slope of the scarp and the vertical joint planes in the lavas, even more because the weight of the lavas is sufficient to have overcome the resistance of the underlying material to shear, the east coast of Skye affords the finest and most extensive landslips in Britain. Not all of these reach the present coast, but it is fair to include all the landslip topography in an account of the coastal scenery. The most spectacular is The Storr; it extends seawards about 5,000 feet (1,524 m). Its nature is shown in Fig. 26. It is not moving at the present time, but further erosion of its toe may well lead to renewed movement in the future. In Staffin Bay, and to the north of it, is the most extensive slip, that of the Quirang. It extends 7,000 feet (2,132 m) from Meall nan Suireamach to the coast. For the most part it is stable, but near Flodigarry the sea is eroding its toe, and there is more or less continuous movement. Fig. 27 indicates its general nature, and it will be seen that it has considerably disturbed the Jurassic sediments in Staffin Bay. Farther south, on the south side of Portree Harbour, there is an unstable slip on the east side of Ben Tianavaig.

At the close of the main glaciation the eastern face of this mountain must have presented a cliff face of about 500 ft [152 m] largely determined by small faults. The lower part consisted of Jurassic sediments . . . In the present slip 2000 ft [610 m] of sediments separate a thick dolerite sill from the base of the lavas. Thus the margin of the slip should be almost 10,000 ft [3,048 m] from the cliff face. It is in fact about 3500 [1,067 m] to sea-level and the toe of the slip is being actively eroded (Memoir).

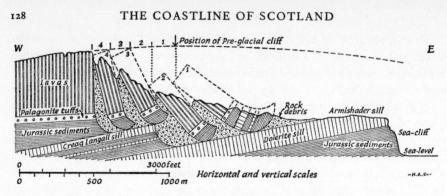

Fig. 26. Section of the Storr Landslip (based on Geological Survey)

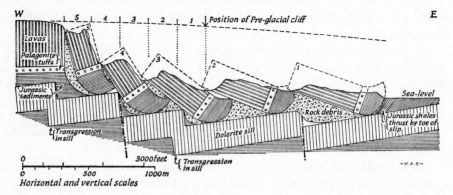

Fig. 27. Section of the Quirang Landslip (based on Geological Survey)

Richards (1969) discussed in some detail the evolution of the coastline of Skye between Staffin and Portree. This coast coincides with the largest exposure of Mesozoic rocks in the island. That the undercliff is a fossil cliff is indicated in three ways: (1) several parts of it are not affected by modern marine erosion; (2) much of the remainder is obscured by waste; and (3) there are several occurrences of cemented beach material. Richards argues that the third point is significant. Near Invertote the cave deposits appear to be overlain by boulder clay, suggesting that the deposit, the beach, is earlier than a glaciation. About one mile south of Staffin Bay pebble accumulations in a cave seem to have been eroded after cementation and 'since the cave is unaffected by present day sea-level it seems necessary to invoke two periods of sea-level slightly higher than the modern'. Near Rigg there is a small platform, about 200 yards (183 m) in maximum width. The height at the back of the platform is 85 feet (26 m); seawards it is cut by the undercliff, 40–50 feet high (12–15 m), at the base of which is a narrow surface which Richards thinks is of present-day origin; it is certainly within the zone of modern wave activity. Two streams cross the higher platform, and in the more southerly of them there is a small re-entrant in the cliff. At the base of this cliff there are patches of cemented

2. The Isle of May, a dolerite sill inclined downwards to the north-east

1. St Abb's Head: cliffs in lavas of Lower Old Red Sandstone age

3. Cliffs, near Heathery Carr, in much folded Silurian rocks

4. The Bass Rock, a plug of phonolithic trachyte

5. Panorama near Newburgh, Firth of Tay, looking to the north-east

6. Old Red Sandstone cliffs at Dunnottar Castle. The rock platform and storm beaches are at the foot of mainly fossil cliffs

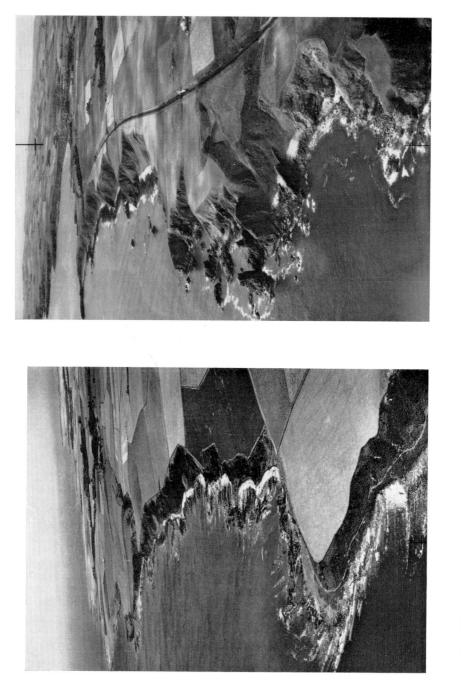

8. Cliffs in Dalradian rocks at Hall Bay, north of Stonehaven

7. The Fife coast near Kinkell Ness. The fossil cliffs and platform are cut in rocks of the Calciferous Sandstone Series

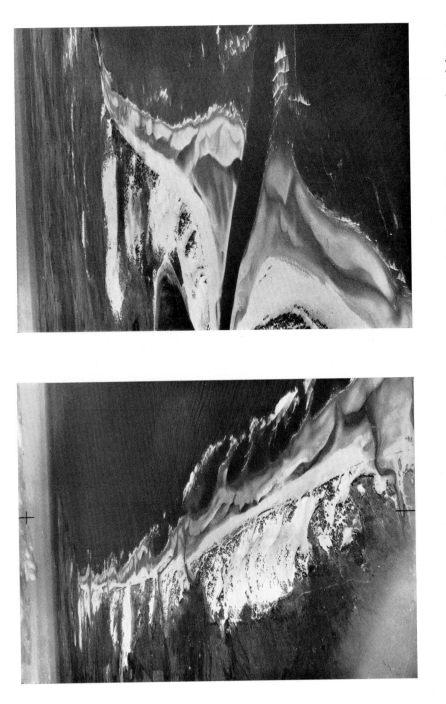

10. The mouth of the River Ythan. Forvie sands and the great transverse dunes are beyond (i.e. north of) the mouth. The more distant dunes can be seen to rest on a rock surface

9. Coastal dunes north of Aberdeen

12. The coast between Peterhead and Rattray Head

11. The northern end of the Sands of Forvie:
cliffs in Dalradian rocks

13. The coast from Macduff to Troup Head. The cliffs nearer
Macduff and at Troup Head are cut in metamorphic rocks of
Dalradian age: they are separated by a narrow band of
Middle Old Red Sandstone

14. The mouth of the River Spey, looking eastwards

16. The Bar off the Culbin sands. The white arc in the forest is the parabolic dune near Maviston

15. Findhorn, the mouth of the Findhorn river and the Culbin sands before the present afforestation. The Bar (Plate 16) is visible in the distance

17. The Carse of Delnies, west of Nairn. Fossil cliffs are on the right (i.e. south) of the marsh

18. The compound spit and ancient cliffs at Ardersier

19. The '25-ft' and '100-ft' beaches and cliffs near Helmsdale

20. Berriedale and the joint mouth of the Berriedale and Langwill burns. Scaraben in the background

21. Duncansby Head and Duncansby Geo and Stacks. Flat-topped cliffs of Middle Old Red Sandstone

22. Old Red Sandstone cliffs at the Kame of Hoy, Orkney

23. The Broch of Rennibister, about three miles West of Kirkwall, Orkney. The rock platform is well-developed, in relatively sheltered water, and is probably to be associated with the ? '25-ft' beach

24. Panorama near Brough Head, Orkney. Cliffs cut in rocks of Middle Old Red Sandstone age

25. Shetland: St Ninian's Ayre; Mainland in the background

26. Shetland: panorama near Aith

28. Shetland: Unst, cliffs cut in Dalradian metamorphic rocks, looking south-west over Herma Ness

27. Fair Isle: North Haven. Old Red Sandstone cliffs

29. Cape Wrath: cliffs cut in Lewisian Gneiss

30. St Kilda: Dunn, Eucrite cliffs

31. Entrance to Sandwood Loch in foreground. Cliffs of Lewisian Gneiss on north side of the loch, and of Torridon Sandstone beyond

32. Enard Bay: cliffs and islets in Lewisian Gneiss

33   Lewis: Loch Grimshader, Lewisian Gneiss

34   Lewis: Leurbost, Lewisian Gneiss

35 Lewis: panorama near Cromore. Lewisian Gneiss country

36. Lewis: the head of Loch Seaforth

37. Lewis: west coast, cliffs in Lewisian Gneiss at Beinn Dhubh

38. North Uist: panorama near Baleshare looking north-west along Kirkibost island

40. Canna and Sanday: fossil cliffs and beach platform

39. The Monach Islands: low-lying Lewisian Gneiss

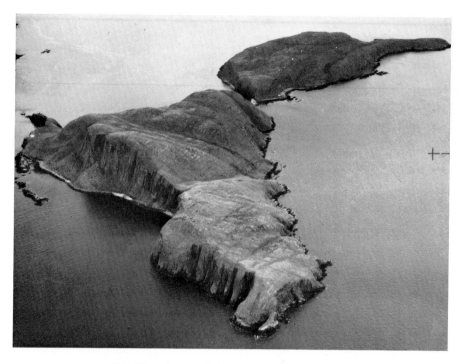

41. The Shiant Islands: dolerite sills and screes at cliff foot

42. Skye: view to north-east from a little north of the lighthouse near Waterstein Head;
Dunvegan Head in middle distance. Tertiary basaltic lavas

43. Kyle Rhea between Skye (left) and the mainland; view to the north

44. Rhum; Askival and Allival. The cliffs, showing several small geos, are cut in Torridon Sandstone. The high ground consists of the Plutonic Complex

45. Eigg: lavas overlying Mesozoic sediments, Tolain and Bay of Laig

46. Panorama near Arisaig: Eigg in background. The skerry-guard is conspicuous
in the middle distance

47. Ardnamurchan; the coastal rocks are mainly gabbros

48. Loch Sunart, view to the west-south-west

49. Staffa: basalt columns and caves

50. Panorama near Port Appin. The northern end of Lismore island in middle distance; the Morvern hills in the background

51. Mull; stratified lavas and scree cliffs, Fionn Aoineach

52. The coast of Argyll north-east from Seil island. Fossil cliffs and rock bench in sheltered waters and the drowned nature of the coast are clearly visible

53. The coast near Oban; the entrance to Loch Etive and Ardmucknish Bay

54. Jura; Glendebadel Bay; raised beach and fossil cliffs in quartzite

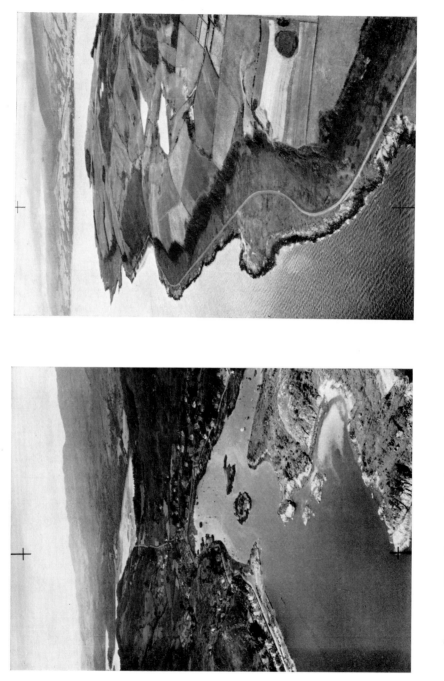

56. The raised beach and fossil cliffs on the west coast of
Great Cumbrae island

55. Panorama of East (foreground) and West (background)
Lochs Tarbert

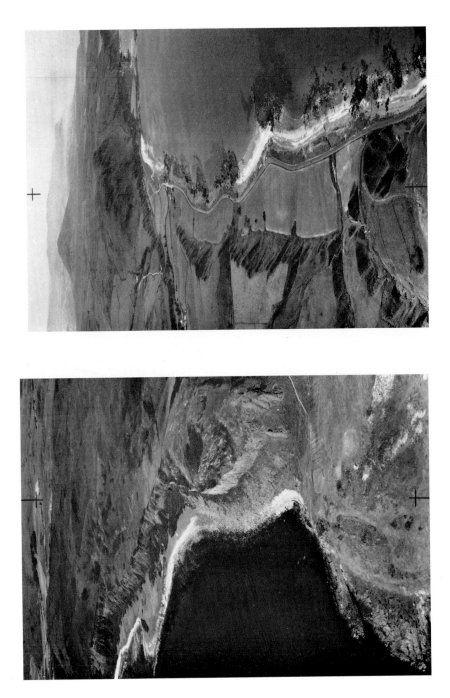

57. Arran: fossil cliffs north of Drumadoon Point; columnar jointing in sill in middle distance

58. The coast, raised beach and fossil cliffs, north of Lendalfoot, Ayrshire

59. Loch Crinan looking to the south-south-east from near the end of Rubha Garbh-Ard to the straight line of the Crinan Fault

60. Arran: coast near Kildonan. Fossil cliff, raised beach and dykes

61. Ailsa Craig, Trammins and Foreland Point. A plug of micro-granite

62. The estuary of the River Cree and Wigtown Bay

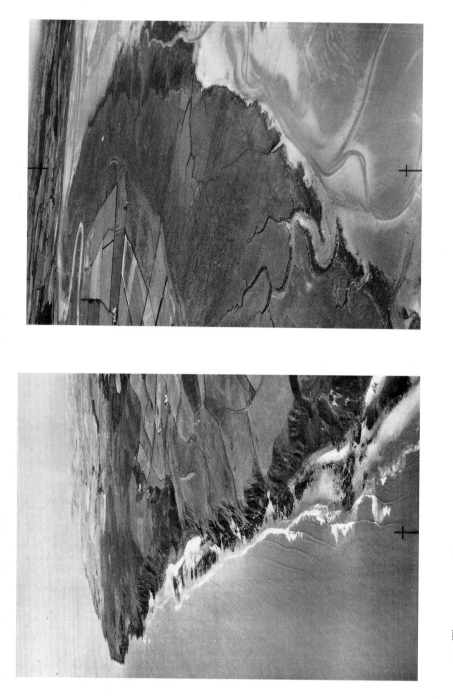

64. Caerlaverock Marsh, Dumfries

63. The west coast of the northern part of The Rhinns, Wigtownshire. Fossil cliffs and pocket beaches

pebbles; between them and the sea is a band of pebbles and rubble about 10 feet wide (3.5 m) which protects the cliff from erosion. Also in this re-entrant is a small wedge of raised beach material projecting from the cliff and resting on rock.

the cemented deposits contrast markedly with the wedge of raised beach deposits despite the fact that they are only separated by a matter of yards. Even though the cementation of the deposits is in itself no proof of considerable age in an area of calcareous rocks, this and their other qualifying features suggest that the consolidated beach materials long pre-date the post-glacial raised beach. To suggest two low-level transgressions, widely separated in time, facilitates an explanation of [these] features.

Richards thinks that the remnants along this coast of the 'Pre-glacial' platform imply the initiation of the seaward-facing slope. The relict cliff means a long still stand, and the low-level beach (= transgression) in its turn implies erosion of the 'Pre-glacial' beach.

This careful analysis of a few miles of coast in sheltered waters is of much interest. It focuses attention on the importance of studying all details of a cliffed coast and for that very reason emphasizes the value of similar studies in many other places.

Portree Harbour is a fine feature and different from others on the east coast. It has been suggested (Hossack, 1930) that Loch Portree was an original subsequent loch later captured by the sea. It has a fairly narrow entrance, and is shut in by high and steep cliffs. The loch is shallow and is filled with sediment, partly deltaic from the Varragil river and partly marine. Part of Portree is built on raised beach. The land between Loch Portree and Loch Snizort is low; a submergence of little more than 100 feet (30 m) would unite the two lochs, which are about six miles (10 km) apart.

The eastern coast of Skye south of Portree is for the most part a protected coast, and wave action is limited. Cliffing is not pronounced and a road follows the coast. In many places, especially between Broadford and Kyle Akin, the road is on raised beaches which are broad and well developed, and form a wide strath, especially near Lower Breakish.

In northern Skye, especially where the Mesozoic rocks underlie lavas, there are several smaller landslips. There is one in Moonen Bay close to Waterstein Head and there are others at Loch Losait in Vaternish and on the west side of Dunvegan Head in Duirinish. There are several small slips in Lùb Score. In many places there are also rock falls. At Rubha Garbhaig the sea-cliff behind the raised beach is a sill of columnar dolerite which has in part disintegrated into a mass of large cuboidal blocks. The fall is clearly later than the '25-ft' beach.

The steep cliffs around much of the lava country of Skye are also responsible for another interesting coastal feature, the waterfalls. One of the best known is that of the stream which drains Loch Mealt. It falls vertically over what appears to be a sill 195 feet thick (59 m). Even if the face under the fall cannot be regarded as a vertical section of the sill, it is nevertheless the finest fall on the coast of Skye.

The Bearreraig falls have been spoiled as a result of hydro-electric schemes. The falls were formerly impressive, and cut in steep gorges. In this case the Inferior Oolite is the surface rock over which the water plunged which drained Lochs Leathan and Fada. There is a fall where the Lealt river reaches the sea at Invertote.

There are some beautiful cliff waterfalls on the north-west coast. Some are permanent, others are mainly wet-weather falls. There are several on the west-facing side of Loch Pooltiel. The cliff scenery from Dunvegan Head to Waterstein Head is magnificent. The fall from Loch Eishort in Moonen Bay is a fine feature as seen from Hoe Rape. There are several falls between Moonen Bay and Idrigill Point, and a well-known one is Talisker Bay. These are but examples, and a traverse around the coast will show many more. The tabular nature of the lavas, the steep cliffs, the numerous small streams, only a few of which cut down their valleys to, or near to, sea-level, and the abundant rainfall provide ideal conditions for coastal waterfalls. In places where the lavas dip seawards falls are less conspicuous, and they are scarcer in the south-eastern part of Skye (see Plate 42).

Beaches and blown sand are found in favourable places. There are beaches in Staffin Bay and Uig Bay, and some dunes at An Corran in Staffin Bay. There is an extensive beach in Lùb Score and smaller ones at Lùb a'Sgiathain and Port Gobhlaig (Kilmaluag Bay). There are also some small beaches of white nullipore (coral) sand. The best-known examples are in Vaternish of which Camas Bàn is the largest (Haldane, 1931–8). There is another forming the spit of sand which connects Lampay island to the mainland. A situation of this type seems to favour the growth of *Lithothamnium calcareum* Aresch, which is the organism which gives rise to the sand. In a small bay rather more than a quarter-of-a-mile south of Duntulm castle the beach sand is dull green in colour and is composed of 80–90 per cent of olivine. Other sands in Trotternish also contain a fairly high proportion of olivine (Walker, 1924–31).

Small islands near the coast of Skye

Trodday (see p. 127) is part of a single sill. The dolerite is columnar, and dips to the south. There are several small geos cut along planes of weakness, and there is a fine stack on the north side of the island. The Skerries, about three miles (5 km) to the north-west, are also parts of a dolerite sill. To their west is the Fladda group. Fladdachuain, about one mile long and a quarter-of-a-mile broad, is a single sill dipping to the south-west at 15°–20°. On the north-east the sill shows an abrupt cliff, reaching 60 feet (18 m) in height. The columnar structure is rather irregular. There are many small geos cut along joint cracks. On the gently sloping south-western side, there are several hollows 'representing the continuation of the geos, so deep that they seem to divide the south of the island into several slices . . . ' All the associated islets, Gaeilivore, the Cleats, Gearran and Lord Macdonald's Table are similar; locally the columnar jointing is almost vertical and well developed.

The Shiants (Walker, 1930a) show dolerite sills intruded into Mesozoic sediments (see Plate 41). There are two main islands connected by a beach. Garbh Eilean reaches 525 feet (160 m), and has impressive cliffs on the north and east. The cliff foot is often buried in scree which is partly overgrown. Eilean an Tighe is 410 feet high (125 m) and shows similar cliffs and screes. Both islands appear to consist of one thick sill dipping south-westward at 10° to 15°. It follows that the south-western sides of the islets are fairly gentle. Eilean Mhuire has high cliffs all round it. It is made of several sills separated by sediments. Galta Mór and Galta Beag are also doleritic. Columnar structure is often remarkable; the columns being large in diameter and often long and straight, sometimes showing curvature at their summits. On the north face of Garbh Eilean, west of Glaic na Crotha, some columns reach 350 feet (107 m) and have a diameter of 5 feet (1.5 m). They are usually hexagonal. There are also major vertical joints trending north-north-east and south-south-west. Near Glaic na Crotha these have been responsible for the cliff breaking off in great parallel slabs. The same phenomenon occurs on the south-west of the island, thus allowing the opening out of major joints so that they form gaping fissures. There are many caves and fissures on the south-western side of Garbh Eilean, presumably the work of waves along joints and other planes of weakness. The arch at Bidean a'Rhoim is eroded along a major joint. In Eilean Mhuire and the rocks named Galtachean there are many examples of columnar structure and minor coastal features; the individual islets are steep on their northern sides.

Raasay-Rona, Scalpay, Pabbay

Raasay and the sounds on either side of it present several points of interest. Since the 1939–45 war new soundings have been made, and off Caol Rona a depth of 1,062 feet (177 fathoms: 323 m) has been recorded. This is the deepest sounding so far recorded on the continental shelf around Britain. The western slope of this deep is steeper than the eastern one, and the bottom is more or less flat. In Linne Crowlin a depth of 828 feet (138 fathoms: 252 m) was found. Robinson (1949) states that between these two deeps the sound is much shallower 'particularly in two patches which probably represent the submarine continuations of the Inferior Oolite sandstones so prominent in Beinn na' Leac in south Raasay'. There is little doubt that the Inner Sound is crossed by several faults, notably the continuation of the great fault which crosses Raasay from south-west to north-east, from near Holoman to Screapadal (Scapadal). That wave action in the Inner Sound was at one time more formidable is shown by the range of cliffs (p. 72) behind the raised beach platforms north of Applecross. Relatively little change seems to be taking place at the present time.

The Holoman–Screapadal fault divides Raasay into two distinct parts, a southern part formed largely of Mesozoic rocks and a large mass of intrusive granite, and a northern part formed of Torridonian rocks which extend to Loch Arnish and include Eilean Fladday, and a northern tip of Lewisian Gneiss which also forms the island of Rona north of Caol Rona. On the east coast south of Screapadal there are great precipices in the Jurassic rocks. The cliffs rise locally to 800 or 900 feet (243 or 274 m), and at Dùn Caan the slope, not strictly a sea-cliff, reaches 1,445 feet (440 m). This hill is topped by an outlier of a basalt sill. Along much of the foreshore there are often fallen blocks of considerable size which obscure the true bedding in the cliffs, and from Hallaig to Screapadal slips and talus obscure the junction of the Upper and Middle Lias. Occasionally these blocks may form a small islet. There are many landslips. Farther south, where a road along the coast follows a terrace, the cliffs are broken. The lower part is in the Middle Lias, but is inaccessible. The upper slope rises to Beinn na' Leac, 1,017 feet (310 m). The ground hereabouts is unstable and further landslips may occur. The landslip at Hallaig is rather more than one mile long. A fault cuts the coast just south of Rubha na' Leac, beyond which cliffs of Middle Lias occur. The east coast north of Screapadal is steep and slightly irregular in the gneiss both of Raasay and Rona; in Rona this is accentuated by numerous epidioritic dykes which commonly trend south-east to north-west. It is suggested (Anderson and Dunham, Memoir, 1966) that the Screapadal fault continues north-north-eastwards parallel to the coast of Rona.

The western coasts of both islands stand in strong contrast with their eastern coasts. South of Manish Point the dip of the Torridonian is to the west, and in Holoman Bay and farther south the coast is not steep. The westerly dip also prevails in Eilean Fladday. In the gneiss areas of Raasay, and in Rona the west coast is very irregular; the several headlands and inlets have, in general, a north-west to south-east trend which is roughly parallel with the epidioritic dykes. This is in no sense to suggest cause and effect; only occasionally do the trends of headlands and dykes coincide. As in so many parts of the coast of Scotland a great deal more research is necessary before even a reasonable understanding of many coastal details is obtained. Before, however, leaving Raasay-Rona it is necessary to call attention to the movements of the ice. The Memoir of 1966 gives a very suggestive map which shows that whereas ice generally crossed Raasay-Rona in a north-westerly direction (although it was more nearly parallel to the coast south of Screapadal), in the Sound of Raasay the direction of movement was directly along the sound. The scouring effect of ice in a narrow sound must have been similar to ice in an alpine valley, and the steepness of the cliffs, especially north of Portree, must be attributed in part, possibly in large part, to this cause. Certainly between Skye and Raasay wave action is limited.

Scalpay, Longay and adjacent islets consist mainly of sandstones of Torridonian age. In general they are well-bedded and show a series of small escarpments. At the north-western corner of Scalpay a patch of Triassic rock forms the hill (142 feet: 43 m) above Rubh' a' Chinn Mhóir and also the islet named Eilean Leac na Gainimh. The small bay at Camas na Geadaig corresponds with a fault trending south-eastwards. In the south of the island a curved fault brings down Jurassic strata near Scalpay House.

Pabbay, on the other hand, is formed of Liassic rocks and is separated by about a mile of sea from the belt of similar rocks in Skye. The highest point of the island is 91 feet (28 m), and there is a remarkably well-developed rock platform on its western and northern shores. A number of dykes cut this platform and also the beach in the south-east. The eastern face of the island is parallel to a dyke.

THE SMALL ISLES OF INVERNESS-SHIRE
Canna

Canna is simpler in structure than Rhum, Eigg and Muck (Fig. 28). It is entirely formed of rocks of volcanic origin. Sanday strictly is a separate island, but is united with Canna at low water. The most conspicuous cliffs on Canna are at Compass Head, about 450 feet high (137 m). The cliff is made up of a variety of rocks (see Fig. 29). The coarsest agglomerate is at the bottom. The name, Compass Head, is given to it because the compass is considerably disturbed in its vicinity, a not infrequent phenomenon on exposed dolerite hills. The cliffs continue along the northern coast; the agglomerates thin out and basalts (dolerites) form the cliffs. Harker (1908) argued that sills give rise to the terraced appearance of the inland hills; in the cliffs they do not make noticeable features, but may make broad reefs on a shelving part of the coast. The sills are often well jointed. The agglomerates also die out to the south and west. There are two conspicuous stacks, Dùn Beag and Dùn Mór which, with the neighbouring cliffs, show a volcanic conglomerate between two sills of dolerite. The columnar nature of the dolerite is locally striking, as near Geodha na Nighinn Duibhe, where the tops of a sill form a regular causeway (see Plate 40).

Canna is nearly separated into two parts by the depression at Tarbert: during the time of the '100-ft' beach there must have been two islands. Traces of a high beach are conspicuous near Tarbert, where it is 70–80 feet (21–4 m) above sea-level. At Coroghon it is more than 100 feet (30 m). Elsewhere in the island there are traces of a beach at about 20 feet (6 m), but the steep cliffs are not favourable to the preservation of beaches. The hollow representing Canna Harbour is eroded in volcanic conglomerates. The steep cliffs which surround much of Canna are best seen from the sea, and in that way also the minor but picturesque details of their sculpture are appreciated.

Rhum

This beautiful island presents many coastal features of great interest. The island is almost entirely surrounded by cliffs which vary in nature with the rocks of which they are made and, of course, with the structure of the island. (Much of this account is based on McCann and Richards' comprehensive paper (1969).)

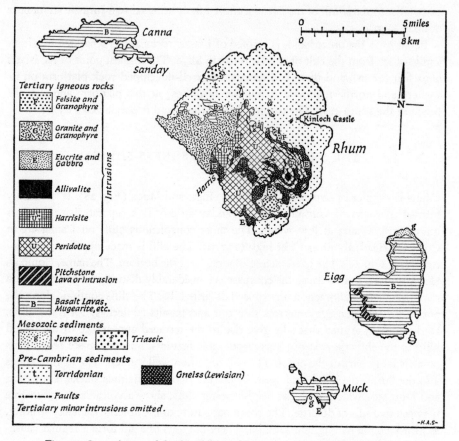

Fig. 28. General map of the Small Isles of Inverness-shire (Geological Survey)

In the north and east of the island Torridonian rocks form not only the coastline, but also much of the interior. There is a small patch of Triassic rock in the north-west around A'Mharagach. The Torridon rocks generally dip to west and north-west at about 20°–30°, and are separated from the high ground of the Tertiary igneous rocks which form a truly mountainous area reaching 2,500 feet (762 m) in Sgurr nan Gillean. Between A'Bhrideanach and Harris the coastal rocks are grano-phyre, and between Harris and Papadil they are mainly basic in type. Despite the boldness of much of the cliff scenery, it is necessary to realize that the work of the present sea is limited to relatively slight modification of existing forms rather than the creation of major new ones.

McCann notes five phases of marine activity in Rhum: (1) a high level rock platform, 60–100 feet (18–30 m) above high-water mark; (2) the dissection and partial removal of this by a sea approximately at the present level; (3) Late-glacial raised beach deposits and minor features of erosion between 40 and 98 feet (12 and

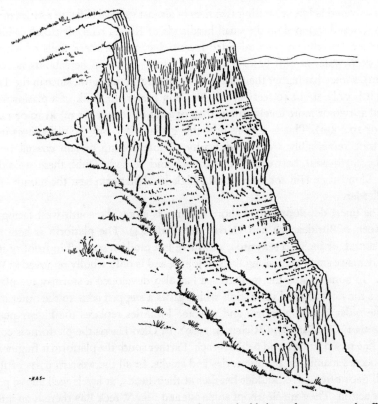

Fig. 29. View of cliff at Compass Hill, Canna, showing bedded conglomerates and tuffs with sheets of columnar 'Dolerite' (rep. from *The Geology of the Small Isles of Inverness-shire* (Mem. Geol. Surv., 1908), Fig. 10)

30 m) above high-water mark; (4) Post-glacial raised beach deposits and minor features produced by erosion up to 27 feet (8.2 m) above high-water mark; and (5) modern beach deposits and some erosion. These five phases of marine activity must also have affected neighbouring islands and the mainland coast. In Rhum their effects are particularly clear and have been worked out more fully than elsewhere. It should not be thought that the whole of the great precipices at the south-western corner of the island are wholly produced by marine action. McCann thinks marine activity has made itself felt up to 900 feet (290 m), and even more in Bloodstone Hill. However, the slopes are seldom simple in form, and it is difficult to assess the precise effect of sea erosion. The northern and eastern coasts of the island are less spectacular. In the Torridon Sandstone there are many places in which cliffs range from 50 to 250 feet (15 to 76 m) and, as in all similar coasts, there are many stacks, caves, and small geos cut out along lines of weakness. South of Loch Scresort and near Dibidil and Rubha nam Meirleach there are several discontinuous patches of the high-level platform. The form of the platform varies from a narrow

scree-covered ledge, separating two tiers of almost vertical cliffs, to a series of wider ledges, best developed on the small headlands of Rubha nam Meirleach, which are separated laterally by vertical sided recent chasms, the sheer walls of which fall into deep water at cliff base. The height of the platform hereabouts is 100 feet (30 m), somewhat higher than in the west and suggesting a tilt, since in the Triassic area it is only about 50 feet (15 m). McCann prefers to think of a platform developed at two or more levels, rather than of a tilt of 50 feet (15 m) in 10 or 12 miles (16 or 19.3 km). This is a more probable explanation, but the differences in level illustrate remarkably well the difficulty of interpreting certain coastal features. In the north-west, between A'Bhrideanach and Monadh Dubh there are old cliffs near Bloodstone Hill and traces of the high-level platform near the mouth of Glen Shellesder.

The finest development of shore features is along the south-west facing coast between A'Bhrideanach and Harris (see Plate 44). The platform is here nearly horizontal, or inclined seawards, and there is a modern sea-cliff in front of it. It is cut by many small geos along lines of joints, and is only wholly removed at Wreck Bay. 'In some places the modern sea has also developed a storm-wave platform above the normal level of erosion which gives a stepped edge to the outer margin of the older higher platform and at some localities replaces the higher platform altogether' (McCann and Richards, 1969). In Glen Harris the platform is covered by a fine raised shingle and pebble beach. Farther south the platform is fragmentary, and occurs mainly on promontories and stacks. In all the western parts of Rhum, small geos often have boulder beaches at their heads, at levels well above present wave action. They are clearly of some age and near Wreck Bay there is an interesting piece of evidence. Immediately south-east of the bay, Late-glacial marine gravels rest on the platform and are covered by steeply dipping scree-like material. The gravels and scree are continuous across a drift-filled cleft, which is an old geo cut into the platform by a sea at about the present level. Other clefts indicate similar events. This low-level erosion, earlier than the Late-glacial, accords with what is known elsewhere in western Scotland, and especially in Colonsay and Oronsay. There are many traces of Late- and Post-glacial beach deposits, including Harris, Schooner Point and Kilmory; that at Harris is regarded as the best example, apart from those on Jura, in western Scotland. There are several curved ridges, mainly without vegetation, and the ridges are in two parts, on either side of the Abhainn Rangail. The remains of these beaches, in gullies cut along joints, to the west of Kilmory are noteworthy.

Modern beaches occur in many small inlets. They are usually steep. There is a larger beach near Bloodstone Hill, and another at Harris. Sand and dunes occur at Kilmory, and there is a mixed beach at the head of Loch Scresort.

The general shape of Rhum has been ascribed to faulting, and it has been suggested that Rhum, Canna, Eigg and Muck 'are arranged in a reticulated manner, conformable with the coastlines of Rhum, and the four islands can be regarded as upstanding elements in a pattern of blocks and basins' (Ryder, 1968). Faulting may be partly responsible for the line Kilmory Glen and bay, and also for that of Glen Shellesder and Kinloch Glen, and perhaps the trend of Loch Scresort.

Eigg

Eigg is about twelve square miles (31 sq. km) in area. It is divided into two parts physically by a depression which runs between Laig Bay on the north-west and Poll nam Partan on the south-east. The north-east is a hilly plateau which reaches more than 1,000 feet (305 m) in Dùnan Thalasgair. This high plateau is formed of Tertiary lavas resting on Jurassic strata which form a prominent outcrop all round the northern coast. The Lower Estuarine shales in the north first form (in the north-west) the lower part of the cliff, and then the whole undercliff. They begin to sink downwards on the east coast, and have disappeared a little north of Kildonnan. They are penetrated by numerous very thin sills of basalt. The overlying Great Estuarine Sandstone is, in total, about 200 feet thick (61 m). Two interesting features are associated with it. Calcite is often present in sufficient quantity to bind the sand grains and form concretions up to 20 feet (6 m) in diameter. These weather out of the cliff and shore on the north-west of the island. 'A group of these concretions has been developed beneath a hard, less calcareous, band which projects from the cliff face like a shelf. Beneath this band a number of these concretions have been developed, and have been weathered out, and these now present the appearance of huge wasps' nests suspended from the lower side of the shelf'. The other feature is the fairly uniform size and shape of the quartz grains. When they form a beach or layer they give out a curious note when walked on. This is very noticeable in Camas Sgiotaig. A band of limestone rests on the sandstone. It is seen in the Laig cliffs. It is covered by shales and then the Oxfordian beds, but these concern the coast but little. A very small outcrop of Upper Cretaceous rocks can be seen at the end of the scars at Clach Alasdair, to the west of Laig Bay (see Plate 45).

Apart from the Sgurr of Eigg (below) the rest of the island is almost entirely built of lava flows. These are beautifully seen in the sea-cliffs. Locally agglomerates and breccias are intercalated. In Eigg the lavas dip gently to south and south-west. But there are one or two faults of some magnitude on Eigg, and one or more of these has helped in the formation of the Laig-Kildonnan depression. The junction of the Jurassic and volcanic rocks can be seen in the north of the island. In the south-east of the island, intrusions of andesite make a considerable effect on the coast between Kildonnan and Galmisdale Point. The sheets intrude the Oolitic rocks which outcrop on the shore at Poll nam Partan; they also form the reefs and islets north of Eilean Chathastail. They are all thin, but conspicuous because they are usually coated with black glass.

Although it does not reach the coast, no account of Eigg can omit reference to the Sgurr. Even if it is not coastal in the strict sense of the word, it is so close to it and so dominant a feature that it seems, especially at its western end, to be a true part of the lava cliffs underlying it. The Sgurr is a columnar pitchstone reaching 1,291 feet (394 m). It has been interpreted in various ways. Geikie (1897) thought it represented a series of subaerial lava flows flowing down valleys in the lava plateau; Harker (1908) thought that the whole mass was intrusive. Bailey (1914) supported Geikie, but Harker later 'directed attention to the smooth nature of the contact between the base of the pitchstone and the supposed eroded side of the valley as being unlike the terraced hill-slopes of the plateau country of the present day' (Richey, 1961) (see Plate 46).

The '100-ft' beach is well developed around Cleadale; it is mainly a terrace; it is also seen on the east coast near Rubha nan Tri Chlach. The '50-ft' level is seen as a terrace in Camas Sgiotaig; it is very small. On the extreme north, the '25-ft' is represented; it is protected by Eilean Thuilm. In general, the steep cliffs of Eigg do not offer good conditions for beaches. The blown sand in Laig Bay is but a thin layer resting on older sands, possibly Pre-glacial.

Landslips are common and extensive. The surface of the lavas is often greasy and affords an easy sliding surface. Near Cleadale the outward dip facilitates this. In the ice age, pressures of the ice would increase this tendency. In the north-east where the lavas and the Great Estuarine Sandstone rests on shales there is abundant cause for slipping, but since the dip is into the hill, the volume of slip is not great. There is also a marked undercliff. The question has been raised whether this inland cliff is an old marine cliff or a true scarp rising from an old and formerly extensive land area. It is possible that this low rim of Eigg may be a remnant of a much larger area. It is Pre-glacial in Barrow's (1908) view. He concludes, however, that to regard the undercliff as a remnant of a former marine plain, ending in an old cliff, is doubtful. There is considerable scope for further research on the cliffs of Eigg. The eastern side of the island, where the undercliff is well developed, is relatively sheltered. The west is more exposed. It would be interesting to know the reason for the remarkably straight part of the western coast which runs almost due north and south from Rubha an Fhasaidh (see p. 136). Around the coast there are several waterfalls draining from the plateau.

Muck

Muck covers about two-and-a-half square miles (*c.* 11 sq. km); most of it is less than 250 feet (76 m) above sea-level, but Beinn Airein reaches 451 feet (137 m). Eilean nan Each (Horse island) is joined to it at low water. There is a high percentage of pasture and arable land on the island. Although small, its geology is by no means simple. The Great Estuarine Series crops out in Camas Mór where it makes a succession of inter-tidal reefs. Locally they are metamorphosed by an intrusion of gabbro. Tertiary volcanic rocks cover the Jurassics and are nearly horizontal. Occasionally pyroclastic rocks are seen at the base of the lavas; in

Fig. 30. Dykes on the coast of Muck (after A. Harker)

Camas Mór they rest on the Estuarines. The gabbro intrusion is not large, but it makes a range of fine cliffs on the eastern side of Camas Mór.

The most striking features of the island are the great numbers and the significance of the basic dykes which cut through the lavas and sills (Harker, 1908). They are closely packed in nearly all parts of the island. Between the two bays on the south coast, Camas Mór and Port Mór, 40 dykes are visible. Along the whole southern and eastern shores 134 dykes were mapped; this implies an average of one dyke for every 28 yards (26 m). The dykes usually stand salient on the coast, but a few which cut a resistant agglomerate form small trenches since they weather more quickly than the agglomerate. Only a few of the dykes so prominent on the coast can be traced far inland, and these are usually on the lower ground. All the dykes are remarkably parallel to one another; the average trend is N 30° W. On the coast they often form walls 20 to 30 feet high (6 to 9 m), and they are nearly vertical; some hade at a slight angle. Sometimes part of a dyke is displaced relative to the remainder; in Camas Mór two neighbouring dykes show this, the parts of the dykes higher up the beach are displaced westward. Some of the dykes continue as far as Eilean nan Each. Although individual dykes make significant shoreline features it can scarcely be said that they control the shape of the island (Fig. 30).

There are several small beaches, but dykes usually interrupt them. Sand, largely calcareous, is blown inland. This, together with the drift deposits of the interior, make Muck a very fertile island.

COLL AND TIREE
Coll

Coll is about twelve miles (19 km) long and is wholly composed of crystalline metamorphics which are cut by much later minor intrusions. Most of the island is less than 200 feet high (61 m), but Ben Hogh reaches 339 feet (103 m). The crystallines are regarded as part of the Lewisian, and they include metamorphosed sediments. All the rocks are folded isoclinally, so that in a traverse of the island from north-east to south-west we find first a large area of grey orthogneiss which covers about two-thirds of the island. On the coast this rock is often shattered and veined. The dip is also seaward. Then comes a narrow belt of paragneiss followed by pinkish granite or granodiorite, two belts of paragneiss which correspond on the north-east with Hogh Bay and Feall Bay, and on the south coast with Loch Breachacha and Crossapol Bay, and finally a belt of grey orthogneiss. The later dykes are frequent in the grey orthogneiss of the north-east, and where they out-crop on the coast they form minor features.

There are many raised beaches in Coll; the beach material is usually found in rock hollows, but along the north-west the many areas of recent blown sand hide the raised beaches. The beaches have, however, been traced up to about 100 feet (30 m), but terraced features are negligible except in Loch Eatharna at Arivirig and farther south-west at Port na h-eitheir. The best exposure of the '100-ft' beach is on the north coast near Grishipoll Farm. In '100-ft' times Coll would have been divided into four. The low ground between Bagh an Traillich and Loch Eatharna, and also that between Hogh Bay and Loch Breachacha is well below that level. There was also a narrow gap along the line of Lochs Cliad and an Dùin. Occasionally there are small raised shingle spits, for example, near Rudh a'Bhinnein on the north coast. There are also extensive peat deposits which cover much of the low ground, including the raised beaches. Much of this peat has been cut by people from Tiree on which island there is almost a complete absence of peat, a fact not satisfactorily explained. There are many fine areas of blown sand, especially on the north coast. The sand is often very calcareous, and locally is cemented to form a soft limestone. These nearly all face good beaches, and at Feall and Crossapol Bays sand drifts from north and south have united. Coll is partly surrounded by many small islets and rocks, some of which are tied to the main island by sands covered at high tide. The island of Gunna, between Coll and Tiree, is nearly one mile long but at the most a quarter-of-a-mile broad. Its structure is by no means simple, and is made up of at least five different rock types. There is blown sand at its eastern end, where at low water sand flats unite several islets and skerries with Gunna (see Richey and Thomas, 1930, and Fig. 31(a)).

Tiree

This very irregularly shaped island (see Fig. 31(b), is about twelve miles (19 km) long and its greatest breadth is about six miles (10 km). It is low lying; a large part of it is below the 50-foot contour, and the three western hills – Ben Hough 388 feet (118 m), Ceann a' Mhara about 250 feet (76 m), and Ben Hynish 460 feet (140 m) – appear almost as isolated small islands.

The solid rocks are all ancient, apart from occasional dykes. The base is Lewisian Gneiss in which are small areas of dark hornblende gneiss and pink granite gneiss. These rocks form a low platform which must have been almost wholly submerged in higher raised beach times. Today large areas are covered with blown sand. Sand extends from Tràigh Mhór, in Gott Bay, to the two smaller bays on the north coast, Vaul and Sulum. A wider mass covers the central part of the island between Tràigh Bhagh and Balephetrish Bay. It is thought that this over-

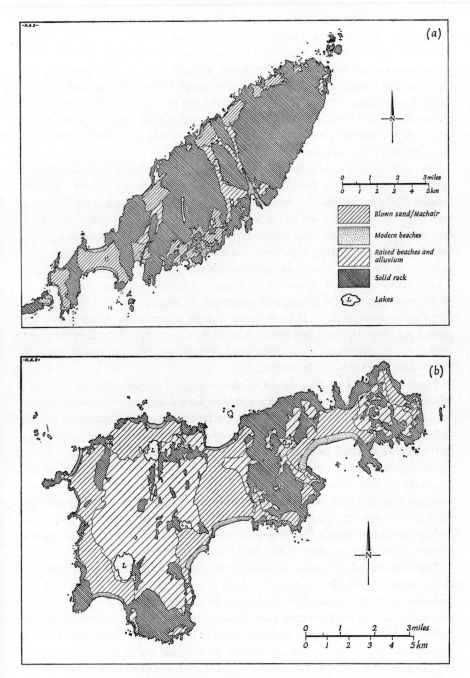

Fig. 31. (a) Geological sketch map of Coll; and (b) geological sketch map of Tiree
(based on Geological Survey)

lies paragneiss. There are other extensive areas of sand in the north-west, and along nearly all the western coast. The coast in the bays shows a succession of magnificent beaches. They are enclosed usually by low and irregular headlands of gneiss which shows its characteristic forms. The three western hills are conspicuous, and Balephuil Bay is enclosed by two of them. The details of the rocky parts of the coast are for the most part such as would be expected as a result of the submergence of a glaciated and low-lying area of gneiss. Occasionally local detail is given by the dykes, as, for example, to the south-east of Ben Hynish. There are a number of off-lying rocks and skerries.

Since the island is so low, it follows that much of its surface is covered by raised beach deposits. This is particularly true of the western part where extensive remains of the higher beaches are found. The lower '25-ft' beach is visible on the west of Balephetrish Bay and to the west of Vaul and Gott Bays. Many other parts are covered with blown sand which forms a true machair. The machair (Vose, Powell and Spence, 1959) is in Tiree the term given 'to the region immediately adjacent to the dunes where the psammophilous grasses have been eliminated'. Two types of soil are recognized in the island, a relatively siliceous drift produced by the weathering of the gneiss, and a younger calcareous soil. The most detailed investigation was made at Balinoe inside Tràigh Shorobàidh. Along a line 550 yards long (503 m), extending inland from high-water mark, the following stages were noted: mobile dune, stable dune, fixed dune, young machair, mature machair, old machair and transition to marsh.

There is no need to give details of the vegetation in these stages; there is a relatively normal transition from foreshore to fixed dunes, at which stage mosses have arrived. The vegetation of the young machair consists of a Festucetum rubrae, with *Trifolium repens*, *Plantago lanceolata* and *Lotus corniculatus* as sub-dominants. In the mature machair bare ground is more or less absent, and in addition to the plants characteristic of the young machair there is 'an increase in the number and frequency of the least valuable grazing species, particularly the mosses and euphrasic species'. The old machair is usually damper, and the transition areas are liable to flooding after heavy rains and consequent rise in stream level. In the more or less permanent marsh the machair species remain, but their proportions and relative importance are altered; *Iris pseudocorus* is common.

There are several other machair areas on the island, and there are local variations in the nature of the plant cover. The sand is nearly all of shell origin, and the carbonate content is about 66–70 per cent. In the mature machair the carbonate percentage can fall to 38 per cent. (On the surveyed line, the percentage of calcium carbonate decreased to 6.1 per cent as a result of leaching and the accumulation of organic matter.) Under natural conditions the machair soil is firm and compact, but easily destroyed by plough or wheels. It is then liable to serious erosion.

Around both Coll and Tiree there is usually a well-marked rock platform bordering the solid rocks. Many small islands and skerries are part of this.

Skerryvore

About eleven miles (18 km) south-west of Tiree is the Skerryvore lighthouse which is built on an isolated rock of Lewisian Gneiss. It is the only true islet in a group of rocks which rise from a submarine bank elongated north-east and south-west.

Dubh Artach, about eight-and-a-half miles (13 km) south-east of Skerryvore, is 47 feet (14 m) above high-water mark. It consists of an intrusive dolerite showing irregular columns. There is no land vegetation.

THE TRESHNISH ISLANDS, STAFFA, MULL AND IONA

The Treshnish Islands

This archipelago lies west of Loch Tuath and Gometra (Mull). It is elongated north-east and south-west and consists of about ten islands or islets and a number of skerries awash at high water. Some of the islands are tied or almost tied in pairs by tombolos, or reefs, for example, Bac Beag and Bac Mór, Lunga and Sgeir a' Chaistell. Fladda is a double island. Lunga is the largest and reaches 337 feet (103 m); Bac Mór (The Dutchman's Cap) reaches 284 feet (87 m). They are wholly basaltic and show raised beach phenomena to perfection. On nearly all there is a platform at about 100 feet (30 m) which has been ascribed to Pre-glacial marine erosion. It is very distinct in Bac Mór and Lunga. The brim of The Dutchman's Cap is the platform, which is a little higher in the north part of the island. It is ice moulded and shows moutonnée forms. Three lavas can be recognized in it; in the peak six can be counted. Similar features can be seen in the other islands, but usually only two or three lavas are present. The islands are fairly closely encircled by the 10-fathom (18.3 m) line.

Staffa

Staffa is a flat-topped island rather more than half-a-mile (800 m) long in a north–south direction, and about a quarter-of-a-mile (400 m) wide where it is broadest. It is wholly basaltic. The highest point is 135 feet (41 m). There are three caves: Goat Cave, Clamshell Cave and Fingal's Cave. The lava rests on red ash which is apparent on the western coast, but farther north disappears under lava flows. At Fingal's Cave three lava zones may be seen. The lower zone consists of massive regular columns, a middle zone of narrow and wavy columns, and a top zone mainly formed of slag. The slaggy zone runs for about 300 yards (274 m) along the east coast south from Goat Cave (see Plate 49).

The close-spaced and wavy columns were probably the result of rapid and irregular cooling, whereas the straighter ones have developed by slow regular cooling of the lower surface. The spacing of the columns may be controlled by the rate of cooling (Bailey and Anderson, 1925).

Mull

Mull is in some respects similar to Skye. In the east and south is a great plutonic complex. As in Ardnamurchan the centres of intrusion moved to the north-west, and extend beyond Ben More, the highest peak in the island. It is very irregular in shape. Loch na Keal and the low ground to Salen; Loch Scridain and Glen More leading to Duart Bay; and Loch Buie, Loch Uisg, Loch Spelve and the low ground to Loch Don and Duart Bay divide the island into four parts. Glen More makes a tract of low ground, but it does not make a geological break; the ring dykes and features of the complex are not interrupted by it. Loch Frisa and the Bellart valley, and also Loch Tuath divide the north-western part of the main island also into four parts. Gometra and Ulva islands are completely separated by the narrow strait of the Sound of Ulva. This irregularity, like that of Skye, must largely be attributed to fluvial action where the land stood higher relative to the sea, and when the island formed part of the mainland. These corridors of low ground are in some way connected with the early drainage system. Several writers (see Chap. 11) have speculated on this, but since Mull itself is only a remnant isolated by erosion and subsidence, it is not easy to be specific. Bailey suggests that the original stream from the direction of Loch Scridain turned near Torness into Glen Forsa. This assumes a continuous course from Loch Scridain to Ishriff. It is suggested (Bailey) that its tributaries, Gleann Seilisdeir from the north, and the Beach river, Glen Leidle and the Abhainn Loch Fhuaran from the south, are beheaded streams, and that they have been restricted by the retreat of the Gribun and Carsaig cliffs.

North and west of the plutonic complex, most of the rest of the island is composed of great lava flows which dominate the scenery (Fig. 32, cf. Fig. 24). In the far west of the southern peninsula, near Bunessan, there is a small tract of Moine Schist, and the Ross of Mull is a granite mass. Iona (see p. 149) consists mainly of Lewisian Gneiss. Around some of the coastal parts of Mull – on the west of Ardmeanach, on the south coast, on either side of Loch Buie and in Loch Spelve and Loch Don and in Duart Bay and one or two places in the Sound of Mull – there are, just as in Ardnamurchan and Skye, fringes of Mesozoic rocks showing under the lavas. The lavas were poured out on an eroded lowland of pre-Cretaceous rocks, some greensands of Cenomanian age, and occasional patches of chalk. The successive flows built great plateaux: 'In Skye and Mull, the lava piles exceeded 6000 ft [1,828 m] in full thickness, and were not less and probably much more than 2000 ft [609 m] in the Small Isles' (George, 1966). Since then there has been an enormous amount of erosion, so that parts of the floor are revealed. We have seen (p. 78) that in Ardnamurchan the intrusions of cone-sheets and ring-dykes led to a considerable tilting of the surrounding Mesozoic strata; the same was true in Mull, especially in the Loch Don area where folding and some overthrusting occurred.

Most of the coast of northern Mull consists of basaltic cliffs and the country immediately adjacent shows a stepped appearance caused by the several lava flows.

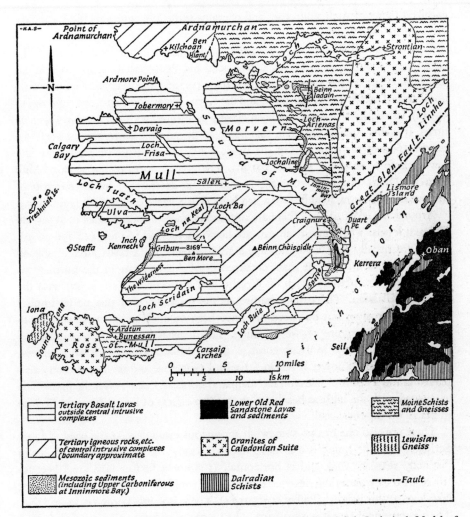

Fig. 32. Mull and the adjacent coastline of Scotland (based on the *Geological Model of Ardnamurchan* (Mem. Geol. Surv., 1934), Fig. 5)

Many dykes cut the coast and are responsible for minor features. The coast between Ardmore Point and Caliach Point is perhaps best appreciated from the sea or, on a clear day, from the Ardnamurchan peninsula. This area is divided into two parts by the narrow hollow of Loch a'Chumhainn. To the south-west of this line 'the featuring of the lavas is irregular in plan; while north-eastwards it is conspicuously lineated in a north-west and south-east direction. The difference is due to the fact that, north east of Loch a'Chumhainn, the lavas lie in the course of the great Mull swarm of north-west dykes, and that many of these latter have served as guides to erosion' (Richey and Thomas, 1930). The north coast is attractive; it affords excellent views of Ardnamurchan, and contains some fine bays, including Calgary Bay and its sand beach.

It is thought (Bailey *et al.*, 1924) that Mull was an island long before glacial times and, as will be shown later, Late-glacial and Post-glacial marine erosion has had little effect on the coast. In the Tertiary period there seems to have been a good deal of oscillation of sea-level, and the Lorne plateau, 800 to 1,000 feet (244 to 305 m), is held by some to have been closely connected with sea-level in Pliocene times. It seems equally probable that some of the south-eastern part of Mull is a continuation of the plateau. On the other hand, on the Ross of Mull there is a lower platform at about 300 feet (91 m). The arcuate form of the coast of Mull between Duart Point and Carsaig is said to be determined by an outcrop of Mesozoic rocks, now mainly below sea level (see Sheet 44, One Inch Geol. Sur.).

Around the coast of Mull there are numerous traces of raised beaches at several different levels. The highest remains are between 100 and 160 feet (30 and 49 m). These are usually notches indicating a submergence on a coast already shaped much as that of today. The main occurrences of this level are in the north-west. This is the 'Pre-glacial' beach of Wright, and corresponds (see p. 80) with the notch at Kilchoan. The patchy nature of the occurrence of this notch implies subsequent erosion, but this need not have been on such an extensive scale as to alter the general nature of the coastline. This level is usually conspicuous, and particularly so on the Mornish coast as far as Caliach Point. It is also present, usually at a slightly lower level, in Ulva and Gometra, and is magnificently displayed in the Treshnish Isles: on the mainland north of Loch Tuath the fossil cliffs associated with the highest beach are steep. The effects of ice erosion can be seen in places on the rock-shelf.

There are also one or two somewhat anomalous examples. At the head of Loch Scridain, near the hotel, is a level of approximately 160 feet (49 m). It is cut in distinctly resistant rock and is horizontal. It is not a former lake level because 'there is no corresponding deserted outlet which could have determined a persistent lake-level at this height' (Bailey *et al.*, 1924). There is a well-known cave on Ulva. There is no doubt that it is a sea-cave. The height of its entrance is 155 feet (47 m) above high-water mark. There is a less striking cave on Little Colonsay; it has not been accurately levelled, but is thought to correspond with that on Ulva.

Beaches of a later date are also numerous on Mull. One of these is similar to the 'Pre-glacial' beach in height, but it can be shown to be Late-glacial in age because it was excluded from the more important glens by tongues of ice. It may occur at different heights in different places. There is no need to do more than indicate some of the places where it is found. In Loch na Keal deltas at Derryguaig and Scarisdale correspond with it, as do also terraces near Killichronan House. There are also traces in the low pass leading to Salen Bay. Traces are frequent in the Sound of Mull, for example in Fishnish Bay and peninsula, in the Java peninsula, and at the head of Loch Aline in Morvern, and particularly at Duart Point. Other occurrences are in the Croggan peninsula, Loch Scridain, Ross of Mull, Iona, the Gribun peninsula, Ulva, and near Quinish Point in northern Mull.

The lowest beach, the one usually referred to as the '25-ft' beach, is well-developed. But here, as elsewhere in Scotland, we can no longer assume that the wide rock platform so often associated with it, is of entirely Post-glacial age. The width of the platform varies greatly. We have seen (Chapter III) that the beach is tilted; this is noticeable in limited areas. In Oban it is approximately 30 feet (9 m), and is only at 20 feet (6 m) in western Mull and Iona. At the head of Loch na Keal there is today, and there was at the time this lower beach was formed, a shingle spit which deflects the river Ba for more than a mile to the north. In '25-ft' times it did not extend beyond Drumlang cottage. At a later date the river broke through the spit about half-a-mile farther south. Other examples of this beach are found in Loch Scridain, in the Ross and Iona, in Gribun and in northern Mull.

The distribution of, and the ways in which, the remnants of all these beaches occur clearly imply no great change in recent times in the shape of the coast of Mull. Many of the conspicuous cliffs today, especially in the Carsaig and Gribun districts owe little, perhaps nothing, to modern erosion. They often stand back of a bench cut, at least in part, at an earlier time. The steepness of the cliffs depends much on their structure, but in Mull as in Skye they have locally been modified by landslips; at Gribun one at least of these slips is Pre-glacial, and consists of lavas and Mesozoic sediments, 'Its Pre-Glacial age seems certain from the fact that it has lost all trace of landslip-featuring, although its tilted and somewhat broken constituents afford a striking enough scenic contrast to the escarpment from which they have broken away' (Bailey et al., 1924). In almost the same place there is a later, Post-glacial, slip about half-a-mile long, and parallel to the cliffs. There are some good modern slips in and near the Wilderness. The north-western face of the Ardmeanach peninsula is wild and interesting. In several places lavas overlie Mesozoic sediments, which in turn rest on Moine Gneiss.

There is a pronounced difference in the coast of the Ross of Mull. The outer part of this is granitic; the inner part on the south side is Moinian. In both the coast is irregular and there are numerous small headlands, re-entrants, and islets and stacks. These are nearly all parallel with the strike of the rocks. The Moinian rocks are cut off by a fault running from Bunessan to near Port nan Droigheann, so that on the north coast they only appear on the south shore of Loch na Làthaich which, at any rate in part, may be determined by the fault. The Moine rocks make the coast between Rubh' Ardalanish and the junction with the lavas. The Moines are repeatedly folded along axial planes which strike north-east. The folds are vertical or steeply inclined to the south-east. It is largely to this structural pattern that the coast in the Moinian area owes its characteristic form, and the large number of small geo-like inlets trending approximately north-east.

The coast in the granitic area is also indented, but its pattern is unlike that of the Moines. All the area is of moderate height; several hills are between 200 and 300 feet high (61 m and 91 m); one reaches 411 feet (125 m). Raised beaches are frequent. Rock shelves are well developed on Rubh' Ardalanish; the upper rugged surface is about 100 feet (30 m), and the lower at about 50 feet (15 m). The main road from Bunessan to Iona ferry traverses the granite area where it is often well below 100 feet. In all this part there are widespread relics of the '100-ft' and sometimes of the '50-ft' beaches. Erraid Island and its smaller neighbours are separated from the Ross by narrow channels with a reticulate pattern. There is an intricate mixture of low rock platforms, shallow channels and salt marsh. The south coast of the Ross is much indented, and several of the larger bays possess good beaches, for example, Ardalanish Bay and Port Uisken (in the schists). It would be interesting to have more information about the formation of these numerous inlets. The Torran Rocks are nearly all granite; they are bare of vegetation and swept over by waves in all storms. Eilean a' Chalmain is a mass of diorite.

We have so far been concerned with details of the coast. No account of Mull should omit the magnificent views that can be seen in all parts of the island. It is a high, mountainous island and nearly all the coastal views are dominated by this high ground. The view up Loch Scridain includes the coast and the great peak of Ben More. Both are in a full sense part of the coast. On the north of the loch are the great terraces of lava forming the Ardmeanach peninsula (see Plate 51); to the south is the lower and completely different country of the granite and schist of the Ross of Mull. The view from Iona is similar, but brings out also the contrast in colour, as well as of structure, between the dark lavas and the pink granite of the Ross. Along the Sound of Mull there are beautiful views, made more impressive by occasional glimpses of high mountains. The view from a few miles south of Tobermory, and looking to the high ground of Ardnamurchan above Kilchoan, and to the entrance to Loch Sunart is outstanding. The striking nature of the plateau country of the north-west (p. 144) is quite different; its terraced and lineated nature can be appreciated if one looks down to Dervaig and Loch a'Chumhainn. The glen in which this loch lies, the smaller Loch Mingary and Glen Gorm and Laorin Bay emphasize the trend of the country. In the south of the island the scenery around Carsaig Bay is distinct. In the cliffs of the bay Liassic sediments covered by lavas give steep cliffs over which waterfalls plunge from the west. This great line of cliffs is continued westwards along Aoineadh Mór and Malcolm's Point and beyond to Tràigh Codh an Easa and Aoineadh Beag. These are for the most part old cliffs, relatively little affected by the present sea. The section at Carsaig shows lava resting on Tertiary sandstone, Cenomanian sands, and Upper, Middle and Lower Lias shales. The sediments almost disappear about two miles (3.2 km) farther west. There are landslips on the east side of the bay and also about halfway between the bay and Malcolm's Point. At the point is a dolerite sill over basaltic tuff and below a conglomerate of rolled flint pebbles. The general line of the coast is smooth; a dolerite sill makes a small protuberance near Nun's Pass. The Carsaig Arches are well known and reached by the sea.

No description of the coasts of Mull can omit reference to Macculloch's Tree situated near Rubha na h-Uamha on the west of the Ardmeanach peninsula. This stands upright, a coniferous trunk 40 feet high (12 m), which is preserved in lava. It is in a recess in the cliff and the tree is a cast and resembles a pipe. 'The lower part of the pipe encloses a partially silicified semi-cylinder of wood glittering with quartz crystals. The diameter of the cylinder is about 3 ft [0.9 m], and outside is a hollow where soft black coniferous wood, a couple of inches thick, may be dug out with a hammer' (Bailey *et al.*, 1924). Although it can be reached on foot, it is most desirable that the visitor should approach the tree *on a falling tide*.

Iona

Iona is separated from Mull by a channel about one mile wide. It was suggested in the Geological Survey Memoir (Colonsay, Oronsay and The Ross of Mull, 1911) that the disposition of the rocks in Iona and the mainland indicate that the Moine thrust passes along the Sound. In this somewhat remote sense the sound may have a tectonic origin, but it is more probable that it owes its present form to fluvial erosion along a line of weakness possibly formed by the Moine thrust, and later to submergence. Whatever the true origin may be, the separation of the Lewisian and possibly some Torridonian rock on the island from the granite of the Ross is profound.

Most of the island is Archaean; along the east coast is a belt of Torridonian rocks; in the north-east they are flagstones; in the middle part of the coast shales; the remainder is conglomeratic. In the Archaean area there are some interesting features including bands of marble. These, however, need not concern us. There are two low peaks; in the north Dùn is 332 feet (101 m), and in the south-west corner there is a knob reaching 243 feet (74 m). The coast is somewhat irregular, especially on the west and south. A hollow runs roughly east and west through the central part of the island. In it are several raised beach terraces at about 65 to 70 feet (20–1 m) above high-water mark. Some less defined remnants reach about 95 feet (29 m). Lower beaches of deposition are common on all coasts. Some of them are shingle, and at Calva in the north and Port an Fhir-Bhreige in the south the old shingle lying above the modern shingle indicates a vertical movement of about 20 feet (6 m). Blown sand occurs in the north and west, and there is a small area of machair more or less corresponding with the golf course. Eilean Chalbha and other small islets close to it are all Torridonian. To the north-west of Iona Rèidh Eilean and its neighbour, Stac Mhic Mhurchaidh, are basaltic and show good columnar structure, especially on the Stac where the columns are about 100 feet high (30 m). Soa, about two to three miles (3.2–4.8 km) to the south-west, and the group of islets named Eilean na h'Aon Chaorach are Lewisian. Eilean Mór and Eilean Carrach on the east are granite.

Gillham (1957) has discussed certain aspects of the coastal vegetation of Mull and Iona. There is much accumulation of silt in the sheltered bays and inlets, and some development of salt marsh which is grazed by various animals. Grazing leads to a dominance of dwarf rosette forms of e.g. *Plantago maritima* and *Armeria maritima* and also keeps down the growth of edible grasses.

In the dunes and machair up to 81 per cent of $CaCO_3$ was recorded some distance from the sea in north-eastern Iona. The pH of Mull and Iona sands was 8.3 to 8.4, causing a rapid decay of vegetable matter in the sands. Since organic matter accumulates but slowly, there is little humus in the machair sands. Nevertheless they make a good plant habitat. Calcicoles are not found in non-calcareous soils. There are species excluded from acid soils but 'indifferent as to whether acidity is counteracted by $CaCO_3$ or sea salt'. In addition there are neutral grassland species and dune and sand species on mobile sand. At Calgary (Mull) erosion occurs along the edge of the machair pasture, and Calgary is now a favourite camping place. There are also erosion and blow-outs in Iona. Iona shell sand often rests on red granite and produces an acid soil. But the red granite of the Ross of Mull, where not sand covered, included only five out of the 130 species found on the adjacent coast of Iona; none of these occurred on the dunes or machair.

In the sheltered inlets the transition from acid moor to salt marsh took place within a few metres. Some moorland species, e.g. *Molinia caerulea* and *Eriophorum angustifolium*, penetrated below high water; halophytes, including *Armeria maritima* and *Plantago maritima*, extended upwards into the Molinietum of the moor zone. 'Except on the most sheltered shores where wave action was negligible, and the most exposed shores where all silt was scoured away, organic matter was at a minimum at high-water mark, increasing both upshore and downshore.'

THE GARVELLACHS, COLONSAY, LUNGA, SCARBA, JURA AND ISLAY

The Garvellachs or Isles of the Sea

This group of three main and several smaller islands trends north-east to south-west to the west of Luing at the entrance to the Firth of Lorne. They consist of a central fold of limestone, but the bulk of the islands are made of quartzite and quartzitic conglomerates. They are situated in an exposed position, and their north-west side receives the full force of the Atlantic waves.

The main fold, which is recumbent, runs parallel with the trend of the islands. The fine, mainly fossil, cliffs on the north-western coast have been cut largely in the limestone. The surface slope of the islands is usually to the south-east, away from the western cliffs. As a result of differential erosion, the rocks have been eroded into cuestas. There are a number of basalt dykes; these and the slightly faulted rocks alongside them, are more easily weathered and so form furrows of low ground, including the Bealach an Tarabairt on the largest island, Garbh Eileach. These features all depend primarily on the solid geology.

Apart from the northern and western cliffs, marine erosion has produced caves, blow holes and natural arches. There are also many remains of raised beaches, especially at about 30 feet (9 m) above OD. The best development of beach is in the south-east of the islands, and may reach 50 + yards (46 m) in width. There is also an inter-tidal platform which clearly shows the effect of ice-moulding, and also striae which show that the ice moved approximately in the direction of the trend of the whole group.

The islands are interesting also in other ways. One of the seminaries of the Early Columban Church was situated on the largest island. The islands are fertile; soil alkalinity gives rise to a black mull which supports a rich vegetation. The islands are consequently green and verdant, and not of the brown-purplish colour which is so common elsewhere (Hunter and Muir, 1952–6; Peach and Kynaston, 1909).

Colonsay

From the point of view of the student of shoreline features Colonsay is perhaps the most interesting island on the west coast of Scotland. Together with Oronsay, from which it is separated by a narrow and shallow channel dry at low water, it is about nine miles long (14.5 km), and little more than two miles wide (3.2 km). It is a rocky island, but the interior is fertile and supports an interesting vegetation. Since the island is so wind-swept, trees are limited to sheltered localities and to parts of the east coast.

Most of the rocks of which the islands are formed are Lower Torridonian in age, and include limestone, phyllites, mudstones, flags, grits, and conglomerates. There is also a tiny outcrop of Lewisian Gneiss in the bay just north of Rubh' a' Geadha. There are also outcrops of plutonic intrusive rocks, the largest being that of the diorite at Scalasaig. In the north there are numerous lamprophyre dykes and sills, and some Tertiary dolerite dykes. The sediments are much folded and cleaved. Faulting is often conspicuous, some of the more important faults running approximately south-west to north-east across the central part of the island. Fig. 33 (a and b) shows sections through the island and indicates the disposition of the beds. The general structure of the islands is fairly simple but the individual beds often show intense puckering and contortion. Locally, there may be minor overfolding or even inversion (Cunningham Craig et al., 1911).

Colonsay reaches only moderate heights; several peaks exceed 400 feet (123 m), the highest being Càrnan Eoin, 470 feet (143 m). Between the higher hills are low valleys or straths, some of which are noticeably straight, and run with the faults mentioned above. The most remarkable of the straths is that in which Loch Fada rests. Others are easily traced on the Ordnance Map. On the west coast Port Sgibinis, Port nam Fliuchan, Port Mór, Port Lobh and Tràigh nam Bàrc are the drowned ends of these valleys. Some of the straths have been formed along lines of crush or fault, so that in that sense much of the detail of the west coast is indirectly

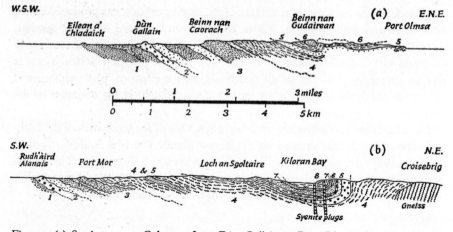

Fig. 33. (a) Section across Colonsay from Dùn Gallain to Port Olmsa; (b) section across Colonsay from Port Mór to Croisebrig. 8, Staosnaig Phyllite group; 7, Colonsay Limestone group; 6, Kiloran Flag group (very uniform and constant); 5, Milbuie group (largely epidotic grits and phyllites); 4, Kilchattan group (phyllites and sandstones); 3, Machrins group (grits and mudstones); 2, Dùn Gallain group (epidotic grits); 1, Oronsay group (sandstones below, mudstones above) (based on Geological Survey)

of tectonic origin. It is thought that Kiloran Bay owes its shape to the erosion of softer phyllites in the centre of a synclinal basin. The big inlet, a little farther to the north, which was occupied by the sea in higher raised beach times, was probably formed by the erosion of the kentallenite. The passage between Colonsay and Oronsay may have an origin similar to the more northerly straths.

Despite the interesting features presented by the present coast, the traces of former sea-levels in Colonsay are of outstanding importance. There is first of all a plain of marine erosion, probably of Pre-glacial age. It has, of course, suffered much since it was formed. It is finely developed in Uragaig, to the west of Kiloran Bay, where it is about half-a-mile wide. At its inner edge, beneath the old cliffs of Torrnach Mór, it is 135 feet (41 m) above sea-level. Fig. 34(b) shows the distribution of the remnants of this platform; it will be clear from an inspection of this map that at the time of the formation of this platform the higher parts of Colonsay formed separate islands. On the east the islands would have been exposed to the main impact of the ice, and this can be seen in the smoothing and rounding of the old cliff, especially in the less resistant rocks. The suggestion is made in the Memoir (Cunningham Craig, et al., 1911) that either as part of this planation or during a second Pre-glacial level, the numerous islands and skerries off the west coast were formed. They are ice-worn and covered at high water. They call to mind the similar skerries off Arisaig.

In addition to these early levels, Colonsay affords remnants of later beaches, many of which are represented by great banks of shingle (cf. Jura). The best are at Lower Kilchattan. Some shingle ridges are now found well away from the sea.

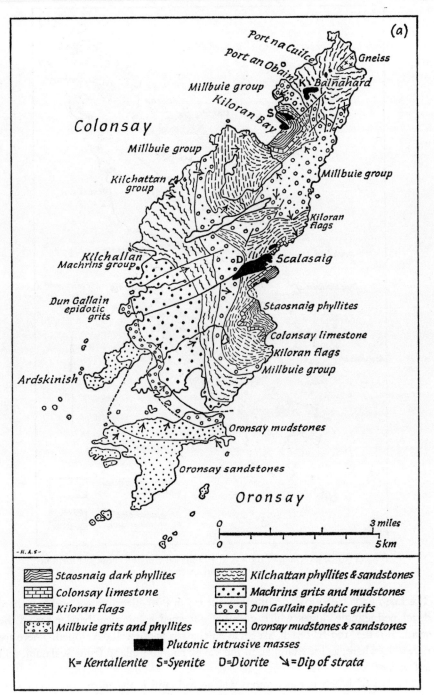

Fig. 34. (a) Sketch map of the geology of Colonsay and Oronsay

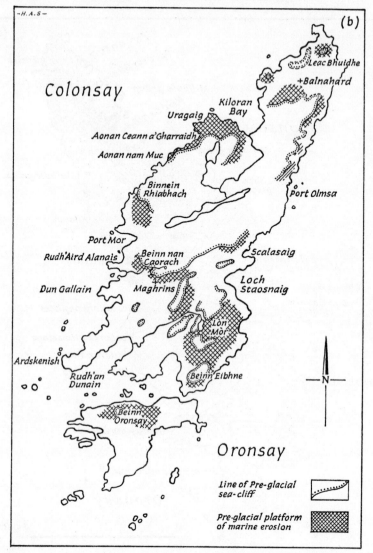

Fig. 34 (b). Sketch map of the 135-ft Pre-glacial cliff and shore-platform in
Colonsay and Oronsay

The Loch Fada valley was occupied by the sea in Post-glacial times, and in it there
developed shingle ridges which eventually formed a complete barrier across the
strath. But the '100-ft' beach, except at Kilchattan, is not well developed. There are
traces elsewhere, as for example, in the valley running inland from Scalasaig, and
at Garvard.

The '50-ft' beach is much more widespread, and is the most conspicuous. It
often occurs in the form of great gravel embankments as at Balnahard, Uragaig,

and several other places. The former cliffs associated with the beach can be followed, and near Uragaig its coastline is cut by several geos. Since the gravel deposits which are assumed to be associated with this beach vary considerably in height, any precise evaluation of the sea-level at which they were formed is, even in Colonsay itself, almost impossible.

The lowest, the '25-ft' beach, is not well developed in Colonsay. There are several places where it can be seen either as shingle formations or as a bench. The estimation of the level of the sea at which the beach was formed is just as difficult as that of the higher beach. A large part of Oronsay is composed of the gravel of this beach, including some cuspate shingle formations. Shingle formations are also present on the neighbouring smaller islands, Eilean Ghaoideamal and Eilean nan Ròn. There are several caves associated with this level. They are large and indicate long-continued action of the sea. There are four caves at Port Easdail; eight more are listed in the Memoir. It is also remarked in the Memoir that the caves are in striking contrast to the otherwise trifling signs of erosion at this level. This fact may be consistent with the view, now generally held, that the '25-ft' beach is not wholly Post-glacial.

Lunga and Scarba

Lunga and Scarba, together with many smaller neighbours, should perhaps be considered together with Jura. Since, however, Jura is so much larger it will be treated separately.

In these islands there is abundant evidence of isoclinal folding. The major, western, parts of the two main islands are quartzite, but there is nevertheless a good deal of difference in the resistance offered to erosion by the various beds. In the eastern part of Scarba schists and slates form a wide belt; in Lunga they are also present to a much smaller extent, and mainly in the south. There are also a number of nearly north–south dykes of epidiorite. These, in turn, are sometimes cut by dykes of Lower Old Red Sandstone age. On the west of Lunga, a porphyrite dyke can be traced for about half-a-mile southwards before it passes out to sea. Dykes of this type are conspicuous on the platform at the north-western part of Scarba. At the northern end of Lunga the island is cut by several channels which, at high water, turn this part of Lunga into a small archipelago. The small islands, Eilean Dubh Beag and Eilean Dubh Mór, are mainly formed of quartz and traversed by many approximately north–south dykes. In all of these islands there are many remnants of raised beaches.

Scarba is a much larger island. There are some conspicuous north–south gashes, and at about three-quarters-of-a-mile (1.2 km) westwards from the east coast one of these gashes or hollows traverses the whole island. On the south coast each of these is marked by a small bay. The island reaches 1,470 feet (448 m) in Cruach Scarba; it is a detached part of Jura.

These and other islands are all parts of a once continuous land mass which was

sculptured by subaerial erosion, both fluvial and glacial, and then submerged – the
raised beaches alone indicate that this was not one single movement – to give the
present topography. It has been suggested that the line of Crinan Loch, Dorus Mór
and the Gulf of Corryvreckan represents the former extension of the River Add.
Similarly, the Barbreck may once have followed the line of low ground, Stain
Mhór, to Tràigh nam Musgan and thence via Cuan Sound to the Firth of Lorne.
Peach and Kynaston (1909) remark that

The erosion of these valleys dates from a period before the capture of the Add and Barbreck
rivers by the longitudinal tributary of a larger consequent stream, part of whose valley is now
represented by the Sound (of Islay) between Jura and Islay. This capture took place at a time
when the sea-margin lay at the outer edge of the continental shelf.

It is not difficult to reconstruct an extensive river system in this way, but it by
no means follows that the reconstruction is correct. Nevertheless to account for the
intracacies of the coast in this part of Scotland one must in the first place envisage
differential erosion by rivers and ice, working on a varied series of rocks cut by
numerous dykes and a number of faults. The later submergence(s) account(s)
for the extreme irregularity of the present-day coastline.

In all these islands fossil cliffs and a rock platform are conspicuous features,
especially on their exposed coasts. Along the south coast of Scarba there are
several caves associated with the old cliffs.

The Strait of Corryvreckan between Scarba and Jura is well known for the violent
tidal streams that pass through it as a result of the differences in the times of high
water in the open sea and in the Sound of Jura. A west-going stream at about 8.5
knots (15.5 km) extends several miles westwards, and with strong west winds gives
rise to heavy overfalls up to three miles (5 km) from the entrance. There is less
turbulence with the east-going stream. Eddies form on both sides of the main
stream, and especially on the north side of the westerly stream. Over the inequal-
ities of the bottom of Camas nam Bairneach the eddy is often violent and dangerous.
Heavy overfalls indicate the boundary between eddy and main stream; the one
may be running at about 8.5 knots to the west and the other nearly as fast to the
east. The eddies on the south are less dangerous, but navigation is hazardous when
the streams are setting through the strait.

Jura

This beautiful island is relatively simple in structure. In the main it consists of a
mass of quartzite elongated north-east to south-west. The dip, except locally
along the Sound of Islay, is persistently to the south-east at angles of 20° to 30°, and
occasionally of 35° to 40° or more. The slates and phyllites on the east are conformable with this dip. The highest points are the Paps of Jura; Beinn an Òir is 2,751
feet (839 m), Beinn Shiantaidh 2,477 feet (755 m) and Beinn a' Chaolais 2,407 feet

(734 m). There are several peaks between 1,500 and 2,000 feet (437 and 610 m), and the island is, throughout, mountainous and rough. Along the east coast, especially at the southern end, there are bands of phyllites of the same type as the Port Ellen phyllites in Islay. In these rocks there are also bands of epidiorite. Thus, along much of the east coast all these rocks give a striped effect, and the phyllites usually form low ground and the epidiorites crags which may reach 400 + feet (123 m). As a result of erosion and submergence the phyllites and epidiorites disappear in part. North of Craighouse there begins a wide embayment partly enclosed by islands (The Small Isles) formed mainly of epidiorite and some phyllites. Farther north the same conditions hold, and Lowlandman's Bay is almost enclosed by a low peninsula of epidiorite. The phyllites continue to follow the coast to the north; Lagg Bay is fringed on its seaward side by quartzite, the western side is in phyllites. Tarbert Bay is aligned south-east to north-west and is partly fringed by raised beaches. It appears to be related to the general structure in a way similar to the other bays on this coast, but its trend is to the north-west. At Lussa Point the point itself and Eilean an Rubha are quartzite; the narrow bay is in black slates, which extend from the bay northward for about two miles. This is a line of low ground running parallel to and just inside the coast which is formed of the quartzite. Ardlussa and Tramaig Bays cut through a low ridge of quartzite and reach the slates. In both, small islands and reefs of quartzite remain. Camas nam Meanbh-chuileag is a small bay facing northwards; its seaward side is a low quartzite ridge. The small inlet, Camas a' Bhuailte, is but a remnant of a somewhat larger bay dating from the time of the lower raised beach. Port Bàn is similar. For the next four miles (6.4 km) the coast, apart from dykes, is formed of quartzite; a small patch of slates makes Rubh a' Bhacain. Between Rubh'an Truisealaich and Kinuachdrach harbour a zone of quartzite alternating with slates, which reaches a height of 270 feet (82 m), encloses a narrow valley now filled with raised beach deposits. This continues behind the quartzite of the Aird of Kinuachdrach and reappears in Port an Tiobairt. The slates disappear at Port an Droighinn, but reappear in the eastern part of Scarba.

The western coast of Jura is of great interest. The raised beach phenomena are remarkable and cut and built features are both common. The finest display of beaches is, perhaps, south of Loch Shian, but they are more or less continuous as far north as Glendebadel Bay (see Plate 54) and are also found in the northward and north-westward pointing bays up to the Gulf of Corryvreckan. These bays, Glengarrisdale Bay, Bàgh Uamh nan Giall, Glentrosdale Bay, and Bàgh Gleann nam Muc, are structural and are partly associated with dykes. The whole island is nearly separated into two parts by Loch Tarbert. In raised beach times this was somewhat larger, and at the time of the highest raised beach extended as far as Tarbert Bay on the east coast, thus dividing the island into two main parts. Although the general trend of Loch Tarbert is east–west, its upper parts into which Gleann Aoistail drains, and also that part, a mile or so farther west, into which Gleann Dorch flows, are aligned roughly parallel with the east coast, and with the trend of the rocks both in Jura and the mainland near Loch Sween.

The raised beaches on the west coast of Jura have frequently been described, but the only comprehensive account is that by McCann (1961). His thesis refers to a considerable part of the west coast of Scotland, but the details and analyses of the west Jura beaches show not only their range in height, but also in the shingle spreads, their evolution. McCann regards the higher platform as Inter-glacial; it is backed by a cliff, and on the platform there are abundant gravels associated with the Late-glacial sea. Much of the gravel is overgrown by peat and vegetation, but there are numerous shingle ridges and shingle spreads. The front of this higher platform is the cliff associated with the Post-glacial '25-ft' beach. This lower platform is usually narrow, perhaps 100 or 150 yards (91 or 137 m), whereas the width of the higher and older platform may locally reach about half-a-mile (805 m). Since the shingle ridges, almost wholly composed of quartzite pebbles and cobbles, derived in part from boulder clay, are well preserved, it is possible by mapping them to determine the way in which they have been built up. There is a remarkable group in the former embayment which contains Loch an Aoinidh Dhuibh. Fig. 35(b) shows how McCann envisages their evolution. The highest ridges are those nearest the former cliff; the next group, at a somewhat lower level, appear to have been built from the south-west. The third and lowest group have a more northerly trend. Although this group is the most complex, beaches at corresponding heights occur all along the coast from Corpach Bay southwards. On the south shore of Loch Tarbert the beaches are well developed, and Fig. 36 shows how in Glen Batrick they are related both to moraine and solid rock. Beaches extend along most of the Jura coast of the Sound of Islay and in the coast north of Brosdale island they tie what were formerly islands of epidiorite into the mainland coast.

In western Jura the highest shingle is at 115 to 120 feet (35–37 m) above sea-level, and is regarded by McCann as belonging to the Late-glacial sea. In south-western Jura the inner angle of the main terrace varies from 85 to 92 feet (26 to 28 m). This suggests that the maximum of the Late-glacial sea was rather more than 100 feet (30 m), a figure consistent with the maximum height reached by the shingle. The numerous beaches record a fall from this maximum. There is a

change from a relatively steep to a more gentle seaward slope of the surface of the shingle deposit at 75–80 feet [23–4 m] . . . at numerous localities. It is considered that the steeper upper slope reflects a rapid fall in sea-level from a maximum to about 75–80 feet, when there was a short halt, followed by a second fall in sea level at a slower rate, which is reflected in the more gentle slope. (McCann, 1961)

At the south end of Loch Shian the features just described are present, but there is also a wide ridge at 55–63 feet (17–19 m). This is a unique feature (McCann) and may be because the outer part of the Inter-glacial platform is for some reason very

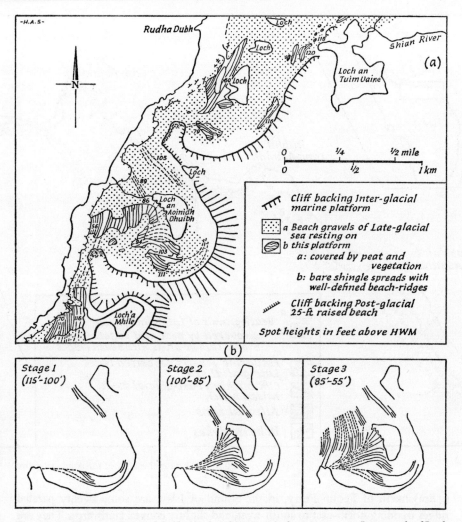

Fig. 35. (a) The central part of the Inter-glacial marine platform in western Jura, north of Loch Tarbert; (b) stages in the formation of the beach ridges in the Loch an Aoinidh Dhuibh embayment of the Inter-glacial marine platform (after S. B. McCann)

low, or possibly because there was a second halt in the fall of sea-level represented here and presumably destroyed elsewhere.

Along the west coast there are numerous caves, arches, stacks or skerries. Many of the caves are in the old cliffs, and now well above sea-level. But here, as elsewhere, the lower caves and other features, which are still in the range of wave action, almost certainly have a long history, and are not wholly formed under present conditions. Many caves and minor indentations correspond either with dykes or lines of weakness, usually joints, in the quartzite. About two-and-a-half miles

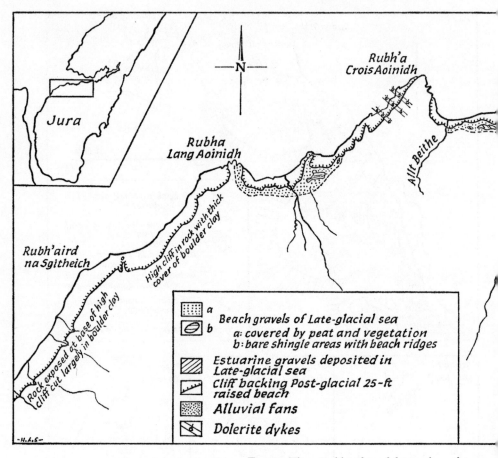

Fig. 36. The raised beaches of the southern shore

(4 km) north of Feolin Ferry, in the Sound of Islay, are some twenty parallel
ridges trending north-east to south-west, and slightly convex landwards. They are
formed of pebbles four to six inches in diameter. The innermost ridge reaches 29
feet (9 m); seawards they merge into the present beach ridge. They are in a small
hollow, and the Abhainn na h'Uainaire drains by a small gorge into Whitefarland
Bay. The river formerly reached the sea where the ridges are now found. Ting
(1936) suggests that this part of the coast is most open to wave attack from the
north and argues that the flood current from the north and the long shore current
(?beach-drift) along western Jura gather up material and take it into the Sound.
In a high sea the south-running current is strong. The ebb is to the north. 'During
the ebb . . . when the tidal current, which is stronger, flows in the opposite direction to the long-shore drift, it checks the latter and gives rise to heavy deposition
at the point where they meet. It is here that the beach ridges are formed.' The

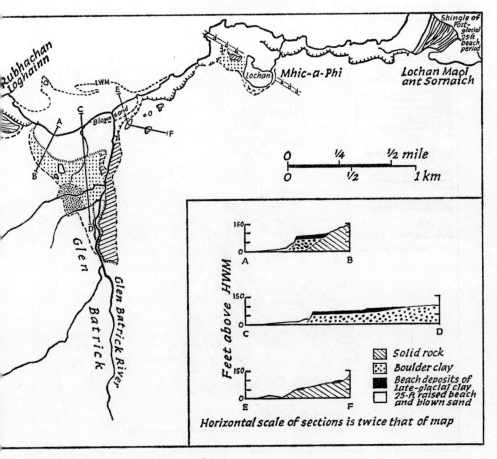

of Loch Tarbert, Jura (after S. B. McCann)

whole of Jura south of Loch Tarbert is traversed by numerous dolerite dykes, the trend of which is persistently north-west. They are seldom more than a few yards in width. Inland they often make conspicuous features, but on the west coast they stand out as great walls, sometimes form arches, and also numerous stacks. They diversify both the present and the older beaches.

The barren and rugged nature of Jura – all the sparse population is in the south and east – and its exposed position are in themselves sufficient to give it a special appeal. When the infinite detail of past and present beach and shore formations are seen in relation to the major features we have a coastal landscape not only of great and wild beauty but also one of infinite interest. It is only equalled in Scotland by the remarkable coast of north-eastern Islay (Fig. 37).

6

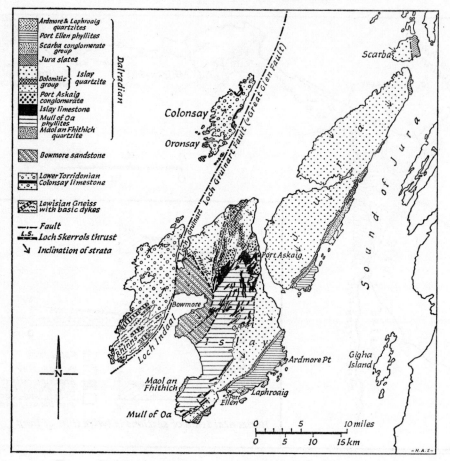

Legend:

Dalradian
- Ardmore & Laphroaig quartzites
- Port Ellen phyllites
- Scarba conglomerate group
- Jura slates
- Dolomitic group } Islay quartzite
- Port Askaig conglomerate
- Islay limestone
- Mull of Oa phyllites
- Maol an Fhithich quartzite

- Bowmore sandstone

- Lower Torridonian Colonsay limestone

- Lewisian Gneiss with basic dykes

--- Fault
L.S. Loch Skerrols thrust
↘ Inclination of strata

Fig. 37. Geological map of Islay and Jura (based on Geological Survey)

Islay

In the eastern part of Islay the rocks are similar to those in Jura. In the north-east there is a mass of quartzite in which there are faulted zones of phyllites which locally reach the north coast. The quartzite is more or less continuous along the east coast as far as Carraig Mhór. It also forms a broad belt, narrowing south-westwards. Between the two main outcrops of quartzite, the central part of the island, east of Loch Indaal, consists mainly of phyllites, which in their north-eastern part contain many bands and patches of metamorphosed limestone. In the south and east of the island the phyllites reappear and together with epidiorites give a broken and picturesque coast. The western part of Islay is quite different. To the south-east of Loch Gruinart and on the north of Loch Indaal there is a mass of Torridon Sandstone which is, on the surface, separated by superficial deposits from the greater mass which forms most of the western peninsula. The southern part of this area, The Rhinns of Islay, consists mainly of Lewisian Gneiss and intrusive igneous rocks. Thus the western peninsula resembles and is aligned with the islands of Colonsay and Oronsay. The highest ground in the island is in the quartzite areas. In the northern area Sgarbh Breac reaches 1,192 feet (363 m), and Giùr-bheinn 1,037 feet (316 m); in the southern area Beinn Bheigeir is 1,609 feet (491 m) high, Glas Bheinn 1,544 (471 m), and Beinn Sholum 1,136 feet (346 m). Over most of its outcrop the phyllite forms lower ground. The Torridonian also makes relatively low country. The Lewisian Gneiss, although not reaching more than 758 feet (231 m), is rough and hummocky ground (Wilkinson *et al.*, 1907).

The variety of rocks implies a coastline of considerable interest, a coast which includes the remarkable development of raised beaches in the north-east of the island, the most spectacular examples in Britain. Moreover, in higher raised beach times Islay was a small archipelago. There was a broad strait connecting Lochs Indaal and Gruinart, both of which were then wider. A narrow strait ran from Sanaigmore Bay to Saligo Bay, Machir Bay, and eastwards to the Indaal–Gruinart strait. A small island remained at Carn Mór. The Oa was also almost completely separated, and there were differences in the south-east coast.

The south-east coast resembles the corresponding coast of Jura. That in Islay is on a bigger scale. There are numerous islands and skerries, and the many indentations and channels, for the most part running north-east to south-west, follow the trend of the rocks and give a beautiful coast. The epidiorites usually form ridges, and with the small masses of quartzite compose the higher ground which, in fact, seldom exceeds 100 feet (30 m) near the coast. This part of the island is also well treed. There is a narrow band of phyllite between Rubha han Leacan and Rubha na Mèise Bàine in the Oa, but between that headland and Port Chubaird it has disappeared and gives place to the almost enclosed bay in which Port Ellen is situated.

In the Oa there is an imposing coast; the higher parts are mainly in the quartzite, although at the Mull of Oa the phyllites give prominent cliffs. Here, as elsewhere, the cliffs are not wholly of present-day origin. There are locally fragments of raised benches in front of them. But the coast is broken and picturesque. Near Bheinn Mhór there are landslips. North of the Mull there are many rocks and skerries and small bays where the Abhainn Ghil and Glen Astle reach the coast. Caves and minor indentations on the south-east and north-west sides of the Oa are nearly all cut out along dolerite dykes which, in the quartzite belt, are less resistant than the country rock.

Between the Oa and Laggan Point is the fine expanse of Laggan Bay. There is a magnificent beach backed by dunes. The blown sand covers the '25-ft' beach, and it may be also assumed that the present beach and the '25-ft' beach grade into one another. Behind the beach and dunes there is a broad belt, well over a mile wide in the south, of Late-glacial marine gravels. These, in turn, pass into fluvio-glacial gravels north of, approximately, Glen Machrie. (McCann (1961) correlates these gravels with the Coir Odhar moraine, and with the Highland re-advance in Scotland.) Laggan Point itself is not at all conspicuous, and owes its formation to a dolerite dyke. There, and all along the coast to Saltpan Point and Bowmore, the '25-ft' beach is present and is backed by Late-glacial marine gravels. Along this stretch of coast rock (phyllite?) outcrops only on the foreshore.

The coast of the Rhinns is, on the exposed south and west sides, picturesque and broken. Near Portnahaven the trend of the rocks is transverse to the coast, and this and the off-lying islands add much to the scenery. On the west coast Lossit and Kilchiaran Bays possess beaches. The part of the coast between Lossit Point and almost as far as Kilchiaran Bay is in intrusive rocks. North of Kilchiaran Bay the Torridon Sandstone makes the coast. It is very broken and there are numerous minor, but picturesque inlets, for example, Port Bàn and Tràigh Bhàn, the form of which is governed by the lines of weakness in the country rock especially in softer slates. Where grits are in mass the cliff-scenery is bold. The folding of these beds is well developed at Tòn Lagain. The inward-dipping rocks at and near Cnoc Uamh nam Fear make the coast bold and precipitous. Between this headland and Tòn Mhór there is a series of geos. Farther south in Saligo Bay, and especially in Machir Bay, there are fine beaches and dunes. North of Sanaigmore the land falls in level; the coast is rather less spectacular, and details depend mainly on the great number of basalt dykes which intersect it. At Ardnave Point there is a considerable spread of blown sand. The islands off the point are of similar formation to the adjacent mainland.

It has already been noticed that a strait formerly connected Lochs Gruinart and Indaal. The flats now dividing the two lochs are partly composed of gravel terraces laid down during the maximum (raised beach) submergence, and partly of deposits of the '25-ft' sea. The eastern entrance to Loch Gruinart is somewhat narrowed by the extensive area of high dunes forming Killinallan Point. To landward the dunes give place to Late-glacial deposits in which a gravel ridge, trending south-west to north-east, can be traced for three or four miles (5–6 km). From a little south of Gortantaoid the cliff of the Interglacial marine platform follows the coast and coincides with it for rather more than a mile north of the Doodilmore river. The well-known raised beaches are along that part of the northern coast between Uamhannan Donna and Rubha A'Mhàil. The old cliff which encloses the Inter-glacial marine platform is remarkably well developed. Near Aonan na Uamh Mhór this cliff is 200 feet high (61 m). The platform reaches a height of 105 feet (32 m), and is quite distinct from that of the '100-ft' beach. Along the western part of the platform, a solifluction deposit follows the cliff foot. The details of erosion in the raised beach platforms are remarkably fine. They represent a series of small geos and can be examined very easily westwards of Rubha Bholsa. Their general trend seems to indicate a structural control by jointing. The lower and present beaches merge at the seaward ends of these features. Farther to the east the solifluction deposit gives place to moraine. Both the solifluction deposit and the moraine are fringed by beach-gravels of the Late-glacial sea. In one or two places the gravels can be seen to rest on boulder clay. The gravels extend seawards to the top of the cliff cut by the '25-ft' sea. Since the upper platform may be more than a quarter-of-a-mile (402 m) wide, and since the cliff behind, and the lower one in front, are well developed, the whole forms a most striking piece of coastal scenery. South of Rubha A'Mhàil, along the west side of the Sound of Islay, there is a long strip of the lower gravels and traces of the corresponding cliff.

Since the formation of the old marine platform is north-eastern Islay, there is, therefore, evidence from the deposits which now rest on the platform, of a period of general glaciation (the boulder clay), followed in turn by a period of local glaciation (the Coir Odhar moraines and the solifluction deposits at the foot of the old cliff line). The local glaciation and the deposition of the high level marine gravels can be related to the general chronology of Late-glacial events in Scotland, but the period of the formation of the rock platform itself remains problematical (McCann, 1961).

The marked structural separation between the Archaean rocks of western Islay and the metamorphics on the east is associated with the southerly prolongation of the great Glen Fault. This is assumed to run close to, and parallel with, the east coast of Colonsay and then to pass through the Gruinart–Indaal depression. The sandstones (? Torridonian) between this fault and the metamorphics are bounded by the Loch Skerrols thrust.

There is no definite reason to suppose that the Sound of Islay is of structural origin; it may well be part of a former river system. The 10-fathom (18.3 m) contour indicates a long narrow trough running midway along the channel. The Islay shore is distinctly higher and steeper than the Jura shore.

Chapter VI

The Outer Hebrides

The long string of islands extending from the Butt of Lewis to Barra Head (Berneray) present some of the most interesting coastal features in Scotland. They are almost entirely made of Lewisian Gneiss except for a limited area of sedimentary rocks near Stornoway and the Eye peninsula. From the structural point of view the most significant feature is a belt of crushed rocks which is almost continuous along the east coasts of the islands from Tolsta Head in Lewis to Sandray. This is a belt of shearing and marks a line of very ancient earth movement directed towards the north-west. The crushed rocks rest on a thrust-plane.

The general topography of the islands varies greatly from place to place. There is, first of all, the profound difference between the east and west coasts. From Barra to the Sound of Harris this is particularly noticeable; the east coasts of the southern islands are hilly, even mountainous, the west coasts for the most part low and flat and carry extensive areas of machair. Harris and South Lewis are high, but nevertheless the west coast is lower and there are extensive areas of machair and sand. Much of the remainder of Lewis is a plateau, but on the east there are prominent cliffs. The Butt of Lewis is cliff-girt; on the north-west coast there are cliffs, but they are lower and often cut in drift, and (see below) traces of ancient cliffs and raised beaches occur. South of Barvas the coast is marked by several impounded lagoons, and from Carloway to West Loch Tarbert it is much indented; cliffs are prominent and locally there are beautiful sandy bays. The interior of North Lewis is largely peat- and boulder clay-covered and very sparsely inhabited.

These land features are closely related to the surrounding submarine topography (Fig. 3 and pull-out map). Along nearly all of the east coast there is deep water. The coasts rise steeply to the hill summits. To the west the level falls at first steeply from the summits to the peat areas and they in turn give way to the machair and wide beaches. The water deepens very gradually and the submerged shelf may reach to St Kilda. The Monach Islands and reefs and shoals nearer inshore rise from this surface. In the north-west of Lewis the contrast between east and west in the submarine topography is less marked. The whole archipelago conveys, even to the most casual observer, the effect of submergence. The coasts of the several islands are extraordinarily irregular and, especially in Benbecula and North Uist, the intricacy of outline and the great number of fresh-water lakes give a landscape not matched elsewhere in Britain. The islands were heavily glaciated. The main ice advanced from the east, from the mainland, but local centres existed on the higher parts of the islands. The contrast between the flat boulder clay and peat plateau of Lewis and the bare mountains of Harris, mountains and valleys which look as if ice had melted from them only yesterday, is profound. The former extent of the ice to the west has undoubtedly furnished the abundant non-calcareous material of the western sea bed and also the cobble beaches which are so well developed in many places of the west coast. In the following pages more details will be given of the inlets, but it may be noted at this point that although ice has certainly played an important part in the modelling of some of the inlets, they are not the exact counterparts of the sea-lochs on the west coast of the mainland. There is scope for a great deal of research work on the Outer Hebrides.

LEWIS AND HARRIS

The Butt of Lewis, and the northern tip of Lewis between Cross Sand on the west and the northern tip of the small bay to the north of Cellar Head on the east is built mainly of banded gneiss. The coastal scenery is imposing; there are several deep clefts or geos. Inland the gneiss is covered by drift. Near the Butt the gneiss is folded along lines trending west-by-north. At Port of Ness there is a good beach. The remainder of the west coast as far as Mealasta island is mapped as undifferentiated gneiss. In and around Galson the cliffs are relatively low and the upper parts are often cut in drift. Seawards there are reefs and shoals. The coast is very exposed and it is perhaps surprising that erosion is not more serious since the 5-fathom (9 m) line is close inshore. The reefs at beach level help in minimizing the effects of waves. Baden-Powell and Elton (1937) describe a raised beach at Galson. It is 10 to 25 feet (3 to 8 m) above present high-water mark, and was formed earlier than an overlying midden which is ascribed to the early part of the Christian era; it is also probably earlier than an adjacent Iron Age earth house. It is probably later than the greenish-coloured boulder clay. Fig. 38 shows the beach in section. The authors were particularly careful to make sure that the beach does not consist of matter thrown up in modern storms; stone size and grading and the blown sand prove this. The coast was examined in some detail as far as Bad an Fhithich but no sign of a low-level beach was found. To the south-west the beach was traced for about half-a-mile. My own impressions from two widely separated visits to the coast suggest that for some distance on either side of Galson the cliffs are partly fossil, and that despite their nature they suffer relatively little erosion. The whole stretch from The Butt to Barvas would repay study in detail. Along most of it there is a rock platform at, approximately, beach level. It would be interesting to know how much of this is of present-day origin; there is no doubt that it protects the cliffs behind. (S. B. McCann, in *Geography at Aberystwyth: Departmental Jubilee* (Cardiff, 1968), expands this point and confirms my views.)

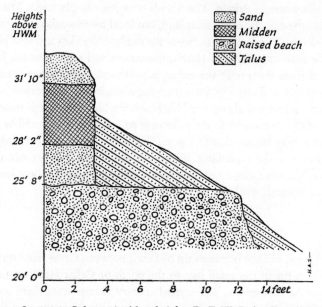

Fig. 38. Section at Galson raised beach (after D. F. W. Baden-Powell and C. Elton)

In another paper, Baden-Powell (1938) discusses the Glacial and Inter-glacial marine beds near the Butt of Lewis. On the west coast from Swanibost sands to the mouth of the Dell river, the cliff section is rather obscure. Baden-Powell interprets it in the following way. The lowest 15–20 feet (4.5–6 m) consist of stiff brown boulder clay; about 10 feet (3 m) of shelly sand and some pebbles follow, and on the sand is boulder clay which may be locally 40–50 feet (12–15 m) thick. The sections, when Baden-Powell investigated them, were most clear at Tràigh Chumil. The shelly sands (Inter-glacial marine beds) were also noted just to the north of Barvas. On the east coast corresponding deposits in the cliffs are found just to the south of Port of Ness. This section was limited and interrupted. The stratigraphical differences between the east and west coasts need not concern us here, and the outcrop on the east soon gives place to the high-cliffed coast.

The beach at Europie, on the west coast, is about a quarter-of-a-mile (402 m) long and is very exposed. There is erosion of the cliffs to the north and south, especially where boulder clay is present. This, in turn, feeds the beach. The general drift of material is to the north. A cobble ridge extends along the head of the beach. Ritchie and Mather, whose account of the beaches of Lewis and Harris for the Countryside Commission of Scotland (1970) is the only authoritative work on the subject, regard the cobble ridge as fossil. Although there is plenty of sand on the beach, there are no continuous dunes. There is a machair zone in rear. The greater abundance of dunes is in the south where there is a considerable growth of marram. There are other beaches at Swanibost and Tràigh Chumil. The former, about 109 yards (100 m) long, is backed by shingle cut through by the Swanibost river. Machair extends inland up to 44 yards (40 m) or more. Tràigh Chumil, just to the south, is broken by outcrops of rock and is backed by cliffs of boulder clay. There is some growth of dunes at the northern end. Ritchie and Mather call attention to exposures of dune sandstone, peat and till which closely recall those mentioned by Baden-Powell (1938).

The coast from Europie to Barvas may perhaps be regarded as a unit and is of somewhat similar appearance throughout its length. It is for the most part a coast of which the upper part is boulder-clay covered, and it is relatively densely settled on account of its fertility. The villages stand on or close to the main road, but several settlements are on the cliff tops. Throughout the whole length there is a marked rock platform at beach level. Several small streams reach the sea in open valleys with wide floors and gentle slopes. At Barvas the nature of the coast changes. From that place to Dalbeg Bay a series of lochs are ponded back by shingle bars. The cliffed headlands between the bays and lochs seldom exceed 100 or 125 feet (30 or 38 m). The rock platform is usually present, but may be hidden by beach deposits. A glance at the One Inch Ordnance Map indicates their appearance better than any description. Seawards of the bars there are often good beaches. They are partly boulder-covered, and often extend along the sides of the inlets and are not limited to their heads. The inlets are wide open, and the lochs and beaches are similar to the oyces and ayres of Orkney.

Loch Mór Barvas is ponded back by a shingle beach which seems to have originated as a spit. It is continuous with the shingle that encloses Loch Ereray to the west; but it is not regarded as a 'free standing form'.

The peninsula at the south end of the outlet of Loch Mór Barvas, which at first sight appears to be a shingle-built feature, is seen, on closer examination, to be composed largely of angular and sub-angular boulders, with only occasional cobbles and pebbles thrown up by storm waves. Thus it may possibly represent a deposit of glacial till left behind on deglaciation, and it has exerted an important influence on the subsequent evolution of the marine-built complex by functioning as a foundation and also as a source of material. Till also outcrops at the south-west end of Loch Ereray, where an actively eroding till cliff backs the beach. (Ritchie and Mather, 1970)

Erosion of the gneiss affords some material for the beach, but most of the sand is organic, derived from shell debris, and is somewhat finer than that of the extensive machair to the north of Loch Barvas. Movement of the coarse material of the beach is to the north. The transition, shingle to machair, is abrupt; there is but little dune along most of the bay.

Dalbeg beach is about 109 yards (100 m) long. It is primarily a shingle beach on which sand and dunes have accumulated. Thin boulder clay spreads over the surrounding slopes, but the cliffs at the north end of the beach consist of deeply-weathered gneiss, and it is the erosion of this which seems to have given rise to the inner bay, although it is sheltered. Deep water generally is close inshore in this part of the coast. Dalmore beach is in a picturesque inlet surrounded by cliffs reaching 150 feet (46 m) in height; these are well jointed and subject to erosion. Behind the sand beach, containing much shell debris, is a cobble ridge which widens northwards. There are some dunes behind it. South of Dalmore the coast, although much indented, is rocky and cliffed and is akin to that of East Loch Roag which it adjoins.

A great change in the nature of the west coast of Lewis takes place near Carloway. The major indentations of East and West Lochs Roag and Loch Roag, and the outline of Great Bernera island are almost certainly associated with tectonic movements. A line of crush runs from Valtos parallel with Little Loch Roag and cuts across the land to reappear in Loch Seaforth. On East Loch Roag intrusions are locally parallel to the coast. Jehu and Craig (pt v, 1931–3) note signs of crushing along the hollow followed by the road from Breasclete to Carloway, a line more or less conformal to the trend of East Loch Roag. The parallelism of the east coast of Loch Ceann Hulavig, the west side of Tòb Valasay, and the valley running from Miavaig to Camas na Clibhe all suggest, even if they do not prove, a tectonic origin. Many other corresponding features can be picked out on the Ordnance maps. Seawards there are numerous small islands usually arranged in lines, and the east–west valleys and depressions are also noteworthy. The cliffs on the outer coast vary a good deal in height, but apart from those near Rubha Mór they are picturesque rather than spectacular.

Although it is not a coastal feature, Glen Valtos is of considerable interest. It is a glacial overflow valley and continues the line of one of the heads of Loch Roag. A little north of its western end Tràigh na Berie is a major beach and is associated with dunes and machair. The beach is rather more than a mile long, and together with the machair impounds two small lakes. The beach sand is probably derived from the offshore zone on which there is much glacial material. At the present time there is some erosion of the beach at the western end; the mid-part is advancing and there is accretion in the east. This is interesting because the nourishment zone at the western end is much wider than that in the east. However, the sand supply is adequate and the dunes behind the beach are easily maintained. Movements along the beach may be variable, the stream at the western end is deflected to the west; that one near Berie is turned to the east. The machair zone is wide, usually well-drained, but there are marshy areas round the lochs. A little to the north is another beach, Tràigh Valtos. It is much smaller and is divided by a rocky outcrop. Although it is sheltered, the dunes are being undercut, but the enclosing cliffs are no longer active. The beach sand is probably derived from off-shore – the boulder-clay cliffs certainly do not now contribute any noticeable amount of debris.

Tràigh na Clibhe is about half-a-mile (805 m) long at the head of a wide inlet to the west of Valtos. The whole inlet is outlined by faults, and a major one corresponds with the western side of the inlet, which is bordered by high cliffs. The slopes on the east are more gentle. There is not much shingle on the beach, and it seems that boulder clay and rock falls provide much of the material of which it is composed, although some is derived from off-shore. There is local undercutting of the dune front.

Uig sands fill the inner parts of a large and complex bay; which has a relatively narrow inlet. Most of the blown sand has gathered to the north and west of Ardroil. In long inlets such as Timsgarry and Crowlista there are sand flats and mud. Around the whole bay there are eskers and other glacial deposits which have produced material for the beaches and flats. The machair in the east is flat, there is rather more relief farther west. Wind erosion today is unimportant; in the past it was significant and associated with overpopulation of rabbits.

The coast from Camas Uig to Gallan Head is wild and exposed, and broken by numerous small inlets. The cliffs seldom exceed 100 feet (30 m). To the south of the bay the coast, as seen from Aird Fenish, is magnificent; the cliffs are nearly vertical and reach 200 + feet (61 + m). It is a much broken coast and includes Mangersta sands which occupy a small inlet in the gneiss cliffs. The line of the inlet is continued eastwards by low ground including Loch Scaslavat; another depression runs northwards. Both may be remains of a Pre-glacial drainage system. Deep water reaches the enclosing cliffs and their erosion forms much of the material of which the beach is built. The cobble ridge behind the beach is built of local material. The sand behind is flat and usually bare, but some dunes have gathered along the steep margins of the depression. A good deal of erosion in the sandy areas is caused by rabbits. Ritchie and Mather remark that there is now little supply of sediment from off-shore, and that this results 'in the starvation and eventual removal by wind erosion of the dune zone, and the blowouts would gradually extend backwards and consume the flat machair behind the dunes'.

The coastal flora of Uig and neighbourhood has been briefly described in a small book edited by Campbell (1945). In the book A. J. Wilmott points out that the cliffs are usually precipitous but with rounded tops. On their faces are found *Ranunculus acris*, *Cochlearia officinalis*, *C. scotica*, *Silene maritima*, *S. acaulis*, *Anthyllis vulneraria*, *Sedum rosea*, *S. anglicum*, *Saxifraga oppositifolia* (in two widely separated places), *Ligusticum scoticum*, *Crithmum maritimum* (at Mangersta only; its northmost limit in Britain), *Angelica sylvestris*, *Tripleurospermum maritimum*, *Primula vulgaris*, *Armeria maritima*, *Plantago maritima*, *Rumex crispus*, *Asplenium marinum*. *Polygonum viviparum* was found on low rocks near the sea. It is pointed out that this is probably an incomplete list, but it will nevertheless indicate the plants typical not only to the cliffs near Uig, but also of those over a much greater part of the west coast of Lewis.

On beaches made of small blocks of rock and shingle *Atriplex glabriuscula* grows in sheltered places. *Cakile maritima*, *Potentilla anserina*, *Galium aparine* (Mangersta), *Tripleurospermum maritimum*, *Glaux maritima* and *Rumex crispus* are common. At the head of the beach, in front of the dunes, *Cakile maritima*, *Honkenya peploides*, *Potentilla anserina*, *Atriplex laciniata*, and *A. glabriuscula* are found. *Ammophila arenaria* is the main dune grass; there is some *Elymus* in Camas na Clibhe. There is no need to list the plants commonly found on the dunes. Behind some dunes there are wide expanses of sandy pasture. The word machair (see Appendix to this chapter) is used to describe them, but the term has not necessarily the exact significance it has in e.g. South Uist. It is noteworthy that at Uig these pastures are characterized by abundant orchids, *Dactylorchis fuchsii* and *Coeloglossum viride*. Two areas of sandy pasture are found in Camas na Clibhe and at Ardroil.

Salt marshes are not well developed. They usually consist of short turf, and the most abundant plants are *Armeria maritima* and *Plantago maritima*. Other plants found include *Cochlearia scotica, Spergularia marina, S. marginata, Glaux maritima, Plantago coronopus, Salicornia* sp., *Triglochin maritima, Juncus gerardii, Blysmus rufus, Eleocharis quinqueflora, Agrostis stolonifera, Puccinellia maritima* and *Festuca rubra*.

Southwards from Aird Fenish the coast is for the most part bold and much indented with numerous geo-like inlets. Camas Islivig and Camas a'Mhoil, facing north and south respectively, point towards Loch Greivat which lies in a hollow behind Aird Brenish. South of Camas a'Mhoil there are several islands, the longest of which, Mealasta, is separated from a bold and high coast by Caolas an Eilein. Here and northwards as far as Rubha Buaile Linnis there are several steep gullies and waterfalls. To the south of Mealasta the nature of the coast changes; large inlets take the place of geos. Loch Tamanavay, Loch Tealasavay, and the much longer Loch Resort, which resembles a fiord, face the large island of Scarp. In all this district heights on or close to the coast are considerable, and the southern slopes near the mouth of Loch Resort are steep. South of Scarp the coast turns inwards to West Loch Tarbert, the immediate north coast of which is lower and less precipitous, but is backed by the Forest of Harris and its high and conspicuous hills. At Tarbert the peninsula between West and East Lochs Tarbert is only about half-a-mile wide, and follows a line of crushing. A similar line a little to the north, runs from Loch Leosavay to Husinish Bay and the narrow neck between that bay and Caolas an Scarp.

In Husinish Bay there is a south-facing beach enclosed by low sloping cliffs. There is a stable machair zone behind the beach. The shell sand content of Husinish is nearly 83 per cent.

The southern shore of West Loch Tarbert is a steep slope, not a true cliff, reaching 1,654 feet (504 m) in Beinn Dhubh (see Plate 37). A short distance beyond Ard Groadnish, Luskentyre Banks, on the north side of the inlet containing Tràigh Seilibost and Tràigh Luskentyre, form a blunt promotory of beach, dune and machair. The complex seems to represent a time when active prograding took place or it may owe its present form to the submergence of a machair surface – the same submergence that has produced the inlet on its southern side. Changing processes of erosion and deposition and shiftings of the drainage channels have made interpretation difficult. In general Ritchie and Mather think submergence is more likely. The banks are fronted by good beaches, especially Tràigh Rosamol, which faces north-west. Landwards the banks consist of high and gently sloping machair built on finger-like ridges of rock running south-east and north-west. Seaward of the ridges and higher machair is a series of curved erosion surfaces, sand ridges and high sand hills. The whole complex is of great interest. 'The height, degree of dissection and variety of land forms found in this comparatively small area make it a unique and fascinating element in the diversity of coastal land forms of Harris.'

Tràigh Seilibost and Tràigh Luskentyre are wide sand flats filling a deep inlet just north of the main road from Tarbert to Leverburgh. They are separated by the northward-pointing sandspit named Corran Seilibost. This is a built feature, but it is suggested that it is the remnant of a former more extensive dune and machair area. 'The triangular shape, the absence of the characteristic fish-hook bend of the true coastal spit and the evidence of the stratified sandy salt marsh-machair which forms the southern "base" of the Corran peninsula suggest the landform is a product of submergence and retreat . . . ' (Ritchie and Mather, 1970). Crago, the north-west pointing ridge to the east of Corran, is a rocky promontory partly covered in blown sand. The wide foreshore to the west of Corran affords an abundant sand supply and accounts for the dune growth on Corran.

The vegetation has been described by Gimingham *et al.* (1949). The inlet is partly protected by the island of Taransay but, as on all this coast, winds are frequently strong. On the other hand, the range of temperature is small (39.2°F (4.0°C) February to 54.6°F (12.6°C) July (means)). Rainfall is not heavy, although rainy days are frequent. The inlet is broken up by ridges running somewhat to the east of north.

On the foreshore *Ammophila arenaria* is common; there is some *Agropyron junceiforme, Cakile maritima, Honkenya peploides* and *Atriplex laciniata*. There are well developed mobile and fixed dunes. On the former *A. arenaria* and the foreshore plants are common, also some *Cerastium tetrandum* and *Festuca rubra*. On the fixed dunes *Trifolium repens, Poa pratensis* and *Cerastium tetrandum* usually begin the stabilization, and are followed by *Senecio jacobaea, Bryum pendulum, Ranunculus repens, Cerastium tetrandum* and *Festuca rubra*. There is also some dune pasture behind the ridges, and when fully established there is a loose turf of *Festuca* and *Poa* with much *Galium verum*.

In general the succession is like that on English dune systems, but the authors emphasize the minor role of *Agropyron*, and also the very subordinate part played by cryptogams.

The line of crush following East and West Lochs Tarbert finds several parallels farther south. An inspection of the One Inch Map of Harris shows several valleys or depressions cutting through the island and joining inlets on either coast. One runs from Tràigh Luskentyre and is followed by A859 as far as Loch Bearasta Mór; it may continue into Loch Grosebay. The high hills of central Harris are separated by bealachs having the same north-west and south-east orientation, and on the eastern coast their continuations into Lochs Stockinish, Geocrab, Flodabay, Finsbay and Lingara Bay are not improbable. There is no doubt that the Lingara Bay, Loch Langavat and Borvemore line is a continuous one. It probably extends into the narrow neck which nearly divides Taransay into two parts. Farther south Loch Rodel, Glen Rodel, Glen Strondeval, the Obbe, Glen Caishletter and Tràigh an Taoibh Thuath mark a distinct feature. It may well be that the lines of islands, for example, Ensay to Dùn-aarin, and Killegray to Groay and the intervening channels in the Sound of Harris indicate directly comparable lines.

Ard Nisabost, a small headland nearly 200 feet high (61 m) and joined to the mainland by blown sand, separates the Luskentyre area from the beaches to the south-west. The headland is, to seaward, fringed by a rock platform and is itself largely sand-covered. Tràigh Nisabost, to the east of the headland, is a continuation of Tràigh Seilibost, and is backed by machair (Horgabost machair). Despite appearances the east beach area is suffering some erosion from the stream and from waves and wind. For the most part, however, a high degree of stability prevails in the area. It is a popular part of the coast.

To the west and south of Ard Nisabost there are several good beaches. Tràigh Iar is the western equivalent of Tràigh Nisabost, and like those farther south, is exposed to wind and sea. Borve Beg is much sub-divided by small rocky outcrops and is dominated by the progradation of the Borve Beg river. Usually the lower part of the beach in this bay is a coarse shell sand, giving place upwards to rock, sand and shingle. In the south at Borve Lodge, the sand is piled up on rocks and forms a high machair. The sand cover continues southwards into Borve Mór but the movement of the sand is definitely to the north, a fact indicated by several pronounced blow-outs. The sand is more stable in the south. The off-shore zone is sandy, but air photographs indicate many rocky outcrops on the lower beach and just offshore.

Sgeir Liath separates Borve Mór from the fine beach of Tràigh Scarasta and the extensive flats of Tràigh an Taoibh Thuath which fill the space between Toe Head and the mainland. In normal weather two distinct areas are notably clear on Tràigh Scarasta; a lower inter-tidal beach more than 109 yards (100 m) wide and a higher convex one from 65 to 208 yards (60 to 190 m) wide. These may be separated by a steep slope which may be almost vertical and up to 8 feet (2.5 m) high in the north of the bay. Ritchie and Mather think that the great flats are the result of drowning, and that at an earlier time the Scarasta dunes extended much farther westwards. They argue against infilling in the absence of large rivers, glacial melt water activity or any other infilling agent. This may be so, but marine action may well have played a part despite the greater exposure of the coast to north and east and to the northward-moving beach material.

The machair at Scarasta is noteworthy. The inner and older part 'consists of a single, scalloped sand ... escarpment which rises to between 26–39 feet [8–12 m]

O.D. . . . The scalloped edge, although stabilized by mature machair vegetation, represents the ancient line of blowout activity . . . In front of the old escarpment is a highly indented topography of dunes, sandhills and ridges with rare areas of an older flatter surface . . . in the central part of the area . . . ' (Ritchie and Mather, 1970).

Beach, dune, and machair also link Chaipaval (Toe Head) to Northton. The link is also partly formed by a small rocky outcrop, reaching just over 100 feet (30 m).

The intricacies of the coast near Leverburgh and the whole length of the east coast are probably to be explained by several factors – the submergence of a heavily glaciated island; the main north-west and south-east valleys referred to above; the different effects of glacial, subaerial and marine erosion on the para-gneisses, anorthosite gneisses, and gabbro-diorite rocks in the south and east, and on the undifferentiated gneisses of the remainder of South Harris; and the dykes and lines of minor faulting or jointing or crush which are generally parallel to the major lines. Most of these features are traceable also in the adjacent islands, such as Scalpay and Taransay. Marine erosion today is rather limited on this coast, and between the major inlets and the infinity of minor ones, which indicate lines of weakness in the rocks, the coastal slopes are rounded rather than cliffed.

We must now glance at the east coast of Lewis south of Port of Ness. From there to Meall Geal the banded gneisses reach the coast which is much broken by numerous geos and similar inlets, and also by the bigger indentation of Port Skigersta. To the south of Meall Geal anorthosite rocks, indented but less deeply so than those to the north, extend to Guiashader where there is a small outcrop of paragneiss. The cliffs for some distance north of Cellar Head are steep and inaccessible and occasionally overhanging. From a little north of Cellar Head as far as the Geiraha river undifferentiated gneiss reaches the coast to form fine cliffs. Along all this coast there are occasional pocket beaches. In Geiraha Bay there are several stacks indicative of past rather than present erosion. There are also traces of benching in the bay. In general the whole line of cliffs appears to suffer but moderate wave erosion. The beach, Tràigh Geiraha, is separated from Tràigh Mhór by a rocky headland, but there can be interchange of materials.

Tràigh Mhór is the largest beach on the east coast of Lewis, and rests at the foot of a cliff which locally reaches about 120 feet (37 m). In the south Tolsta Head exceeds 200 feet (61 m). The amount of sand on the beach is considerable, and its main source appears to be off-shore. The general direction of drift along all this coast is to the south, and Tolsta Head acts as a great groyne. The cliffs behind the almost completely sandy beach are in crushed gneiss; some of the steep streams from these cliffs may help to feed the beach; the more so because some are re-excavating valleys plugged with boulder clay. Everything suggests that a long time has elapsed since the cliffs were cut; there are fossil stacks but the platform on which they stand is not visible. Dunes fringe the beach, but they are usually low. There is a longitudinal hollow between the free dunes and the sand-covered cliffs. On the embryo dunes *Cakile maritima* is the chief pioneer but is soon followed by marram. *Tussilago petasites* (butterbur) is dominant in many of the steep valleys. Long constructional waves from the north-east play a significant role on Tràigh Mhór. On the south side of Tolsta Head is the smaller beach, Giordale sands. This, too, is backed by cliffs, the upper parts of which show intense gullying.

The crushed gneisses cease at Sheilavig Mór, where there is an abrupt change. The headland of Druim Mór is cliffed and to the south the Torridonian rocks and sandy bays give a very different type of coast. Gress Sands are interrupted by Gob Tais, and they rest on a rock platform which is prolonged seawards by reefs. Marine erosion is active on the enclosing headlands and locally on the dunes. The fact that the streams are not graded to sea-level is also indicative of erosion. The conglomerate cliffs are an important source of sediment, both sand and shingle. The Gress river also brings down a certain amount of debris, but also acts as a partial barrier to the southerly drift of shingle along the beach. It is suggested (Ritchie and Mather, 1970) that the shingle is in fact a southerly pointing spit on which the beach and dunes rest. This also implies the possibility of a rather higher sea-level, a factor which seems to be consistent with the old cliff in Tràigh Mhór and elsewhere. In this area *Cakile maritima* is the most important pioneer. The dunes are for the most part close turfed. It is thought that 'the balance between marine erosion and accretion is a delicate one, and little sediment is now available for beach or dune construction'.

Coll Sands, a mile or two farther south, are contained by low and readily eroded headlands of conglomerate. An abrasion platform seems to be present under the sands, which are thin. Since the cliffs are easily destroyed it is curious that the beach is not better fed. Ritchie and Mather think that sand may be swept out into Broad Bay and that it may accumulate in Melbost Sands. It is known that the Coll river brings down fine shingle and sand to the beach. The Coll has been diverted a little to the north by beach-drifting, whereas the Allt-an-t-Sniomh has been partially blocked by a south-growing spit. Both these diversions are caused by shingle, although the main beach consists largely of sand. Despite these constructional forms, erosion is perhaps more noticeable today, especially on the ridge that runs seawards at Upper Coll.

Aird Tunga, a low promontory surrounded by fossil cliffs and a wave-cut bench, divides Coll sands from the much greater and more complicated accumulations at Melbost and Tong. Teanga Tunga is a spit extending southwards for about a mile on the inner side of the main channel in Melbost sands. It is used as a source of gravel for building. The spit is not a simple feature; at its proximal end two short spits at a rather higher level run into it from the north-east. These appear to have been built at a somewhat higher stand of the sea. The main spit is suffering some erosion, although there are places where there is accretion. There is blown sand on the shingle of the spit. This shingle is derived from the conglomerate cliffs to the north. The fossil cliffs give place to active erosion near Aird Tunga. Some of the shingle reaches the small island a little to the south of the spit; very little crosses the deeper channel to reach Steinish. There is a good deal of inter-tidal and generally immobile shingle on the eastern side of the spit; on the west there are sand flats giving place to salt marsh. The streams, the Laxdale river and the Amhuinn à Ghlinne Dhuibh, have changed their courses in the flats from time to time, a matter well-illustrated in different editions of the One Inch Map.

On the south and east of the main drainage channel, there is a sand spit, un-named, which is actively growing to the north-west. It grows from a small conglomerate headland and at this end of the spit there are considerable accumulations of sand which have been undercut by the waves. There is also some shingle here-abouts. The growth of the spit has been much influenced by the air-field and the wall alongside it. Comparison of early maps shows that the southern end of the spit is retreating (to the south-west) and the free end prograding, so that the whole structure is rotating in a clockwise direction. It is suggested that since the orientation of the air-field wall is slightly different from that of the spit as shown on the first edition of the One Inch Map, the wall may have been responsible for the change of orientation of the spit. The spit is regarded as primarily wind-built, and the fine sand has a high shell content. *Cakile maritima* is the pioneer on the embryo dunes. Behind the spit there is a sand flat with an open growth of *Armeria maritima* and *Plantago maritima*. Above high-water mark *Honkenya peploides* replaces *Cakile maritima* as the main pioneer. It may be noted that the whole of Broad Bay is sand covered.

The Eye peninsula is tied to the mainland by a sand and shingle tombolo which is probably resting on a rock platform at no great depth. There are beaches on both sides but the whole feature has been much modified by man. Godard (1965) thought it was a fossil feature and Baden-Powell and Elton (1937) hint at the same view. The coarse sand and shingle are derived from the conglomerate cliffs and the widening off-shore platform. Since there is erosion, sea-walls have been built from time to time (Jehu and Craig, pt v (1931–3), and Ritchie and Mather (1970)) and also groynes to retain the sediment. The drift along both north and south beaches is to the west.

It is difficult to avoid the conclusion that the many fossil cliffs, various extensive off-shore platforms, occasional high-lying shingle spits and related features all point to the presence of a recent fall of sea-level in Lewis and probably in other parts of The Long Island. There may have been a subsequent rise of sea-level leading to the drowning of peat beds and the Dùns (see p. 185), but the evidence is not conclusive. In general the evidence in favour of the presence of low raised beaches seems to be clear; that they are not so prominent as on the adjacent inner islands and mainland may well be explained by a tilt.

Most of the Eye peninsula is formed of crushed gneisses and is cliff-girt all along its irregular coastline. The same type of coast prevails south of Stornoway harbour, except for Arnish Point which is in the Stornoway beds. The east coast is pierced by several major inlets. These sometimes run north-west and south-east, like those in southern Harris. The lower part of Loch Seaforth is continued by a major line of crush right across the island to Little Loch Roag. Loch Claidh, Loch Bhrollum and Loch Leurbost are similarly orientated (see Plates 33, 34 and 35). On the other hand Lochs Shell, Ouirn and Erisort run east and west; they are continued west-wards by valleys which may be traced for long distances; that behind Loch Ouirn corresponds precisely with the upper part of Loch Seaforth. Many of these inlets closely resemble fiords, and have certainly been glaciated. The ice generally approached The Long Island from the south-east. It filled the Minch and, when it reached the islands, the lower layers probably turned north or south (according to location) and followed and presumably deepened the furrow which parallels the east coast of The Long Island. The upper layers advanced over the land and would have followed pre-existing lines of hollow. Nearly all the north-west to south-east lakes rest in true rock basins. But it is difficult to separate completely the parts played by the local ice and the mer-de-glace. The sea-lochs only resemble in general outline those of the mainland. No great depths are found; there is one of 138 feet (42 m) in Loch Seaforth and a hollow in Loch Shell is 60 feet (18 m) deeper than the threshold at its mouth. Lochs Claidh and Bhrollum are somewhat similar. Geikie (1873, 1878) thought that these depressions occur in those places occupied by considerable local glaciers. Most of these lochs lie along lines of weakness of one sort or another and as Geikie pointed out, 'the glaciated outline is super-imposed on an older set of features which in the main appear to have been the result of ordinary atmospheric and aqueous action guided by the structural peculiarities of the rock'. Loch Seaforth, the longest loch, is made up of three distinct parts; the upper part including Tòb Kintaravay runs east and west and is all but connected by a burn to Loch Ouirn (see Plate 36); the middle part runs north-east and south-west, and the lower part north-west to south-east. Each arm follows a structural line. The east coast of Lewis south of Stornoway, between the lochs, is irregular, and locally there are many off-shore islands. There is some cliffing, but steep rounded slopes are more common. Today marine action is much less significant than on the west, and probably much of the shore profile was produced by the ice advancing from the mainland. Many of the smaller inlets, like the bigger ones, follow lines of weakness in the rocks and their trend is often continued inland by burns and small lochs. Scalpay island is separated from Lewis-Harris by the Sound of Scalpay which is a continuation of East Loch Tarbert. The island itself shows many coastal and inland features following the same trend. The numerous islands and rocks in East Loch Tarbert make hazardous navigation.

PABBAY AND BERNERAY

The Sound of Harris is not deep; parts exceed 10 fathoms (18 m) but much of it is considerably less. In one sense it may be comparable to the depression of East and West Lochs Tarbert, but it is much wider. It represents, topographically, a major depression in The Long Island ridge. Pabbay and Berneray are the largest islands in it. Pabbay has been investigated by Elton (1938). It is formed of hornblende schists penetrated by dykes which run generally parallel to Harris Sound. Rock appears at the surface in the centre and north of the island. There are no trees nor shrubs on the island, but birch and hazel formerly grew on it. It is thought that grazing, wind action and lack of shelter explain its present state. The former population of about 200 also demanded fuel and hence the thinness of the peat and the lack of trees. Peat can be found both above and below sea-level. Sinclair, writing in the *Statistical Account* (1794), notes that 'When the sea ebbs out in spring tides to a great distance there are visible, at the very lowest ebb, large trunks of trees, the roots of which spread out widely and variously, and fixed in a black moss, which might be dug for peat to a great depth.' MacRae, who was Minister of the Parish of North Uist, noted in 1837 (published 1845) that sunk forests 'are found under low water mark nearly as low as the water recedes at spring tides . . . the land must have largely extended its [present] bounds . . .'. Elton suggests that moss caught up on anchor flukes off the Outer Hebrides may be compared to that from the Dogger Bank. He adds that if MacRae is correct – that the moss extended some miles from the coast – the banks between the Monach Islands, Haskeir and North Uist may have been part of this sunken land. The lowest recorded submerged woods in the Sound are 11–16 feet (3–5 m) below high-water mark, but because of the movements of the sands on the floor of the Sound any argument based on present-day submarine contours is dangerous. However, an elevation of 30–50 feet (9–15 m) might well convert the Sound into a strait similar to those on either side of Benbecula. Elton argues that the forests imply a subsidence of the order of 16 feet (5 m), and probably more. In support of this he mentions the tradition that the islands of Pabbay and Berneray were separated only by a narrow channel so that people could shout across and even throw things across. Admiralty Chart 2642 shows a well-defined submarine bank connecting the two islands at depths of 12–20 feet (3–6 m) below low water spring tides. Does this imply a real submergence in historical times? Woodland on Pabbay and Berneray seems to have survived into Norse or even later times, but the forest and the submergence may well have been earlier. There is scope for careful pollen analysis in this area. Ritchie (1966) is not convinced that the Pabbay sites are necessarily to be explained by submergence; he argues that coast erosion can explain the facts. This point will be discussed later in connection with South Uist.

There is not much information, relative to the theme of this book, available about the other islands in the Sound. Berneray is the largest, and at least half of it, the west, is an area of extensive beach and dunes. These and Lochs Bhruist and Little Loch Borve recall the machair of the Uists. Boreray is almost split into two parts by the depression in which Loch Mór stands. Around this island, Pabbay and several others there is a rock platform. This may be present on West Berneray, but if so it is obscured by sand. The numerous islands off the north-eastern corner of North Uist are perhaps more correctly regarded as submerged parts of that island. It may be noted that in Hermetray there are cliffs 10 feet high (3 m) in drift (Jehu and Craig, pt III, 1923–6).

NORTH UIST AND BENBECULA

North Uist conforms to the general pattern; along most of the east coast there is a band, a mile or more wide, of crushed gneiss. This belt becomes narrow and discontinuous south of Ronay. Most of the island is made of grey gneiss except for a band of migmatitic gneiss along the west coast from Griminish to Balranald. The highest ground is along the east coast; Eaval reaches 1,139 feet (347 m). The island has a long and complicated coastline. Loch Maddy is the largest of the lochs and is a labyrinth of channels and islands. Loch Eport is a long narrow loch running east and west. The entrance is narrow and the water exceeds 70 feet (18 m) in depth. There is no sill. Ritchie (1968) notes that the cleft through the east coast hills was probably overdeepened by ice from the Minch, but the origin of the loch is more likely to have been caused by faulting. It is not a true fiord. There is a marked difference between the north and south banks.

The east coast of North Uist is steep and is cliffed; off-shore depths fall steeply to the trench which follows nearly the whole east coast of The Long Island. Godard (1965, p. 246) and others have recognized planation surfaces in the island and, as Ritchie remarks,

the coastal morphology of North Uist can be described as the product of submergence operating on these low rock platforms with their highly significant local topographic differences being exploited or modified according to the nature of the incoming marine processes.

(Godard (1965) refers to a site on the south side of Loch Maddy where there is a cliff cut in moraine and shingle a half-a-metre above present high-water mark. J. W. Dougal (*Trans. Edin. Geol. Soc.*, 19, 1924–31, 12) refers to raised beaches in the Outer Hebrides, and J. W. Gregory also makes reference to a possible raised beach on Loch Maddy.)

To describe and try to explain the coastline of North Uist in detail would require a monograph. Along the north, west and south-west facing parts of the coast there are some remarkably interesting sand formations. On the coast facing the Sound of Harris there are several bays, widely open and limited by rocky headlands. The island of Vallay and the strand south of it are important elements. The island is low in the east and reaches to more than 100 feet (30 m) in the west. There is abundant sand, but little smooth machair. In Camas Mór, at the north-west corner of the

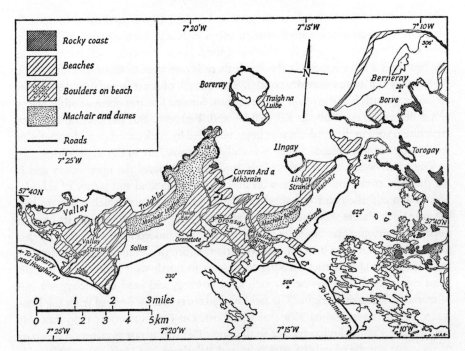

Fig. 39. Sketch map of the coast of North Uist (based on Ordnance Survey;
Crown Copyright Reserved)

island, a rocky outcrop protects a large area of inter-tidal organic deposits which include birch branches, below low-water mark, and embedded in dark brown peat. Valley Strand nearly dries out at low water, and is an area that Ogilvie suggested for reclamation. To the east is the great sweep of Tràigh Iar and Machair Leathann. The sands for the beach and dunes came from the sands exposed at low water, and also from Valley Strand. The beach is held by the small rock outcrops of Veilish Point and Huilish Point and the much larger mass of Ard a'Mhòrain, presumably a former island. Behind Machair Leathann is Tràigh Ear, another extensive sand flat comparable to Valley strand. The most interesting feature is the narrow tongue of Corran Ard a' Mhòrain which runs south-eastwards from Machair Leathann. The tongue is a narrow strip of marram-covered dunes only a few yards wide at high water. At its end a tiny fragment of machair led Ritchie to suggest that Tràigh Ear was partly produced by flooding. On the east of the strand Oronsay island recalls Vallay, but is more protected and has a much more extensive northern shore. The coast here and farther east is sheltered from serious wave attack, and the two areas of machair, M. Robach and M. Suenish are separated by the rocky outcrop at Hornish Point. On account of rabbits and grazing and violent winds these areas of machair are very mobile (Fig. 39).

West of Vallay island, the coast from Griminish Point to near Balranald is rocky. At Griminish Point there are geos and arches. Geo is used more frequently than sloc to describe these small narrow inlets in North Uist, most of which have been cut along the strike in hornblende gneiss. Some coincide with joints and in but few does a dyke appear at the head; there is one west of Caisteal Odair. Sloc Roe nearby is a more impressive outlet. This stretch of coast faces fairly deep water so that wave attack and erosion can be serious. Similar features occur at other places along this coast, especially at Kilphedder and Tigharry. But the general effect of structure is seen in the semi-circular bays, bounded by rock outcrops and sometimes fronted by shingle ridges. There is usually some machair in the depressions.

The south-west coast is particularly interesting. It faces the open ocean and is for the most part a low coastline with much calcareous sand and plenty of shingle. Near Balranald the coast is a machair plain and sand dunes which reach 70 feet (21 m). Loch Paible was once a freshwater loch, but an outlet was cut in 1793 to improve drainage. The drainage of a large area was indeed improved, but the sea filled the loch and gave it its present condition. South of this loch there is extensive machair, but ridges of rock striking south-east to north-west beneath it have had great influence. There has been much sandblowing, and sand often covers a soil of marsh or lacustrine origin. The sand nowadays is mainly derived from the low-water flats between Rubha Mór and Kirkibost. On Reid's maps of 1799 much of the machair was marked as subject to drifting. The two major features on this part of the coast are the islands of Kirkibost and Baleshare (see Plate 38). Kirkibost is the smaller. The coastal dunes are high and often cut by deep blow-outs. There is a central marshy area, and true machair to the south. MacRae (*Statistical Account,* 1837) states that the island was formerly of great value, but in his time (and later) so much sand had been blown away so that the sea now occupies parts which were at one time fertile fields. Baleshare shows hummocky drift, bare rock, marshy areas and lochs and some machair. The southern part, Eachkamish, is a sand peninsula with a fine stone cobble beach on its seaward side. Erosion and deposition proceed at the same time so that the topography is irregular. Steep slopes face the ocean, and the southern end of the island suggests extension by recurved beaches, now sand-covered, and separated by small depressions which have been overrun by salt water from the east. On the west of the island at Ceardach Ruadh, a rock ridge and boulders show at low water on the beach. To the north layers of organically rich sand are known under the beach and extend some way seawards. Their presence implies erosion and a retreat of the dunes. There may also have been submergence. The erosion is also shown by the fact that at Ceadach Ruadh a stone structure and midden, of probably Iron Age date, have been cut into by the sea. In the north Tràigh Leathann is almost shut in by machair, but appears to be working westward against the narrow peninsula of Slugan. This is a potentially dangerous area. Within Kirkibost and Baleshare the coast of the main island consists of a low rock platform covered by drift.

The irregular south coast of North Uist and the north coast of Benbecula may perhaps best be regarded as two sides of a channel or sea-loch that, on the east, recalls Loch Maddy. As a result of submergence the rock-bound eastern part has been extended so that the channel is now continuous. But the western part is in every sense akin to the smaller channel between Kirkibost and Baleshare. It is very shallow and, apart from tidal runnels, dries out at low water. Since the tides of the Atlantic and the Minch do not conform there was always some uncertainty in rough weather in crossing the sands from North Uist to Benbecula. The causeway has successfully overcome this. It should be remarked that the islands of Ronay and Grimsay divide the eastern part of the channel into two distinct east and west sea-lochs, both of which are generally parallel to Loch Eport in North Uist and the main axes of Lochs Uskavagh and Leiravagh in Benbecula and also to the channel between that island and North Uist.

In Benbecula the crushed gneisses which characterize the eastern coasts of North Uist and South Uist are only represented in patches on Maragay Mór and its neighbours, Maaey Glas and Wiay. The whole of the main island is gneissic, and there are areas of machair on the west. The eastern coast is rocky and cliffed, but the cliffs are not high. The highest point on the island, Rueval, reaches 409 feet (125 m). Like North Uist there are numerous lochs, and they, like the sea-lochs on the east, trend south-east and north-west over much of the island. In the lowland on the west their shapes are irregular and resemble the lochs (q.v.) on the western side of South Uist. The open bays, enclosed by low outcrops of rock, on the west coast have fine beaches often backed by cobble ridges. The machair land is behind these. The flat sandy area north of Bailivanish is used as an air-field.

SOUTH UIST AND ERISKAY

The channel between Benbecula and South Uist is like that north of Benbecula and was formed in a similar manner. The eastern end contains many rocky islands; the centre and western end are full of sands, and the curious feature of Gualaan stretches more than half-way across the entrance to the Atlantic. The whole eastern coast of South Uist, except for a part of the south side of Loch Boisdale, is in the zone of crushed gneiss; this is a rocky and mountainous area cut through by sea-lochs which, in general, conform to the south-east to north-west trend. The peninsula of Usinish, elongated north and south, is separated from the main island by a belt of low ground. Islands are scarce on this coast, the only one of any size is Stuley, about two miles (3.2 km) south of Loch Eynort. There are rocky cliffs all along the coast, but seldom do they attain any height. They have been ice-moulded, and modern marine erosion is limited. The hills slope steeply seawards; westwards the slopes are gentler and pass downwards under the peat and machair lands.

The west coast displays the machair lands to advantage. On its seaward side there is usually a dune belt which is narrow and grades gently into the machair behind. The beaches are magnificent; they are not noticeably wide, but are continuous for long distances between low rocky outcrops. Sometimes these outcrops rise a few feet, but several are more or less covered at high water. They represent spurs of rock in conformity with the usual south-east to north-west trend. Since it seems (Ritchie, 1966, 1967) that rock underlies all the beaches at a shallow depth, it may be that there is now only sufficient sand to maintain beach equilibrium without adding to the machair. Wind erosion is active in the dunes and machair. The true machair is low and flat, usually below 20 feet OD (6 m). The flatness is probably related to its age. The landscape of the machair is produced by erosion and re-working; 'the phase of deposition and building from a primary source is over'. The flatness is related to the water table, and the water table is also the base-level for wind erosion. The parallelism of the surface and the water table is significant. But nevertheless small hills and ridges of sand are often piled up against rock outcrops or even less permanent obstacles. The plan and section in Fig. 40 illustrate this. But over a much longer period of time the machair has been subjected to periods of stability and drifting, the evidence for which is found in buried soil horizons. The water table is very important in the evolution of the machair; not only is it the base level of erosion but also sand deposition on a wet surface gives rise to flat spreads or gentle fans. In winter the low-lying parts of the machair are flooded, and around temporary and even more permanent lakes on its surface there may be marked wave erosion. Ritchie rightly points out that, in order to appreciate the role of water in the machair, visits to it must be made in winter.

How old is the machair? The best evidence is obtained from a study of compressed organic deposits below high-water mark. One of these at Borve (Benbecula) contains wood-fragments dated 5,700 ± 170 years BP. At Kilpheder wheelhouse structures are in or on an older sand surface and covered by later sand. They belong to the period AD 100–200. There are a number of similar sites and it seems that at this time the machair was relatively densely settled. The Vikings arrived about the ninth century, but the material evidence refers to a Viking house about 6 feet (1.8 m) below the surface of the machair at Drimore. This suggests sand drifting since the ninth to thirteenth centuries. Thus archaeological evidence indicates that in the last 2,000 years machair has existed, and that there have been periods of severe drifting in that time. The earliest written reference is the place-name Macker-meanache, mentioned by Dean Munro in 1594. This place is identified with the present-day Eochar. There are many later descriptions in the seventeenth and eighteenth centuries.

Combining the evidence of the Clanranald Papers (1668) – and other sources – the general impression is that the machair was over-grazed and over-cultivated, and as a result much of it was laid bare by severe sand erosion and drifting, and that this state continued into the early twentieth century from which period the land has become increasingly stabilised. (Ritchie, 1967)

Certain other features of the west coast of South Uist are noteworthy. The many lochs on the landward side of the machair, that is, those mainly in the black peat land, are orientated east and west; on the other hand the lochs wholly in the

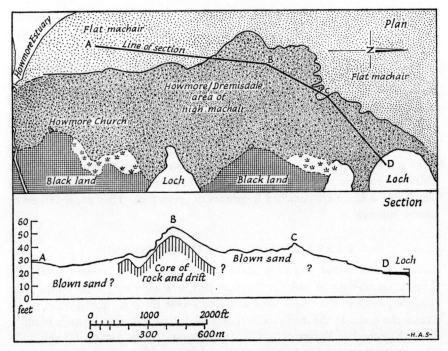

Fig. 40. Howmore–Dremisdale area of high machair (based on W. Ritchie)

machair usually have their long axes parallel to the coast. It is probable that the east–west lochs are related to the trend of the underlying rocks. Mention was made on p. 183 of Gualaan. This narrow spit is part of a former complex eroded by the sea as a result of penetration both north and south of it from South Ford (i.e. the passage between South Uist and Benbecula). Just to its south is the large Loch Bee. Ritchie suggests that the outflow from this may also have helped. Whatever the precise evolution, it seems probable that there was formerly an extensive machair area which stretched across the South Ford.

The evidence for a relatively recent submergence of about 6 feet (1.8 m) is not convincing. It rests on the interpretation of some sites in Vallay; Ritchie thinks that ordinary coast erosion offers a better explanation. The evidence derived from the Dùns, which are connected by causeways to the shores of lochs, is used to show evidence of fall in lake level. The causeways would have been 12 to 18 inches (0.3 to 0.45 m) beneath water level. The change to the present level could well have been the result of drainage. At Borve and Peninerine there is good evidence of severe coast erosion some time before 5,700 BP. True, this would be aided by a rise of sea-level, but the machair must be younger than the organic deposits over which it has developed. There seems to have been a machair surface along the west coast of both Uists before or during Romano-British times. Ritchie holds that the major period of machair formation was between 5,700 and 2,000 BP, that is during

Sub-boreal and early Atlantic times. It may have begun to form even earlier. Lacaille suggests that at that time Scotland was cooler and drier so that sands were easily blown up on to the Post-glacial raised beach and over the organic layers of the Atlantic period. The Sub-Atlantic was cooler and better, and so the sand was more stable.

Eriskay is separated from South Uist by a wide sound, trending in the usual direction. The island is, structurally, a direct continuation of South Uist; the crushed rock is in the east. The north and west of the island is much lower, and there are extensive sands. The island is nearly bisected by the deep inlet of Acairseid Mhór and the low ground which reaches the west coast at Cailleag a'Phrionnsa. This is directly comparable to the sea-lochs in the northern islands. The small stack islands to the south are detached fragments of crush rock. They are separated by narrow channels.

BARRA AND THE SOUTHERN ISLANDS

Barra is a small island, but once again there is the contrast between a hilly and cliffed coast on the east and, for the most part, low and sandy areas on the west. However, both in the north-west and south-west there is high ground which reaches the coast. In the north the spit of Tràigh Eais connects the main island to the high ground of Eoligarry and Scurrival Point. To the east of this spit is the extensive sandy area of Tràigh Scurrival and Tràigh Mhór, part of which is used as an airfield. The belt of crushed gneiss which hugs the eastern shores of the northern islands turns inland in Barra and reaches the west coast in Greian Head and Ard na Gregaig.

The coastal features in the high land owe much to weathering along joints. There are two main sets which roughly follow the dip or strike of the rocks. The north-east coast is much broken and there are several off-lying islands. The sounds between them, which resemble and probably are drowned valleys, trend south-east to north-west. The small bay just north of Ersary on the east coast and the low ground at Borve on the west coast are connected by a bealach (pass) which cuts through the island. It follows lines of vertical jointing. Loch Obe may well have been a fresh-water loch; the channel from it to the sea is shallowest in its mid part. Submergence aided by erosion along joints could account for this. Jehu and Craig (pt 1, 1921–5) compare Loch Obe with Bàgh Beag near Castlebay.

The small islands of Gighay and Hellisay show a narrow band of crushed rock on their east sides. They are high, and separated by a narrow south-east to north-west sound. There is no sand or machair land. The Sound of Hellisay, trending in the usual direction, divides them from the archipelago of small isles which are virtually part of north-east Barra. Fuday is somewhat different. It is wholly formed of gneiss, and on its western side there are extensive sands in Tràigh na Reill. On the northern side, in Tràigh Bhàn, shelly sand has been cemented into a firm sandstone.

MacLeod (1951) has described the vegetation of Barra. That on or near the shore is divisible into five categories. On the unfixed dunes *Ammophila arenaria* is dominant. It is followed by *Senecio, Tussilago farfara, Agropyron junceiforme,* and *Ranunculus.* A few yards beyond the crest *Ammophila* is still dominant, but *Galium verum, Trifolium repens, Achillea millefolium, Festuca rubra, Plantago lanceolata, Carex arenaria,* and *Erodium circutarium* grow freely. At the back of the dunes *Ammophila* decreases, and *Carex* and *Agropyron* have disappeared. On the other hand *Tussilago* has increased, and *Bellis perennis* and *Cerastium vulgatum* have come in. About 100 yards (91 m) back from the dunes conditions are stable. *Trifolium repens, Lolium perenne, Agrostis stolonifera, Holcus lanatus* and *Poa pratensis* abound. There is an absence of bare patches, and there is some moss. This is machair land which is grazed by cattle and covers all the tombolo between the main island and Eoligarry. Sand, blown by the winds, has an effect on the vegetation on the western side of the island up to 300 or more feet (91 m), and for nearly half-a-mile from the shore.

The string of small islands to the south of Barra possess many points of physiographical interest. Vatersay and Sandray show outcrops of the crushed rock. Vatersay is almost bisected by Vatersay Bay and Bàgh Siar in both of which, but especially in the latter, there are wide sands. These two bays and the Sound of Vatersay (north) and Sandray (south) as well as the wider and more open sounds between the southern isles, all conform to the south-east to north-west pattern, a pattern particularly clear in north-eastern Vatersay. This island is prolonged by a series of islets or almost islets (Creag Mhór and Uinessan) to the high island of Muldoanich (505 feet: 154 m) in which there are several slocs. There is also a small islet separated from it by two slocs, one following a dyke, the other a joint plane. Sandray is cut by a deep north-north-west to south-south-east valley, Gleann Mór. It is a high island, reaching 678 feet (207 m). There are dunes somewhat inland on the eastern side, but blown sand is partly responsible for holding Loch na Cuilce. The outline of the island emphasizes in several features the trend of Gleann Mór.

In Pabbay the high cliffs face west. There is a gentle slope to the east and the sands in Bàgh Bàn. The peninsula of Rosinish is almost an island. This and the Rubha Greotach promontory on the west are being separated by erosion along joint planes from the main island. Outer Heisker is pierced (Jehu and Craig, pt 1, 1921–5) in the two narrow necks between the three arms of the E which its shape resembles. Mingulay possesses magnificent cliffs and stacks. That at Aoineig is nearly vertical and 753 feet high (230 m). From it the land slopes eastwards to Mingulay Bay. There is a small stream flowing into this bay, the head waters of which seem to have been beheaded by erosion of the western cliffs. The island was formerly inhabited. It is separated from Berneray by a narrow sound. This island also possesses famous cliffs. On both islands there are narrow inlets, slocs, some of which have been cut along dykes, but more are related to dominant joints. This is seen to perfection in the west of Mingulay where there are several tunnels and caves, stacks and islets. The full force of the waves in stormy weather is great on this coast.

THE HEISKER OR MONACH ISLANDS

The Heisker or Monach Islands are a group of small islands about eight miles (13 km) south-west of Hougharry in North Uist (see Plates 2 and 29). In 1893 the Six Inch Map showed four islands. Recent winter storms have, however, nearly severed the tombolo connecting Ceann Iar and Shivinish, so that at high water there are effectively five islands. They are all formed of Lewisian Gneiss. In recent years they have received a good deal of attention from R. Randall and from the Brathay Exploration Group. The main eastern island, Ceann Ear, has changed considerably in shape since 1893, especially in the dune area on the north-east which has been cut back. On the west coast the northern part of the dunes has receded, the southern part has prograded. Great changes in fact take place on the surface of the islands in storms and heavy swell. The other parts of the coast, of all the islands, are rocky and stable; a low platform covered at high water occurs almost all round the islands.

The plant communities have been investigated but not published, mainly by R. Young of the Brathay Exploration Group. He recognized three shore communities – rocky shores, cobble beaches and sandy shores. In the southern part of Port Ruadh bay (east side of Ceann Ear) considerable modification of the natural vegetation has taken place as a result of the movements of the former inhabitants.

As a general rule the shore vegetation grades directly into the inland types, but in parts of south Ceann Ear a fairly distinct community seems to result from the effect of spray. 'Inland' the pattern is simpler in the north than in the south. There is a marked line running from just north of Rudh'an Thaing and passing along the north side of Loch nam Buadh to the west coast. Young notes that in a few metres typical stable dune or machair to the north gives place to a dark soil with peaty surface horizons in the south. *Carex nigra* is often abundant in the dark soil, but was not seen north of it. The plants of Ceann Ear have been carefully catalogued by Young.

On Ceann Iar, which received much more cursory tretament, various flat areas seemed to correspond with the dark soil of Ceann Ear; *C. nigra* is common. In a short visit neither *Daucus carota* nor *Thalictrum minus* were found; this was odd in view of their abundance on Ceann Ear.

On Stockay (north-east of Ceann Ear) a day's visit revealed a well-developed strand-like vegetation on sand and pebble shores. There was abundant *Cakile maritima* and some good colonies of *Mertensia maritima*. *Secale cereale* and *Arena strigosa* were also found, but not noticed on the main islands.

All the islands are low; there are small rings of the 50-foot (15 m) contour in the dunes of Ceann Ear, and 62 feet (19 m) is marked on a trig. point at Cnoc

Bharr on Ceann Iar. There is a disused lighthouse on Shillay, the island farthest west.

Haskeir Island lies about ten-and-a-half miles (17 km) north-north-east of Stockay. It reaches 120 feet (37 m) in its south-western part, and nearly as much in the north-east. A lowland, nearly divided by a cave, separates the higher parts. In winter spray covers all the surface, on which there are several pools of brackish water. Haskeir Eagach, eight-and-a-quarter cables (1,508 m) to the west-south-west, is really a group of five bare rocks close together, but separated by deep water. There are other rocks and shoals in the neighbourhood. (West Coast of Scotland *Pilot*.)

APPENDIX: THE MACHAIR

Machair is a particular form of sandy pasture that is found in the Outer Hebrides, in Tiree, Coll, and some other islands and in various places on the mainland. It is a calcareous sand, but the proportion of lime varies considerably from place to place, and even along any given traverse (see p. 142, Tiree). It is seen to greatest perfection in South Uist. All along the western side of The Long Island the sea deepens very gradually. The floor is a continuation of the rock plat-form which slopes down from the eastern hills under the peat and pasture lands, and continues several miles out to sea, possibly as far as St Kilda. In glacial times the ice spread over much of this now submerged surface and deposited boulder clay. This is the source of the *non*-calcareous sand, and particularly of the cobbles and shingle which often form long ridges at the back of the beaches. The calcareous sand is derived from the numerous organisms which live on the sea-floor. This shell sand is driven, together with siliceous material, on to the beaches and when it comes under the influence of the wind there must be a pronounced sorting action. The lighter and more flaky material is carried more easily.

The beaches are often long and enclosed between rocky outcrops which are often scarcely higher than the beach; elsewhere they may be in bays shut in by high rocky headlands. At the head of the beach there is usually, but by no means always, a pebble or cobble ridge, and behind this, and partly on it, there are dunes. These may be simple, in the form of a narrow and low ridge, or they may be extensive and show typical erosion forms. Behind the dunes is the true machair. It characteristically takes the form of a low flat sand plain. Locally, the sand may rise up against an outcrop of rock, and in such places there may be more dunes. On its landward side it gives place in South Uist and some other islands to peat over which it has spread. It may, on the other hand, rest directly on the base of the hills.

The machair gives a very fertile sward. The humus content reaches it in various ways. The sand itself sweetens the peat and makes it more fruitful. Animal dung is added and adds greatly to its fertility. In winter great numbers of barnacle geese visit it, and the sea casts up on the beaches great quantities of weed which is taken inland by man and spread over, or ploughed into, the machair. But the machair needs constant care and attention. If it is ploughed or sub-jected to too much wheeled traffic, the vegetation cover may be destroyed and then the wind only too soon regains mastery.

The vegetation of the machair consists largely of grasses and low-growing perennials (see list). The surface is broken by many lochs which if entirely in the machair are usually elongated parallel with the adjacent beach; if they are partly in the peat and partly in the machair their longer axes usually run with the grain or trend of the underlying solid rocks. From the scenic point of view the lochs add much to the appearance of the machair. It is seen at its best in high summer. Then on a clear day and with a bright sun there are few more beautiful sights than these great spreads of sandy pasture covered with wild flowers in full bloom which give it not only colour but also scent. The occasional breaks in the surface show the underlying sand; seawards are the dunes and the dazzlingly white beaches and the intensely blue sea. The lochs, too, are often partly covered with water lilies and are the haunts of birds. Inland the machair gives place to the peat and the hills. The crofts and larger settlements are along the inner edge of the cultivated machair – and cultivation may occupy most of the machair in some localities. When this is so, the crops themselves add variety to the scene.

However, to understand the machair it should also be seen in winter. The water table is, even in summer, but a short distance below the surface; in winter much of the low-lying machair is flooded. The water table, at whatever level it may be, is the base-level of erosion, and during violent storms in winter it is a real advantage that floodwater prevents the drying of the surface and the removal of sand (see p. 184). In summer, when the surface is dry and broken, wind erosion may be serious. The water table also promotes a richer and stronger vegetation. The spreading of seaweed as a surface dressing in winter also gives some protection from wind erosion. Seaweed, by raising the humus content of the soil, also reduces the risk of drought damage.

In any sandy area by the sea in these islands the marram grass (*Ammophila arenaria*) is of prime importance. In the machair islands this is indeed true; if the marram on the dunes is destroyed, then the sand can be easily dispersed and dune protection is lost. Many acres in Tiree have disappeared in this way. In the early stages of colonization the sea purslane and the sea sedge are important. At a slightly later stage the trefoils (*Lotus corniculatus* and *Trifolium procumbens*) come in behind the dunes. These plants fix nitrogen and make it possible for other plants and grasses to take root. In time a shell-sand grassland is created.

Many separate areas of machair have been referred to on pages of this book, and variations of development have been noted. On p. 184 the probable age of the machair of South Uist has been discussed. It is not possible to say that this estimate of age applies to all places in the western isles, but it is probably reasonable to assume that it does.

The following list of plants on the Coll machair is taken from (Sir) Frank Fraser Darling's *Natural History in the Highlands and Islands* (1947).

Buttercup	*Ranunculus bulbosus*
Milkwort	*Polygala vulgaris*
Mouse-eared chickweed	*Cerastium vulgatum*, and *C. semidecandrum* and *C. tetrandrum*
Heartsease	*Viola curtisii*
Cathartic flax	*Linum catharticum*
Blood-red cranesbill	*Geranium sanguineum*
Red clover	*Trifolium pratense*
Wild white clover	*T. repens*
Hop trefoil	*T. procumbens*
Bird's-foot trefoil	*Lotus corniculatus*
Kidney vetch	*Anthyllis vulneraria*
Wild carrot	*Daucus carota*
Ladies' bedstraw	*Galium verum*
Daisy	*Bellis perennis*
Spear thistle	*Cirsium lanceolatum*
Creeping thistle	*C. arvense*
Cat's ear	*Hypochoeris radicata*
Sow thistle	*Sonchus oleraceus*
Bog pimpernel	*Anagallis tenella*
Forget-me-not	*Myosotis versicolor*

Germander speedwell	*Veronica chamaedrys*
Eyebright	*Euphrasia officinalis*
Redrattle	*Bartsia odontites*
Wild thyme	*Thymus serpyllum*
Selfheal	*Prunella vulgaris*
Ribgrass	*Plantago lanceolata*
Frog orchid	*Coeloglossum viride*
Marram	*Ammophila arenaria*
Yorkshire fog	*Holcus lanatus*
Meadow grasses	*Poa* spp.
Woolly-fringed moss	*Rhacomitrium languinosum*

Chapter VII

The mainland coast: Melvich Bay to Berwick

THE COAST OF CAITHNESS; BRORA AND GOLSPIE

At the north-east corner of Scotland the main mass of the Old Red Sandstone coincides closely with the county of Caithness. A narrow strip extends to Melvich Bay, and farther west there are small outliers, one of which forms the coast for a mile east of Strathy Point. In order to appreciate the general nature of the coast of Caithness it will be helpful to refer first and briefly to the interior.

Most of the county is a plain of Old Red Sandstone on which there are extensive Glacial and Post-glacial deposits. The area is for the most part low-lying; it rises to over 1,000 feet (305 m) in the west and south and falls to sea-level in Sinclair's and Dunnet Bays. The slope is not regular, and is interrupted by a ridge of higher ground running east and west from Knockfin Heights to the coast between Ulbster and Sarclet. This ridge is also an important watershed, and includes Ben Alisky (1,142 ft: 348 m) and Stemster Hill (815 ft: 248 m). This and other watersheds in the county are irregular and this, taken in conjunction with the absence of any signs of *recent* (so in the Caithness Memoir, 1914; but see Chapter III above. The Memoir and *The Beaches of Caithness* (1970) by W. Ritchie and A. Mather are frequently referred to in this account of the coast of Caithness) coastal movement, suggest that the drainage system attained maturity some considerable time ago, and also that the plain is not of recent origin. The rocks underlying the Old Red are folded, and the upper beds of the Old Red are faulted in Dunnet Head, and a little south of the county boundary similar faults have affected the Jurassic rocks. It seems probable, therefore, that movements continued until Tertiary times.

Caithness forms part of a structural basin which extends into Orkney and beyond. The lowest Old Red beds are found in the south, and the highest, the John o'Groats Sandstone, in the north. But since a broad anticline cuts the coast near Sarclet, the main trough or basin is subdivided so that the lower beds are brought to the surface. There are also other folds, but it is difficult to date them. The beds are also cut by some important faults, the most significant of which is the Brough Fault which runs almost due south from the east side of Dunnet Head. It is estimated that the total thickness of the Old Red Sandstone in Caithness is between 16,000 and 18,000 feet (4,880 and 5,490 m). The coarser conglomerates, breccias and arkoses, occur near the base and are often very red in colour; the upper beds are mainly fine-grained flagstones of a grey colour. The John o'Groats beds are sandstone, yellow or red in colour.

The rivers cut through the folds, and are therefore either antecedent or super-imposed. But the evolution of the plain remains obscure, particularly in its early stages. At one time it was presumably continuous with the Orkneys; the higher ground on the western side of those islands may have joined the higher area on the western side of the county. At that time the plain probably extended farther west. Morvern, Scaraben, Maiden Pap rose above the plain or plateau, and Ben Griam More, Ben Griam Beg and Ben Armine, west of the county boundary, imply erosion at an early time 'during which this plateau was formed by removal of great masses of rock from above its present surface. These mountain masses alone escaped the processes of erosion, and now stand out boldly with little or no apparent relation to the present drainage system' (Crampton and Carruthers, 1914). When all this erosion occurred is uncertain; it may have been accomplished by the Pre-glacial ancestors of the present rivers, or at a much earlier time. However, before or after the mountains were isolated, rivers formed themselves into two main groups, one flowing to the north and north-west, the other to the east and south-east. River capture and glaciation have produced many changes in their original courses, and in general the northerly flowing streams have gained at the expense of the others.

The ice age had a great effect on Caithness; it largely masked the Pre-glacial topography, and many deeply cut river courses were choked with boulder clay, so that when the ice finally disappeared the streams often had to find new courses, and have often cut Post-glacial rock gorges. The Reisgill burn (Fig. 41) illustrates the nature of these changes very well. Since the ice advanced from the south-east a good deal of the relief of the county is now orientated in that direction, e.g. the line of the Wick river, Loch Watten and Loch Scarmclate. The effect of this advance of the ice from the south-east is occasionally seen in the cliffs where it left a terraced appearance on account of the removal of material from the less resistant outcrops. Marine erosion has locally obliterated these features. The growth of peat has covered extensive areas, and hidden not only the solid rocks but many of the features produced by glaciation. Peat in some respects acts like a sponge, and by retarding the flow of water has the effect of preserving the topography.

This brief outline is sufficient to indicate that not only do the different beds of the Old Red reach the coast, but that their appearance on the coast will depend on their nature and on the folding and faulting, both major and minor, to which they have been subjected. The form of the cliffs will also depend much on the thickness of the boulder clay covering them. Where this exceeds a few feet, the upper part of the cliff nearly always slopes at a much smaller angle than the lower part. Much of the beauty of the cliffs depends upon the nature of the bedding and jointing, and also upon minor faults and the occurrence of dykes of igneous rock. The sea works in all these lines of weakness and produces magnificent stacks and clefts and long narrow inlets called geos. These geos are perhaps the most characteristic feature of Old Red Sandstone cliffs, wherever they occur, but in Caithness and Orkney they are particularly well developed. The height of the cliffs depends mainly on the

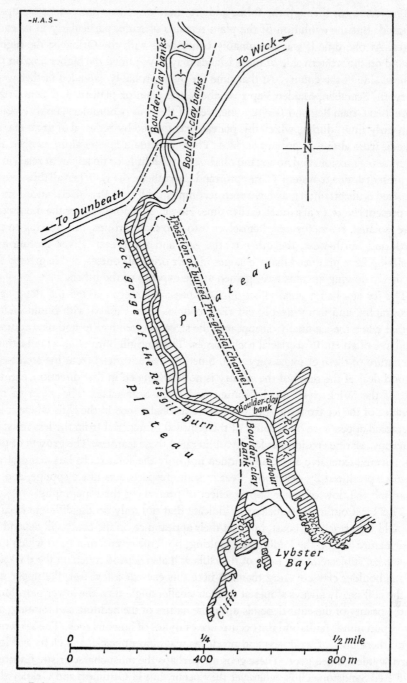

Fig. 41. Relation of the rock gorges of the Reisgill burn to the buried Pre-glacial channel
(based on Geological Survey)

general structure of the area; they reach 300 feet (91 m) at Dunnet Head, and 200 feet (61 m) near Duncansby.

It will be convenient to comment on the coast from Strathy Bay eastwards to Duncansby Head, and then southwards to the Ord of Caithness. In Strathy Bay the junction of the Old Red and older rocks is hidden by sand and alluvium. The Old Red cliffs extend for about a mile, and then give place to an irregular and picturesque stretch of coast, around Portskerra, cut in the Moine series (see p. 56). Cliffs in Old Red Sandstone are, apart from certain bays, continuous from Melvich to just north of Berriedale. Between Melvich and Sandside Bay they locally exceed 100 feet (30 m), and the several geos in this stretch are nearly all associated with faults. Red Point, however, is an outcrop of the old crystalline rocks.

Sandside Bay lies between two headlands of flagstones in front of which are wide rock platforms. The cliffs near Sandside Head are vertical and cut by several geos. Three considerable streams flow into the bay, and they are not fully graded to base level. Since they pass through boulder clay, they bring sand and shingle to the beach. This is augmented by cliff erosion so that there is a thick bed of sand in the bay. The beach is, however, mobile and the mouth of the middle stream, the Reay burn, is sometimes turned as much as 328 yards (300 m) to the west. The dunes behind the beach owe their form and height largely to the fact that the winds are channelled in the depression leading to the bay. Sand is quarried in considerable amounts from the dunes; the sand may contain up to 40 per cent calcareous material. The proximity of Dounreay has had its effect (Ritchie and Mather, 1970). At Brims Ness the bench is about a quarter-of-a-mile wide (402 m). A fault runs out at Port of Brims, and the cliffs for a mile or more are low. But near Ness of Litter they rise to Spear Head and Holborn Head. Here is some of the finest cliff scenery in Caithness. The thin-bedded flagstones are traversed by joints, and other lines of weakness that cut through the headland from top to bottom. The sea working along these lines has cut off a great mass, the Clett, and has so undermined the headland that it stands, as it were, on a number of legs. The sea also works upwards in these caves and sooner or later a vertical passage is made between the inner part of the cave and the surface, and a blow-hole is formed. Holborn Head, like those of Dunnet and Duncansby, is the summit of a hill cut back by erosion during long periods of time to the watershed. It is in these places, too, that there is a tendency for the submarine contours to close into the land.

Thurso Bay is a major indentation. At Scrabster the boulder-clay cliffs are liable to slip and slump, but this is not the direct effect of the sea, although at an earlier time marine erosion may have been more important. The recent movements are probably connected with the extension of house building, sub-surface drainage, and freeze–thaw cycles. The waves during high tides may now reach the base of the slumped material, a change possibly connected with the new sea-wall. Between the Braes of Scrabster and Thurso beach is a flagstone cliff, very much indented and faced by reefs which prevent any interchange of sediment between the two bays. Thurso beach faces north-northwest; the eastern limit is now a wall and breakwater, which, because of wave convergence, is subject to erosion. Beyond is a wide and long rock platform extending as far as Dunnet Bay, except for a short space at Murkle Bay where there is a small beach of shell sand. The abrasion platform on the north side of the bay is wide and backed by a cliff not subject to marine erosion; on the south side of the bay cliff erosion is limited. Thus the beach seems to be mainly derived from slumping of the boulder-clay cover of the cliffs and to a slight extent from the stream draining into the bay. The off-shore zone probably furnishes most of the sand.

Dunnet Bay, enclosed between the relatively low coast east of Thurso and the high headland culminating in Dunnet Head, is a feature of primary importance in the coast of Scotland. Since it is so enclosed the only escape for sand from the bay is landward, to the south-east. It is this which mainly accounts for the height and extent of the dunes. The beach is stable, and the seaward gradient gentle, about 1:124. There is certainly a great reserve of sediment in the bay. Several small streams flow into the bay and they divide the dunes into sections, and also help to make parts of the beach wet. The main road just behind the bay divides the dunes from the more stable links. The dunes exceed 50 feet (15 m) in many places, and there are spectacular blow-outs. Many of these are compound, and residual pinnacles remain. The largest blow-outs are in the south of the bay. Because the bay is at one end of a low corridor extending to Sinclair's Bay, the wind, from north-west or south-east, is a most potent factor in shaping the dunes and probably accounts for their steep easterly slope. Since the Forestry Commission has been interested in parts of the area, the dunes have been largely stabilized in the north-central parts of the bay, but as a commercial venture the planting has not been a great success. In the 'free' areas the dunes may be more than 383 yards (350 m) wide; in some places they are prograding. In places shingle appears under the blow-outs, and may indicate a shingle foundation for the dunes; it is not certain, since no precise measurements are available, if this shingle represents a beach at a somewhat higher level than that of the present. Probably the whole dune barrier is slowly migrating landwards. South-east of the dunes blown sand, the links, spreads widely over peat which rests on boulder clay. Drainage in the area is difficult.

Almost parallel with the dunes, and continuing northwards on the west side of St John's Loch is the Brough Fault which separates the Upper Old Red Sandstone of Dunnet Head from the middle beds. The fault is associated with a belt of low ground which helps to emphasize the summit nature of the headland. The cliffs around the headland are vertical, and at the point are more than 300 feet high (91 m). The somewhat calcareous sandstones of which the headland is formed are less resistant than the flagstones. Many of the geos on the headland are cut along lines of decomposed dykes. The views both of and from the headland are magnificent, especially on a bright day when Hoy and the neighbouring islands stand out clearly.

Eastwards from the promontory of Dunnet Head the coast is formed of low cliffs in the Thurso flags which are cut by numerous small geos. The abrasion platform is more or less continuous and is noticeably wide near the Castle of Mey. At St John's Point it is much cut up and the Men of Mey are fragments of it. The offlying islands of Stroma and Swona are also formed of flags and show similar coastal features. Old (?fossil) cliffs are present on the east side of Stroma; those on the west are vertical and often undercut. Swona is smaller but, especially on the west and south, is bordered by many reefs and skerries. The Brook and The Haven nearly sever the northern part of the island from the main mass.

About a mile south of St John's Point is the interesting inlet known as Scotland's Haven. It is surrounded by high and steep boulder-clay cliffs, and the haven itself follows a line of fault which separates the Thurso Flags to the west from the John o'Groats Sandstone to the east. The depression is an old one since the glacial fill is only partly excavated. It is more than probable that the cliff and abrasion platform, and the caves on the west side of the inlet, were all cut when sea-level was higher. Ritchie and Mather remark that 'in terms of present-day marine erosion processes the whole unit is one which is fossil'. There is a small beach at the head of the inlet, the material of which is probably derived from the till. The mouth is almost enclosed by a fragment of the abrasion platform. Gills Bay is fault-controlled in the sense that faults parallel to its western and southern shores let down a narrow belt of the John o'Groats Sandstone; the western fault extends northward into Scotland's Haven. Throughout the whole stretch there are well-defined bench features, and at Ness of Duncansby two small volcanic necks. A noteworthy feature of the strip of coast between John o'Groats and Sannick Bay is the great amount of shells, broken and comminuted. The finer particles are at the back of the shore over the whole distance, and extend up to 100 yards (91 m) inland. The Memoir (Crampton and Carruthers, 1914) regards them, at least in part, as a raised beach. Although the sand is now fixed by vegetation, it was at one time subject to wind drift. The volume of the materials suggests accumulation over a long period of time, but it is subject nowadays to some erosion in severe storms. The deposit should be considered in relation to recent work on raised beaches on the north coast of Scotland. At Ness of Duncansby this wind-deposited debris of shell fragments is thickest, and dune pasture reaches a quarter-of-a-mile inland, and then merges into the sand deposits of Sannick Bay. In this machair-like surface an ancient midden has been exposed. In the shore reefs between Ness of Duncansby and Ness of Huna patches of shell sand occur between sandstone ridges; one small beach consisting of almost intact shells more than an inch long. Near John o'Groats much sand was removed in the war; what is left is mostly at the foot of a low boulder-clay cliff, and is colonized by many pioneer plants. The Bay of Sannick, despite the fact that its natural beauty has been spoiled by scarring and

dumping, presents some points of interest. It carries a small beach almost wholly
made of shell fragments, but east of the stream it is coarse and formed of cobbles
and slabs of sandstone. The shell material comes from the relatively deep water
species, *Modiola modiola*, *Buccinum undatum*, and *Pectunculus glycimeris*. *Patella
vulgata* from the shore reefs is also a source of supply. At the present time there is
little erosion, and on the west side of the bay the shore platform was almost
certainly cut when sea-level relative to the land was higher. On the beach shell
sand gives place, above low-water mark, to an abrasion platform, and above high-
water mark the shell sand rests on a shingle beach, partly fossil. The vegetation
on the highly calcareous sand is unlike that on more siliceous beaches. *Elymus
arenarius* is present on the cliff-foot dunes; machair is scarce or absent, but
Potentilla anserina and *Honkenya peploides* are present. The machair on the cliff
top has been much eroded.

Duncansby Head (210 feet high: 64 m) may be regarded as that part of the coast
cut off by a fault running from the Bay of Sannick to Thirle Door. It is formed of
Thurso Flags and contains Duncansby Head and the well-known geo just to the
south (see Plate 21). Southwards from Duncansby Head is one of the finest
stretches of cliffed coast in Britain. From Thirle Door almost to Fast Geo the
John o'Groats Sandstone forms the cliffs and some magnificent stacks. One of
these is somewhat higher than the adjacent cliff. The Memoir makes the comment
that this stretch of cliff has retreated somewhat more than the Thurso Flags to
north and south of it, and suggests that this may be because of the larger size of the
sandstone blocks which allowed the waves to attack the cliff base more effectively.
There is also a narrow bench of somewhat recent origin, and probably associated
with a raised beach, referred to in the Memoir as the '5-ft' beach (1.5 m). The
cliff is also higher than that to north and south. In the Thurso Flags which make
the cliffs as far as Skirza Head there are some fine geos; Wife Geo is distinguished
by having two entrances, the one to the south open to the sky, the other a
tunnel. It is a fine feature.

Freswick Bay is to some extent fault-controlled. The Thurso Flags extend from
just north of Fast Geo to include Skirza Head. The John o'Groats Sandstone forms
the inner part of the bay, almost to the Gill Burn mouth, and the Ackergill beds
the remainder. There are traces of the so-called '5-ft' beach in the bay. Behind the
bay there is thick boulder clay. Both shingle and shell sand occur on the beach.
The shingle is derived from the cliffs; the sand comes from the sandstones, the
Freswick burn which traverses till, and from off-shore. The south-east winds
cause maximum deposition at the head of the bay. The exposure to the south-east
has also led to the formation of a marked abrasion platform in the bay. The dunes
have been seriously menaced by sand extraction, and there is no doubt that loss for
this reason is in excess of supply. Southwards from Freswick Bay, flagstones form
the cliffs as far as a quarter-of-a-mile north of Ires Geo. There are numerous geos
north of Sinclair's Bay; some are associated with small faults, e.g. Castle Geo,

Brough Head, Hobbie Geo; others with jointing, minor faults, or other lines of weakness. The raised platform occurs in places; the cliffs fall in height near Sinclair's Bay.

Sinclair's Bay is the largest in Caithness, and occupies a syncline in the Ackergill beds. It lies between Tang Head and Noss Head. As far as the Rough of Stain there is an abrasion platform in the northern part of the bay; near Keiss a higher raised platform, covered with gravels, lies behind the lower one. The south side of the bay runs roughly east and west, and is bordered by cliffs which are well over 100 feet high (30 m) near Castle Girnigoe, and finally disappear near Ackergill Tower. Between these points there is an abrasion platform and several small geos near Noss Head. The depression in which the bay lies extends inland to Dunnet Bay and to the south is paralleled by the Wick river, and to the north by the smaller depressions, the one connecting Freswick and Gills Bays, the other which cuts off the small peninsula of Duncansby Head. These may be parts of an ancient drainage system, but from the present point of view they have played a significant role in coastal evolution since they have been suitable places for the accumulation of marine sediments. In Sinclair's Bay the Loch of Wester has been impounded in this way and, as Ritchie and Mather suggest, probably at more than one level of the sea.

The erosion of the cliffs at the north and south ends of Sinclair's Bay, although not rapid, has supplied the central area of beach. The beach material, coarse and fine, comes mainly from the capping of boulder clay on the cliffs. In the past, when the sea stood higher, the supply would have been greater. Today the older till cliffs are buried under blown sand. Shell debris may reach 63 per cent of the beach contents. It is derived from shallow water creatures, especially *Littorina*, *Patella*, *Tellina* and *Mactra*. The fact that accretion occurs in the central parts of the bay may be helped by the abundance of off-shore material in this region.

The present beach, except in the north, is mainly sand. In the dunes, however, blow-outs cut down to expose shingle and it may well be that originally the embayment was closed by a shingle spit or beach which impounded the Loch of Wester. At this earlier stage the sea would have been eroding the boulder-clay cliffs and so producing much more coarse material. It is on this shingle that the dunes have developed. In the north and south of the bay the dunes form a fairly simple ridge. Since winds from the north-west course down the depression, both inner and outer slopes of these ridges are steep. The dunes attain their maximum development north of the mouth of the river. It is an area containing large blow-outs. At the mouth of the river new foredunes are forming and forcing high-water mark farther seawards. Those on the north of the mouth have continued the deflection of the stream to the south. Changes in the older dunes in this central area have led to the exposure of brochs and other prehistoric structures.

The dune vegetation in general follows the normal pattern. Marram dominates the seaward faces, and lyme grass is also important. The landward slopes carry fescue and other grasses, and *Galium verum* and *Lotus corniculatus* are not un-common. The links behind are much damper; in the wettest places there are masses of *Equisetum palustre* and *Potentilla anserina*. Most of the links are meadows of *Agrostis* and *Festuca* and flowers such as *Parnassia palustris*. Some sand extraction has taken place, but the effects are not serious. The whole area would repay detailed study.

At Noss Head there is a right-angled turn of the coast. Although flagstones occur there is little difference between these cliffs and those farther north. Geos are common. In Staxigoe there is a former stack preserved on the low rock platform. Wick Bay, into which the Wick river drains, makes a relatively large inlet, and affords good sections in the shelly boulder clay. Here, too, as in Lybster, Thurso and Sinclair's Bays peat is found under the sand and is said to be full of tree stumps. Above the harbour at Wick there is a deep channel in the river which is now filled with sand and mud. Difficulty was experienced in finding satisfactory foundations for the bridge. 'All along its site no bottom was found at 70 feet [21 m], and the bridge had to be founded on piles . . . Bottom was not found at 50 feet [15 m] near the Service Bridge and the Lower Weir, where the bed of the river is 6 feet [1.8 m] below high water' (Crampton and Carruthers, Memoir, 1914). These facts imply that since the ice age the river has cut downward at least 60 feet (18 m) below present low water; that sea-level was considerably lower, and at a later stage regained its present level. This change, more complicated than a simple fall and rise, is found in many rivers around our coasts. South of Wick the so-called grey stones, large slabs of sandstone, on the cliff top are interesting. Some weigh several hundredweights, and it seems that they are occasionally moved in storms. They overlap one another seawards. Local geologists thought that they were quarried from the cliffs by wave action. Crampton and Carruthers think it more probable that they were first loosened by ice. One very large one rests on rounded stones and, as the Memoir says, it would be difficult to see how these stones could be so placed by wave action. It is much more probable that these, and other, blocks were lifted by ice, and that drift was later rammed beneath them. Nevertheless, some, if not all, of the grey stones have been moved by waves, but since they are now lichen-covered it seems probable that this action no longer prevails. Once again they may be a feature of a higher sea-level and corresponding beaches. They may, perhaps, be compared with the Grind at Navir in the Shetlands (p. 289), and the large boulder beach near Straenia Water, Stronsay, in Orkney (p. 284). South of the bay the cliffs are fairly low. At The Brough a fault cuts the coast and is partly responsible for the geo and the large stack known as The Brough. A somewhat similar pattern occurs at Girston.

About a mile farther south sandstones reach the coast, and in them are some fine geos, Ires, Ashy, Tod's Gote, Broad Geo and Riera Geo. On the south side of the last a fault brings in the sandstones and conglomerates of Sarclet. This is the part of the coast where the broad anticline emerges; to north and south are synclinal areas – at Ackergill to the north and Latheron to the south. It is also around Sarclet that the ice, advancing to the north-west, had to mount the cliffs and in so doing has left a terraced effect as the result of the removal of the softer material. 'This terraced slope has been in most places subsequently destroyed by sea erosion, and replaced by cliffs. Locally, however, it still remains as a series of scarps, with ledges dipping gently landward, descending almost to sea-level.' The conglomerate cliffs are noteworthy. They extend from Sarclet to the Stack of Ulbster, and are massive, almost resembling granite, and make rugged cliffs. They are well jointed, and the sea working along these lines of weakness has formed caves and buttresses. Often the caves are driven far inland. The Stack of Ulbster and Gearty Head are in the conglomerate and exhibit vertical joints very well indeed. At the Stack of Ulbster a fault separates the Sarclet conglomerate from mudstones and sandstones which make the cliffs nearly as far as Ellens Geo, where there is a local conglomerate which forms the Rowans, a conspicuous headland just to the north of the geo.

From Ellens Geo to Ceann Hilligeo the Helman Head and Clyth beds form the cliffs. In this five miles (8 km) there is little faulting. The cliffs are dark in colour apart from some paler sandstone beds, and on the cliff top there are small sandstone scarps above, alternating with black stone-slates below. These beds are abruptly cut off by a fault at Ceann Hilligeo, a fault which has led to the formation of the inlet, whereas the crush rock in the fault itself is seen in the stack off-shore. Ceann Hilligeo also marks the former mouth of the Clyth burn. This old channel was blocked by boulder clay, and the burn was forced to find another course. This it did by cutting a rock gorge which is crossed by the main coast road at the bridge of Occumster. In this course there are several falls, and the river reached the sea by one about 80 feet high (24 m). There are several other geos in this line of cliffs, including Wester Whale Geo and Hannie Geo. The stacks about one mile to the east of Ceann Hilligeo are peat capped, and the inlet adjacent is cut along a fault. South-westwards from Ceann Hilligeo the coast is more broken and picturesque. The inlet of Lybster Bay, which is attractive, lies on a fault. The harbour is the mouth of the Reisgill burn. This, like the Clyth burn, has been forced to cut a rock gorge (see Fig. 41, p. 194) about 200 yards (183 m) to the west of its old buried course. The new channel reaches the harbour through a boulder-clay bank. The indented character of the coast around Achastle, the mouth of the Forse burn and the precipitous cliffs at and immediately west of Port na Muic are all noteworthy. Behind and roughly parallel to the coast is the Latheron Fault which reaches the coast a little to the north of Janetstown harbour. This same fault may continue south-westwards a little offshore to re-enter the coast near Knockinnon and thence to the north side of Dunbeath Bay. Throughout all this length there are several small crush lines on the coast. These and the burns reaching the sea at Latheron, Janetstown and Dunbeath, not to mention several smaller ones, give the coast an irregular appearance. There is a good deal of faulting on the northern headland enclosing Dunbeath Bay, and a marked shore platform, followed for some distance by a local road, is backed by an old cliff. In fact, all along this coast there are traces not only of benches but of cliffs now seldom reached by the sea. This phenomenon is clearly visible on the south side of Dunbeath Bay where there are low cliffs of Berriedale Flags. About a mile south of the castle, sandstones and later conglomerates reach the coast. Mudstones and sandstones make the cliff between Ceann Badaidh na Muic and Stùrr Ruadh, and a submarine fault follows this same line. Crushes are associated with the fault, and along the fault rocks are turned upon end to form reefs. In the cliffs there are landslips at Tràigh Bhuidhe and Tràigh Fhada where there are also beaches.

Near Lower Newport granite reaches the coast for about half-a-mile; and is followed southwards by the Berriedale conglomerates and sandstones, the former being separated both from the granite and the conglomerate by faults. Berriedale itself is one of the most beautiful parts of the coast (see Plate 20). The joint mouth of the Langwill and Berriedale waters breaks through a narrow rock chasm which deflects the joint stream a little to the north. There is also a small beach. The cliffs on the south rise steeply; the upper slopes are rounded, but at lower levels there is a vertical cliff and a stack. It is really only the lower slope that is a true cliff. Fig. 42 (b) gives a section along the coast between Berriedale and Ceann Ousdale. The Berriedale burn reaches the sea in the flagstones; the sandstones rise up to the south and about half-a mile south at Cnoc na Croiche the Badbea conglomerate outcrops. Near this point there are several landslips which have taken place along joints and small faults running approximately parallel with the cliff face. These give the cliff a step-like appearance. The whole slope is some 400 feet (122 m) from top to bottom where there is a narrow beach. The slips begin along vertical joints which traverse the breccia and allow water to reach the mudstones below. The junction between the two rocks is unconformable, and the sandstones give richly coloured cliffs, red, purple and greenish bands. Where (see Fig. 42 (b)) the Badbea breccia overlies the mudstones in the cliff it is responsible for the steep slope above the true cliff. At the mouth of Ousdale burn there is another arkose which is separated from the granite by a small fault. The two rocks are nevertheless very similar in colour, massiveness, and the way in which they weather. The fault continues to the south-west. The Ord granite forms the highest ground close to the coast of Caithness, but the slopes facing the sea are not the work merely of marine erosion. They are for the most part slopes produced by continental processes of weathering, and erosion along a scarp associated with the fault mentioned above. Only the lower parts have been modified by the sea. The granite coast is limited to the stretch between Ceann Ousdale and Dùn Glas.

At Dùn Glas the fault which brings the Mesozoic rocks against the crystallines the granite and the Old Red Sandstone runs out to sea; it continues southwards to Golspie, and brings about a profound change in the nature of the coast. South of Helmsdale a narrow strip of Old Red Sandstone is faulted between the Jurassic

203

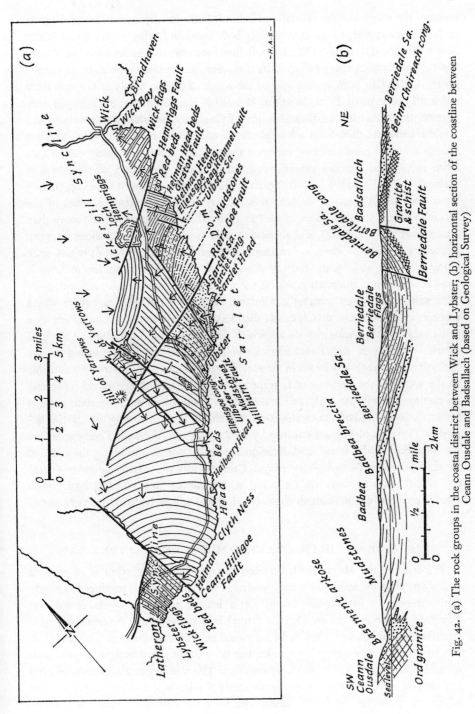

Fig. 42. (a) The rock groups in the coastal district between Wick and Lybster; (b) horizontal section of the coastline between Ceann Ousdale and Badsallach (based on Geological Survey)

rocks of the shore and the crystallines behind. Since the dip in the Mesozoic rocks is, for the most part, to the north, older beds come in farther south. Kimmeridge beds reach from Dùn Glas to Kintradwell; then for rather more than a mile the Corallian occurs between two smaller faults transverse to the main one, to be succeeded by the Oxfordian beds around and to the south of Brora; finally at Golspie there are small outcrops of Trias. North of Helmsdale the Kimmeridge outcrop is very narrow; the beds both north and south of that place are shales, sandstones, and boulder beds and they form a low platform, fringed by reefs. The platform rises southwards and near Lothbeg carries a magnificent series of raised beaches (see Plate 19). These beaches extend northwards, and a great part of the town of Helmsdale is built on the '100-ft' beach. This level also occurs at Crackaig (inland from Loth point), and at or near this locality there are extensive relics of the '50-ft' beach. There are traces of the '25-ft' level near Kilmote (a little more than a mile north of Crackaig), and there also, and at several other localities between Kilmote and the Ord, there are remnants of a level at about 15 feet (5 m). A walk along the coast and by the railway north of the now disused station at Loth is most revealing and interesting. Along the water line there is a rock platform. Near Kintradwell the beach terraces run inland, so that in former times there was a considerable re-entrant at Clynelish distillery north of Brora. Today there is a good sand beach, backed by dunes which now make the foundation of the golf course. The River Brora meanders through terraces which, according to the Memoir (Lee, 1925) may be six in number. Former courses of the stream are visible to the south of the harbour at Inverbrora, the river having shortened its course by cutting through the necks of the meanders. From Brora to Golspie the coast is lower; the coastal platform carries local spreads of gravels resting on glacial deposits. These, in places, are cut into by the '100-ft' raised beach. The beaches are often covered with blown sand. Erosion is usually not serious along the coast of Scotland, but at Golspie The Royal Commission on Coast Erosion (1911) reported that the erosion was estimated at two yards (1.8 m) a year, a very high rate for any part of the Scottish coast. Erosion was still active with the early 1950s.

THE MORAY FIRTH COAST; LOCH FLEET TO SPEY BAY

The coast of the Moray Firth between Golspie and Portgordon is of especial interest to physiographers because it contains so many features produced not only at the present level of the sea, but also at at least three former levels. It was first described in some detail by Ogilvie (1923) in a paper which deserves far more attention than it has had. Ogilvie by no means answered all the questions that even a brief visit to the coast will provoke, but his mapping and analysis are the base on which all future studies must be founded. The whole stretch consists of sand and shingle spits and barriers, extensive sandy forelands and sandy marshes, all of which are associated with the present beach or the so-called '25-ft' and '15-ft' beaches. These are backed in many places by extensive glacial deposits and in others by lines of former cliffs, some of which are cut in the shingle of a higher beach. The coast is markedly irregular today, and at the time of some of the raised beaches was even more so. Loch Fleet and the Dornoch Firth are partially obstructed by spits; the entrance to the Cromarty Firth on the other hand lies between two prominent headlands, The North and South Sutors. Inverness Firth is shut off from the Moray Firth by the forelands of Fortrose and Fort George, and in '25-ft' times the high ground along the coast between Burghead and Branderburgh was an island. There are also long stretches of high coast in the Black Isle and the Tarbat peninsula. These are broken by deep valleys as at Munlochy and, of course, the entrance to the Cromarty Firth. The general nature of these valleys, and the structural control of the coastal outline are discussed on pp. 210–12; here we shall be concerned with the shore features themselves, and we shall follow Ogilvie's account for the most part.

Loch Fleet was originally a wide-open bay leading to the narrow and fiord-like valley farther inland. The entrance to the loch has contracted in course of time by the growth of bars from the north near Golspie. On both sides of the loch there are patches of raised beaches the upper margins of which occur at 80, 70, 50 and 25 feet (24, 21, 15 and 8 m) (see Fig. 43). Ogilvie mapped the area in great detail, and the figure shows clearly the way the ridges running south from Golspie have developed. The figure also shows their heights, and also that most of them are associated with the '25-ft' level. The earliest ridges are those labelled GG; the last are at HH. It will be seen that the bay was almost in its present condition at the end of the '25-ft' episode. The south side, opposite Little Ferry, is obscured by blown sand, but the ridges at Q have grown from the south. There is a deep channel, with a strong tidal scour, which limits the growth of the ridges on either side of it.

This opposed growth of spits at the mouth of an inlet is not uncommon; well-known examples in England and Wales include Newborough Warren and Morfa Dinlle at the southern entrance of the Menai Straits, Morfa Bychan and Morfa Harlech in Tremadoc Bay, and the spits enclosing Poole Harbour. The explanations are not by any means necessarily the same in each case. In the Moray Firth the prevalent winds from a westerly direction are off-shore or, along parts of the south coast of that firth, along-shore. The winds that matter most come from between north and east. These set up waves travelling in similar directions, and it is a reasonable assumption that beach material will, in general, travel southwards along the northern shore, and westwards along the southern shore of the firth. But the intricate configuration of the coast will cause irregularities, and local reversals of movement. Reasoning of this nature may explain the north-pointing ridges near Embo, but a more detailed explanation is required.

The beach at about 70 feet (21 m) is prominent on the south of Loch Fleet between the Skelbo Burn and Embo and beyond, and from Embo to the Evelix mouth the cliffs, cut in glacial drift, belong to the '25-ft' episode. Extending

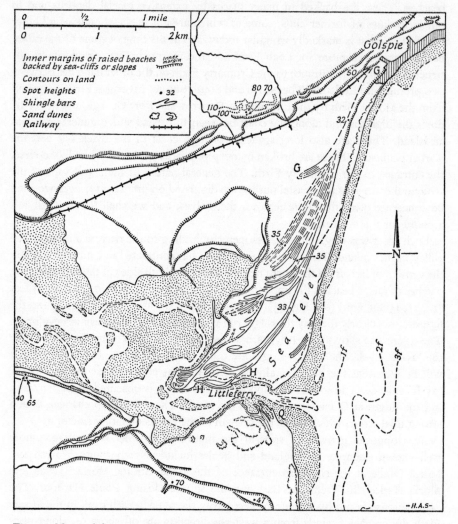

Fig. 43. Marginal features of Loch Fleet. The area between high- and low-water marks is stippled (after A. G. Ogilvie)

southwards from Dornoch, is a level area, Dornoch Links, which was built mainly in '15-ft' times, but the erosion around Dornoch itself is in part responsible for the modern growth of Dornoch Point. Dornoch Point lies seaward of two other important features, Cuthill Links on the north, and Ferry Point, with Ardjachie Point, on the south. Cuthill is a complicated structure, partly covered with sand. Fig. 44 shows the general run of the ridges, and indicates the southern growth of the point, almost all of which belongs to the '25-ft' and '15-ft' episodes. The small recurves about half way between the point and Ferrytown need further investiga-

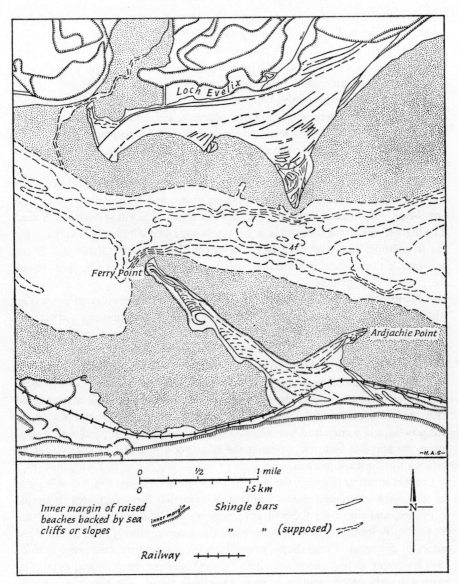

Fig. 44. Central part of Dornoch Firth. The area between high- and low-water marks is stippled. Isobaths at 1-fathom intervals are drawn from soundings of 1837 (after A. G. Ogilvie)

tion. On the opposite coast the long and narrow spit running to Ferry Point is a shingle structure of similar age to Cuthill; the higher part is at about 25 feet (8 m), and the lower part, including Ardjachie Point, is at 15 feet (5 m). Ogilvie (1923) remarks, 'The existence of this spit, built apparently of smaller gravel materials eroded from the foreland nearer its point and carried south-eastward, shows that shingle is no longer being transported hither from the east, a fact probably connected with the building of the sandy Morrich Mhór, east of Tain. The smaller spit is probably due to eddying tidal currents.' Even if currents have played a part, they have not built the ridges, and a great deal more research is required before we have an adequate explanation of all these features.

In the middle stretch of the Dornoch Firth there are many traces of former beach levels. They take the form of low plateaux, and since they grade into, and are largely formed by the waste of, glacial deposits it is by no means easy to make hard and fast distinctions. In places, for example near Ardmore, there are groups of recurved ridges.

Morrich Mhór is a sandy foreland about nine square miles (23 sq. km) in area. It has a remarkably straight north-west-facing shore, and is extending toward the north-east. All round its landward margin there is a sea-cliff cut in red boulder clay. The highest part of the foreland is in the south-west where it clearly belongs to the '25-ft' episode. The level declines north-eastwards, suggesting that much of the mid-part of the foreland is associated with a sea-level of about 15 feet (5 m) above the present. The gradual slope to the north-east is continuous, so that it is not possible to define a break between the 15-foot level and present conditions. The whole feature has grown forward with a falling sea-level, and in many ways it may be compared with the Culbin area and also with Tents Muir (see p. 256). Study of the Ordnance Map suggests that the feature has grown seawards by the successive additions of sandy barriers formed by wave action on Whiteness Sands. Innis Mhór and Innis Bheag are offshore ridges of this type. These barriers have been to some degree driven south-westwards, but undoubtedly silting has taken place behind them, and so the level of the foreland has gradually risen. Blown sand has helped this process, and so dry land has extended seawards. The lines of parallel elongated lochs, the spit-like features enclosing the Blue Pool, and the offshore ridges all indicate this forward growth.

Erosion is taking place on the straight north-western shore. On this side there are also some prominent dunes. Ogilvie thought that they originated as simple foredunes, and that their form in 1913 depended on the erosion caused by the prevailing westerlies. The oldest ones are the more northerly, and are at right angles to the wind. These dunes, assuming their position on the survey map of 1873 is correct, moved about 200 yards (183 m) between that year and 1913. The more southerly group includes some fine parabolic dunes. The rest of the area is almost flat, and falls on the south-east to a tidal creek. The cliff of the '25-ft' beach continues along the east side of the creek and can be easily traced together with both lower and higher beaches as far as Tarbat Ness. From east of Portmahomack around Tarbat Ness and all along the eastern side of the peninsula the fossil cliff is well preserved. In many places there are raised beaches, the best development is perhaps at Hilton of Cadboll where the cliffs behind the '25-ft' beach make a noticeable re-entrant. Along the shores, too, are some small traces of Mesozoic rocks, mainly Jurassic. They are separated from the Middle Old Red Sandstone by a major fault which is directly on the line of, and also a continuation of, the Great Glen Fault.

It is convenient at this stage to refer to recent work on the Great Glen Fault. On pp. 82ff. the importance of lateral movement along this fault on the matching of rock types on either side of Loch Linnhe and the Firth of Lorne is discussed. But a lateral movement of 83 miles (134 km) must have had a considerable effect in the Moray Firth. Holgate (1969) shows that such a movement caused a great displacement of the Mesozoic beds in that area. The small patches of Upper Corallian and Kimmeridgian, Lower Oolite and Oxfordian near Balintore and Port an Righ in the Tarbat Ness peninsula, and the Kimmeridgian at Ethie in the Black Isle are now isolated. Holgate (see p. 83) finds good evidence for a further 18-mile (29 km) movement along the Great Glen Fault. Before this second movement took place the Mesozoic sediments in the Tarbat peninsula and the Black Isle would have been approximately on the same east–west strike as those near Burghead and Lossiemouth. There is no doubt about a post-Kimmeridgian movement. That 'this movement [can] be identified with a major dextral wrench on the Great Glen Fault seems highly probable'. If this is allowed, then the Dornoch and Moray Firths make a single great embayment, broken only by the Tarbat Ness peninsula which 'is determined by the strongly upturned beds margining the Northern Highlands block at the fault-line, and would suggest that this upturning was associated in origin with the post-Mesozoic movements on the Great Glen line' (Holgate, 1969). (See Fig. 45a and b.)

When the submergence was at a maximum there was a sea connection between the outer Dornoch Firth and Nigg Bay in Cromarty Firth. This is proved by a study of the levels; no traces of erosion or deposition confirm it and, since the area is boulder-clay covered, has been much cultivated, and its slopes subject to soil creep, there is little reason to expect evidence of this nature. Over the whole district glacial features such as kames, drumlins, and kettleholes are common, and sometimes glacial ridges of one type of another can be confused with marine features, especially in places where, e.g., esker ridges run down towards sea-level.

In the Cromarty Firth there are also many traces of raised beaches and some features associated with present conditions. There are two small cuspate forelands at Dunskeath and at Cromarty itself; both are situated just inside the entrance in positions where such features are to be expected. Ogilvie thought that both were in part produced by wave action in piling up current-borne material. The foreland on which Invergordon stands is wave-built. Since there are several remnants of glacial deposits hereabouts its formation may have been complicated by them as well as by the fossil delta at Tomich. At Alness and Evanton there are two magnificent fans, the origin of which is ascribed to glacier-fed rivers. The Alness fan shows three dissected cones in addition to the modern delta. The higher ones correspond to beaches at 85, 28, and 15 feet (26, 9 and 5 m). The delta at Evanton is more

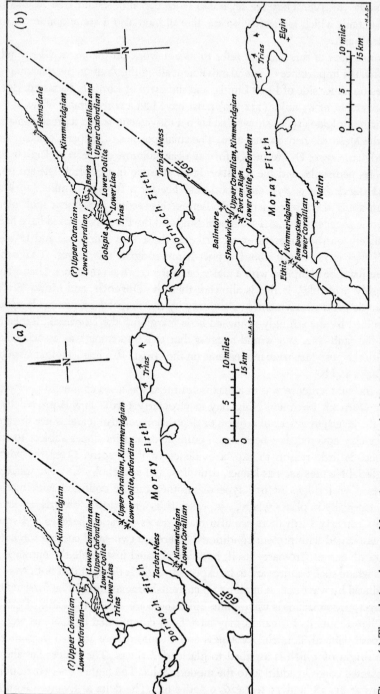

Fig. 45. (a) The relations between Mesozoic outcrops in the Moray Firth–Dornoch Firth area, prior to a post-Kimmeridgian dextral wrench dislocation of 18 miles on the Great Glen Fault. Note how the Dornoch Firth was formerly integrated with the Moray Firth as an inshore extension of the latter. The headland of Tarbat Ness represents a later feature originating in the Cainozoic fault movements; (b) the present distribution of Mesozoic outcrops on either side of the Great Glen Fault (GGF) in the Moray Firth area (after Norman Holgate)

complex since it appears to have been built by two streams. There are levels corresponding to those at Alness. At Alness Point and Balconie Point (Evanton) there are small shingle bars growing to the east and north. On the southern shore the only marine feature, apart from raised beaches, is the minor foreland at Ferryton Point. Between Jemimaville and Cromarty, Smith (1963) calls attention to the fact that the higher raised beach thereabouts is much gullied. The gullies are graded to the lower beach, but all the debris they brought down was incorporated in the lower beach, or possibly aided in the formation of the Cromarty cuspate foreland. The two beaches may be correlated with the so-called '100-' and '25-ft' beaches, and Smith concludes that the relation of the gullies to the two beaches suggests that the gullying occurred in the pluvial part of the early Atlantic climatic phase.

At the Sutors of Cromarty, both north and south, there is a small outcrop of Moine Schists. On the north side the schists form the coast more or less corresponding with the Hill of Nigg; on the south the outcrop is smaller and narrower and disappears near Navity. Part of the higher ground, notably at the south Sutor, is in the Old Red Sandstone. It was suggested by Hugh Miller that the River Conon at an earlier period flowed over the Old Red Sandstone plain and extended its course farther eastwards along the line of the Cromarty Firth. The opening between the Sutors may have been eroded by the river as it cut its way downward through the relatively soft sedimentary rocks which rest on the Moines. In *Terrain Analysis of the Northern Coastal Zone of the Cromarty Firth*, a report for the Highlands and Islands Development Board, 1967, K. Walton discusses fully the coastal features in the Cromarty Firth. He suggests that the morphology of the Firth is structural. The Udale and Nigg areas are synclinal; the Sutors and the deeps running to Invergordon are fault-line features. He also suggests that Pre-glacial drainage and ice may also have played important parts. The deep channel leading to Invergordon and the direct fetch through the Sutors have allowed waves to make cliffs at Invergordon. Walton also suggests that the original drainage along the line of the Firth found its exit through the depression between Nigg Bay and Morrich Mhór. The schists reappear on the coast near Ethie and are continuous as a narrow coastal strip as far as Rosemarkie. At Ethie, however, is a fragment of Kimmeridgian which corresponds to the similar fragments noted above near Balintore. The Black Isle coast corresponds with the Great Glen Fault. Much of this coast is a steep slope with the fossil cliff and a boulder-strewn beach at its foot. South of Chanonry Point (see p. 213) the coast is broken by small streams at Avoch and Kilmuir, and by the much larger inlet of Munlochy Bay. Horne and Hinxman (1914) suggest that this bay is but a remnant of an old valley which drained from the mountains of Ross-shire. This river was later diverted along the Cromarty Firth. The bay is now largely silted up. The narrow strait between North and South Kessock probably marks the old course of the Farrar, Glass and Beauly 'which still preserves more or less its easterly course, but the lower part of the valley has been modified by subsequent events, and in earlier times the river probably flowed over a plain of Old Red Sandstone that occupies the position of the Beauly Firth, discharging its waters into the Moray Firth far to the eastward of the present shores of the Black Isle' (Hinxman, 1907). The Beauly Firth is shallow, and the narrow connection between it and the Inner Moray Firth, which lies between Craig Phadrig and the southern part of Ord Hill, can, if Hinxman's and Miller's views are acceptable, be directly compared with the entry to Cromarty Firth between the Sutors.

The upper parts of the Cromarty and Beauly Firths grade into extensive carse-lands, and between them is the low-lying area of the Muir of Ord. Strathpeffer is a glacial trough the floor of which consists of alluvium. The terraces alongside it are taken to be river terraces since their gradients are relatively steep and, as Ogilvie (1923) comments, 'their upward gradient towards the land would be evidence of differential uplift after recession of the ice'. The lower part of Strath Conon was occupied by the sea, and the steep bluff just to the south of Dingwall is regarded as a sea-cliff. Just below the lowest part of the strath is an extensive former beach corresponding to a level of about 80 feet (24 m). On its southern border it gives place to what Ogilvie called the isthmus of Ord, an area of fluvio-glacial features, the whole of which remained above sea-level even in the maximum submergence. If this was so, then the locally steep surrounding slopes are not sea-cliffs, except at the 80-foot level just mentioned. The Beauly carse is divided by the Kirkhill ridge, the smaller part being to the south. Former beach levels can be traced, but the '25-' and '30-ft' levels seem not to have extended beyond the Moniack fan.

In the Beauly Firth there are several features of interest. On the north shore there are three widely-opened scallops, to use Ogilvie's word. In them there are traces of raised beaches which are often but narrow strips, and where beach and glacial deposits both occur their separation is not always easy. 'The hummocks and hollows may be described as either (1) kames deposited on the 50-ft or other lower beach; or (2) kettleholes, the result of the melting of "dead" ice buried in the 80-ft beach.' On the southern shore the features of interest have been produced by running water more than by marine action. In the strait of Kessock the River Ness has built a magnificent delta which falls from rather more than 30 feet (9 m) to the floor of the strait without a break. On the north shore there appears what at first sight is usually taken for a fine raised beach. Certainly a cliff corresponding to the 15-foot (5 m) level is cut in a terrace, but since the surface of the terrace is uneven, and its seaward edge varies in height from 50 to 84 feet (15 to 26 m), and since, too, its landward margin is at approximately 150 feet (46 m) in places, it is more than probable that it is all that is left of an alluvial fan, and that it is not a marine terrace.

Craigton Point and the Ness delta separate the Beauly from Inverness Firth, that part of the Moray Firth enclosed by Chanonry Point and Fortrose. The south shore of Inverness Firth is part of the low coast which extends from Inverness to the Spey. It is a rich agricultural area and its surface is made up of glacial features, kames, osar, drumlins, old lake bed and carses with wide wave- and wind-built land nearer the sea. In Inverness Firth the most conspicuous feature is Alturlie Point, a fluvio-glacial remnant which has been sloped on the north and west by the waves of the '15-ft' sea. Just to the east of it is a small creek which was deflected by a spit of the '25-ft' sea; this is now being eroded.

At the entrance to Inverness Firth there are two major features, Chanonry Point on the north side and the Ardersier foreland on the south. These have been analysed by Ogilvie (1914) (see Plate 18). The northern shore, apart from Chanonry Point itself, is remarkably straight, and is the continuation of the Great Glen Fault. The south-east side is less regular and this is largely on account of the extensive glacial deposits. The two forelands are built features, consisting of raised beaches belonging to at least four episodes. Levels at 100, 50, 25, and 15 feet are all found, and also a line of what appears to be an old sea-cliff corresponding to a sea-level of 75 feet (23 m) is found in Chanonry. Ogilvie carried out some detailed and careful mapping of the two forelands and of the channel between them (see Fig. 46). There is today a deep reaching 143 feet (44 m) between the two and, as the map shows, the slopes down from the tips of the two forelands are notably steep. The currents in this part of the firth are indicated by arrows on Fig. 46. The more important waves approach from the north-east, and occasional effective waves may come from the south-west. The combined effect of the two factors, waves and currents, is shown by erosion on the north-west-facing side of Ardersier, and mainly on the southern side of Chanonry, although there is also some on its northern side. Deposition takes place mainly on the southern side of Ardersier, and at the tip of Chanonry. There is also deposition along the east–west part of the north side of Ardersier. The present outline of the foreland has evolved over a long period of time. At the time the earliest beach formed the north coast was almost straight; on the south there was a kink, produced largely by glacial deposits; but the 'solid' shore was relatively straight. Ogilvie's figures show how each new beach was added, and also that parts of a former one were eroded before the new one was completed. The beaches now make well-defined terraces and much of the land is cultivated.

Ogilvie maintained that waves and currents carried beach material to this area, and then that currents and conditions similar to those at work today began to build the transported material into forelands. To quote his own words, 'the forms . . . are the product of processes such as those which are at work now; but the processes have acted through at least four episodes in the topographic cycle, each of which was introduced by uplift of the land relative to the sea'. This explanation however, does not give a reason for the precise position of the forelands. Why should they not have been a few miles farther in or farther out? It is doubtful if any exact answer can be given to this. It is, therefore, of interest to turn to Zenkovich's (1967) view on the development of accumulation forms in a straight bay with parallel shores and a continuous submarine slope. He contends that

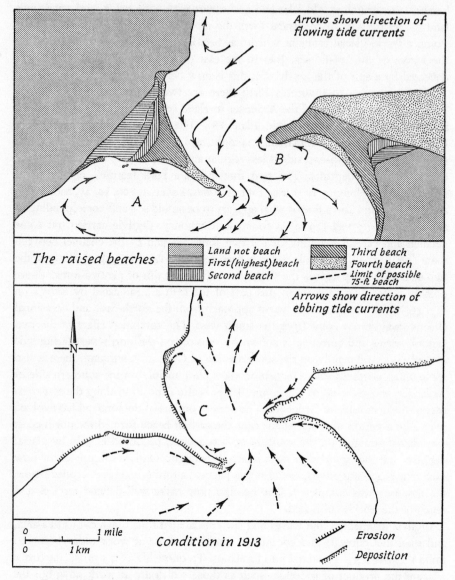

Fig. 46. The raised beaches of Chanonry Point and Ardersier, Inverness Firth
(after A. G. Ogilvie)

Supplies of material fed by abrasion of the outer coast extend along both sides [of a bay or
inlet] towards the bay head. Waves advancing along the axis of the bay undergo refraction
on both shores and their parameters are gradually reduced. As a result the saturation of both
flows increases towards the bay head and its limit is reached at some point. The whole of the
material can no longer be carried beyond this point and an accumulation *unrelated to any shelter*
will form. A sheltering effect could be manifested only if, owing to a difference in flow rates,

a projection developed on one shore earlier than the other. In that case, the reduction in the
water area of the bay would compel a deposit to develop at the corresponding point on the
opposite shore (facing the first) even if the internal properties of the flow would have determined its formation nearer the bay head. The second form would be classified as an induced
form.

I have italicized four words in order to emphasize that if a bay is almost straight,
it is only excessive load which will lead to deposition. In the case of the Inverness
Firth the second point may apply, namely that at Ardersier there was a projection
of glacial material at the highest beach level. This seems likely, and if so it was, on
this hypothesis, responsible for initiating the deposition on the northern shore.
It is pertinent to add that, at any rate on the southern shore, there was no shortage
of detritus.

There are some dunes on each foreland. Those near to the point of Chanonry
are small and of no particular interest. On Ardersier there are similar dunes on the
north shore, and also fixed dunes at some distance from the sea, produced by north-
east winds. They remain just seaward of the '25-ft' level. Their form has been
modified by south-west winds. To the west of Fort George there is a particularly
good development of an old cliff.

It is interesting to note, as Ogilvie points out, that the names of Rosemarkie
and Fortrose, both on Chanonry, have in common the Gaelic root ros, meaning
promontory. Markie refers to the stream which reaches the sea at that place. Fort
is not the English word, but 'probably derived from a Gaelic adjective indicating
the position of the place at the base of the "ros".' In 'Ardersier', the syllable ard
also means promontory.

On the low south coast of the Moray Firth the town of Nairn stands in a key
position. It is there that the River Nairn breaks through the glacial deposits and
reaches the sea, and it is there also that there is some erosion. Along all the low
coast beach material travels to the west, and it will be convenient first to describe
that part of the coast to the west of Nairn. The harbour works prevent a good deal
of material passing to the west of that place, but beyond the harbour the erosion
of the foreshore, consisting of sand covering a platform cut in the Old Red
Sandstone, undoubtedly supplies the coast to the west and helps in the building
of the long spit fringing the Carse of Delnies (see Plate 17). Just how much
erosion has taken place at Nairn is not known, but Wallace (1896) says that the
castle stood in what is now the sea. Ogilvie described Delnies briefly in 1923; some
detailed mapping was done by W. G. V. Balchin in 1937 (Steers, 1938). Individual
shingle ridges are not shown on Ordnance Maps. Wallace (1896) states that in
1823 there was no shingle there. Precisely what this means is difficult to say;
there was certainly no foreland comparable to that of today, but it seems most
unlikely that some shingle had not accumulated along the eastern part of Delnies
spit. In 1937 the spit consisted of a series of well-defined shingle ridges, each one of
which was cut off by erosion at its eastern end, whereas its western termination was

a true shingle end. Since each ridge showed the same phenomenon, it is clear that the shingle is all moving to the west (cf. The Bar, p. 222). In the 69 years, 1868 to 1937, the spit had extended 900 yards (823 m). Behind the spit there is a tidal area of sandy saltmarsh. At the back of the carse is a prominent cliff which seems to correspond with the '25-ft' episode. At Hilton, at the foot of this cliff there is a well-defined terrace and then another, smaller, drop of about six feet (1.8 m). This lower cliff may well have been that produced by the '15-ft' sea.

The section of coast between Nairn and Burghead, although but a part of the low coast, may well be treated as a distinct unit. The occurrence of the Bay of Findhorn, east of the Culbin foreland, at first sight suggests that there are two units to be considered, and in a sense it is true. On the other hand the probable history of the Bar off Culbin, the known deflections of the narrow mouth of the bay, and the view held by Ogilvie (1923) that some of the older shingle bars east of the bay once extended across it into Culbin, make it more convenient to treat the whole stretch as one unit. Moreover, there is no outcrop of solid rock on the coast between Nairn and Burghead. For our present purpose we will take the line of the '25-ft' cliff as the landward boundary, a line roughly approximating to the 50-foot contour on the One Inch Map. At that time there was a wide, open, and shallow bay interrupted by islands of glacial detritus as at Grange Hill and Upper Hempriggs. The Findhorn drained into this bay, which was gradually shallowed by detritus brought down by the river and that swept in by waves from the east, and also by the growth of salt marsh. We do not know the course of the Lower Findhorn at that time; it may have run south of the great fan of shingle on the Culbin foreland, but of this there is no proof.

The Culbin foreland and adjacent parts of the coasts originated as sand flats, produced at least in part by the smoothing out of glacial sands. On the Culbin side of Findhorn Bay a complex system of shingle ridges was built and, with the shallowing of the sea, more and more dry land appeared, and blown sand began to cover much of the area (see Plate 15). Culbin had long been habitable. A great number of finds have been made, many by amateurs, so that essential details of location are not recorded. There has, however, been another and more important difficulty. Remains of occupation of different ages are all too often found at the same level. The sand on which settlement took place has been eroded, and articles have dropped down to the surface of the beaches beneath the dunes. There are also shell heaps and kitchen middens, consisting almost entirely of edible molluscs. Among these the presence of the oyster is the most interesting. Up till fairly recent times it was much sought after; now it has completely disappeared from this district. It is probable that, since it likes a somewhat muddy habitat, the increasing amount of sand has made this part of the coast unsuitable for it.

In historical times, in about 1240, the first known owners of the area were the de Moravia family. The Morays remained until the early part of the fifteenth century, when, by marriage, the Kinnairds became owners, and remained so until the final inundation at the end of the seventeenth century. No map or plan is known of the estate before it was overwhelmed. It covered about 3,600 acres (1,457 ha) and even in 1694 the rental was £2,720 Scots, 640 bolls of wheat, 640 bolls of bear, 640 bolls of oats, and 640 bolls of oatmeal. In 1733 the 'annual net produce in money' was £494. 4s. 4d. These figures include Earnhill as well as Culbin; they nevertheless throw a vivid light on the degree of destruction. The sand continued to accumulate, and it was not until 1839 that Grant of Kincorth began the successful planting of conifers. This example was followed by other landlords, and in course of time the Low Wood (see the One Inch O.S., Popular Edition, Sheet 28, 1929) covered a large area. This wood was felled at the end of the First World War. In 1921 the Forestry Commission began to acquire land at Culbin, and in 1937 owned some 5,000 acres (2,023 ha). Low Wood and additional areas (see current One Inch Map) were planted, and the big sand hills began to be reclaimed. Scots Pine, Corsican Pine and the Lodge Pole Pine are the species mainly used. This new planting has profoundly changed the appearance of the district. In 1937 it was possible, when many of the trees were small, and the big sandhills were still almost untouched, to follow the trend of shingle ridges and map them. The map (Fig. 47) was made in 1937; today it would be impossible to make it.

The Culbin dunes, i.e., those in the central area, were probably the finest in these islands. Their height remains, but since they are now almost entirely covered with vegetation it is no longer possible to see them in their natural condition. They were mapped by Ogilvie just before the First World War, and C. A. Fisher (Steers, 1937) re-mapped them just before the Second World War. Since their lower parts so frequently merged, it was only practicable to make the 30-foot (9 m) form line the lowest shown on the map. The upper parts, at that time nearly all bare sand, were subject to a good deal of change. The windward slopes were gentle; the lee slopes often very steep. Several dunes reached approximately a height of 100 feet (30 m). On the northern shore there are long ridges of fore dunes, either parallel to, or slightly oblique to, the sea-line. Locally they have developed into parabolic dunes. In many places there are relict dunes, and near Findhornhill and Binsness there are some particularly interesting features. Near Findhornhill a former dune area had been planted, but the trees were later cut down, and in 1937 wind erosion was uprooting the stumps, and had produced many bizarre forms. Nearer Binsness there were many relict dunes which resembled small buttes or zeugen. On their surfaces were small caps of closely turfed blown sand; this sand rested on a layer of nearly black sand, and this in turn on a layer of indurated sand. At the bottom there were shingle ridges. All these buttes had been exposed by erosion of a former flat, and if natural conditions prevailed would sooner or later be recovered by the advance eastwards of the great dunes.

Sand is still accumulating on this shore today. Mackie (1893–8) analysed the dune sands, and noted that quartz accounted for 78 per cent and felspar 18 per cent. In the river sands of the Nairn and Findhorn, the felspar content is about 24 per

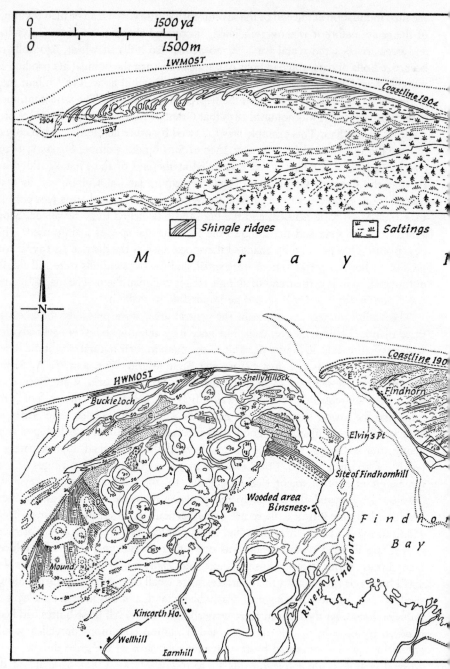

Fig. 47. The Culbin sands and Burghead B

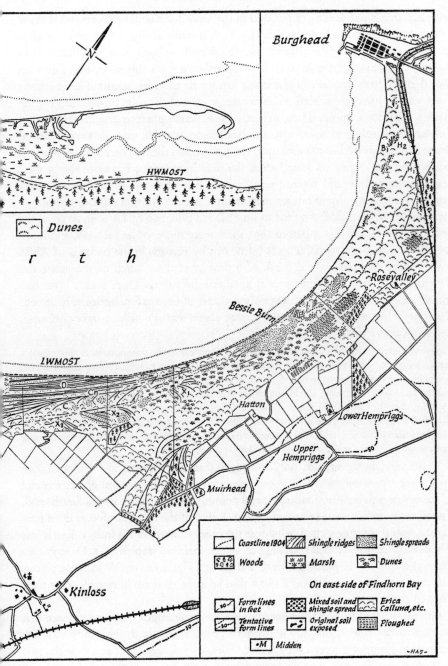

Burghead

Roseyalley

Bessie Burn

HWMOST

Dunes

r t h

LWMOST

Hatton

Lower Hemprigs

Upper Hemprigs

Muirhead

Kinloss

Coastline 1904		Shingle ridges		Shingle spreads
Woods		Marsh		Dunes

On east side of Findhorn Bay

Form lines in feet	Mixed soil and shingle spread	Erica Calluna, etc.
Tentative form lines	Original soil exposed	Ploughed

•M Midden

et: The Bar (after J. A. Steers)

cent; and this is reduced to 18 per cent in the Culbins. The felspars resemble those in the Ardclach and Kinsteary granites. It is usually assumed that much sand is carried down by the rivers, which are occasionally subject to severe floods as in August 1829. Since the coastal drift is towards the west this sand, together with that derived from erosion of the coast farther to the east, accumulates between Burghead and Inverness Firth. The numerous glacial deposits on and fringing the coast all add their quota. These extensive sand accumulations and the downward movement of sea-level relative to the land, exposed wide sand flats and allowed the prevailing westerlies to build up great dunes upon them. There is no doubt that much sand had blown up long before the final inundation; but since the process was relatively slow and intermittent, it does not seem to have attracted much notice. Records are absent but we may assume that the western sands continued to increase. Storms would from time to time blow a thin, patchy layer on some of the fields, but in the years just before 1695 there were more violent storms, and more sand was carried eastwards. This is borne out by records in the parishes of Alves and Kinloss. We may, then, think of a thin layer over much of the estate the owners of which at that time were in financial difficulties. Once such a layer had spread and begun to thicken it was only a matter of time before agriculture ceased. The one thing we may be sure of is that there was no sudden overwhelming. Bagnold (1937) wrote

Experiments have clearly shown that any given wind can only transport in a given time a certain weight of sand, and no more. As a result of a calculation I made, I found that if a wind was to blow over the area for twenty-four hours at an average speed of 40 miles [64 km] an hour and the sand which the wind carried on to the area was deposited over a depth downwind of one mile, then the greatest thickness of sand which could be deposited would be only one inch. Moreover the wind at 40 miles an hour would have to blow against such high resistance due to sand movement that it would be equivalent, upwind in open country where there was not any sand, to a continuous wind of nearly 70 miles [112 km] an hour.

Eventually, however, not only the estate, but the cottages and Culbin house itself were overwhelmed. The house reappeared on one occasion about the end of the eighteenth century, and then, and possibly earlier before it was first buried, a good many stones were removed from it. Relics of the estate are found from time to time, and this suggests that, whatever the precise nature of the inundation, it was effective enough to stop the salvage of all the farming implements. In 1937 (see Fig. 47) traces of old ploughed land, and plough rigs were visible in some places. The removal or accumulation of but a thin layer of sand could quickly alter their size and appearance.

On the Culbin foreland there are numerous shingle ridges. Mapping showed that they are in two main groups. The western group is clearly in the nature of a fan; the eastern group suggests a prograding shore. It is likely that at one time the two were connected in some such way as indicated in the figure. In the writer's view all these ridges appear to be independent of those to the east of Findhorn Bay. Ogilvie (1923) suggested that the eastern group, or some of them, might be the western ends of some of the older Burghead Bay ridges. The 1937 map, which shows with reasonable accuracy the trends of all the ridges, does not suggest that Ogilvie's explanation is likely. However, the ridges in the eastern group are truncated, and this may have been the work of waves in Findhorn Bay, which may then have been more open, or it may have been on account of some change in the course of the river. Whether the river ever flowed to the south of the ridges is a moot point. There were some low-lying damp hollows which may indicate an early course, but a shingle fan could well develop on a sand flat, and there is no particular reason why it should have deflected the river.

On the east of Findhorn Bay the ridge system is even more complex, and in 1937 most of it was easily accessible. Along its whole length the bay is bordered by ridges belonging to the '25-' and '15-ft' stages. Some of them are dune covered, but since the westerly winds caused sand to gather near Burghead, the western part is relatively free. Within the ridges is carse or flat-lying shingle which is now under afforestation. The oldest ridges (A on map) are near Muirhead, and they seem to be a quite independent group. The ridges H1, H2 near Burghead are high and massive and appear to belong to the '25-ft' or even to a somewhat higher level. When the A ridges were built Findhorn Bay would have had a much wider mouth, and possibly these A ridges formed as spits running from the higher ground of Rose Isle and Hempriggs. The next episode seems to have produced the ridges, only the ends of which remain, at B and B1. Their mid-parts have long been destroyed by erosion in Burghead Bay. The great series, C, then followed. Their north-eastern ends are truncated, but this does not necessarily imply that at one time they were connected with the high ground at Burghead. Almost certainly the erosion of the B ridges helped to build the C group, just as the erosion of the C group has fed those of group D, the north-eastern ends of which are also cut away. The modern ridges, E, have all formed at present sea-level. A walk along the beach will show that the older groups belong to higher sea-levels; the beach is often backed by a true shingle cliff cut in the higher ridges. The E series once extended much farther west (see below). The evolution here suggested by the several series of ridges in Burghead Bay does not necessarily imply a great deal of erosion. Earlier writers, e.g. Wallace (1896), say that people only a few centuries ago could walk nearly in a 'right line' from Findhorn to Burghead, part of the way being across a submerged forest. At its face value this implies that two to two-and-a-half miles (3.2–4 km) have disappeared from the middle of the bay. We need not doubt that a path or track existed across the bay, but this does not necessarily mean a great loss of land, and the careful mapping of the trends of the remaining shingle ridges agrees with this view. Findhorn is now a popular resort and various buildings obscure some of the ridges.

The ridges at X1, 2, 3 are puzzling; they are not recurved ends of any of the other groups. Their nature and appearance suggests that at one time they ran much farther to the west, and it is they which may have had some connection

(Ogilvie, 1923) with ridges on the Culbin side of Findhorn Bay. The ridges at Y are also puzzling. It is possible that they had some connection with those at X3. Since agricultural operations on the various farms have upset the disposition of the shingle, we cannot be certain that some ridges which might have given a key to the problem have not been destroyed.

We must now return to the modern ridges marked E. Although there are no early maps, we are fortunate in that in 1758 Peter May made a careful survey of the mouth of the river as a result of a dispute concerning fishing rights. His map shows clearly the old course of the river, and that it ran through the low ground now occupied in part by Buckie Loch. The old town of Findhorn stood about one mile to the north-west of the present town, and during a storm on 11 October 1702 a breach was made and the river formed its present mouth, an outlet which has fluctuated from time to time. These conclusions are supported by Avery's survey of 1730 which shows what is called on May's map The Old Bar as a long spit. There seems no doubt that the part of the channel within The Old Bar gradually silted up and became part of Culbin. There is no evidence available about the date of origin of the gap marked The Breach on May's map. The 1730 map makes no mention of The Bar as we know it today; it presumably did not exist in its present form, or perhaps it was in such an immature state that it was not thought necessary to survey it. But in 1835 The Bar was a major feature as is proved by its insertion on Admiralty Surveys. It was about five-sixths of a mile (1,340 m) shorter then, but of much the same shape as it is today (see Plate 16).

The Bar in 1937 was a most interesting structure (see inset, Fig. 47). The northern part was made almost entirely of sand, and its seaward side had been widened by dune ridges. There was, however, some shingle at the northern end, and a little along the inner side. The dunes may rest on shingle. Near the middle of The Bar were some widely spaced recurved ridges, and in that part The Bar was narrow and seemed to have been overrun in storms. A little farther south The Bar is a fine shingle foreland, with sometimes 18 ridges running nearly parallel to one another. The inner ridges are thickly covered with vegetation, including gorse, heather and crowberry. The inner side consists of a fine series of recurves enclosing narrow saltings with *Salicornia* spp., *Suaeda maritima*, *Puccinellia* and other plants. On the seaward side the northern ends of the ridges are all truncated by erosion. It is quite evident, even in a short visit, that the shingle is moving to the south-west; the erosion in the north is balanced by accretion in the south. This also means that the shingle as a whole was once farther north. Comparison of charts shows that between 1835 and 1937 The Bar had advanced to the south-west about five-sixths of a mile. If for the moment we assume that this rate held in earlier times, it means that the old mouth, as shown in May's map, was in use about three and a half centuries ago. Any estimate of this sort is very rough, but it is sufficient to indicate the probability that the present Bar is the direct descendant of the Old Bar; the whole structure has moved along the coast just as it is moving at present. This does not explain why the northern part of the present bar is so different from the south end, but if, as I think we should, we regard the shingle as the main part of The Bar, the great difference between its two ends is less significant than the continuous movement of the shingle. As in all such structures, there has probably been some shoreward movement along the whole length of The Bar, the evolution of which may thus be explained as shown in the inset of Fig. 47.

The salt marshes within The Bar are sandy; it is rare to find (1937) even six inches of silt. The tidal range at springs is 13 to 15 feet (4–5 m), and would be much greater in a surge such as that in 1953. The central parts of the marsh are only covered by big tides. There are salt-pans, but they are not prominent, and like many of the creeks are frequently overgrown with plants so that they appear only as shallow depressions. In general the marsh resembles those on the Cardigan Bay coast, and not those on the East Anglian coast. The main colonizer is *Puccinellia maritima. Zostera* occurs locally, and the sward in addition to *Puccinellia* includes *Glaux maritima, Armeria maritima, Plantago* spp., *Spergularia media, Aster tripolium* and *Triglochin maritima. Juncus* spp. is found at higher levels. The marsh is extending, and has had some effect in stabilizing The Bar. The landward side of the marsh is the flat on which is situated the Low Wood. At Maviston, when the sand in this neighbourhood was moving more freely, large dunes formed, and one or two became magnificent examples of parabolic dunes. They advanced over a forest and, in 1937, the dead trees over which they had passed stood gauntly in their rear. The dunes are still burying woodland in their advance.

The whole system of sand flats, shingle ridges, dunes and marshes between Nairn and Burghead may be regarded as a link between two areas of solid rock. We shall see later that there is a corresponding, but less well-known one between Branderburgh and Portgordon. The high ground between Burghead and Branderburgh was formerly an island or, at times, a group of islands of Permian and Triassic rocks. The map based on Ogilvie (1923) shows that at the time of the maximum submergence there were two major islands separated by a narrow strait. As the sea-level fell relative to the land, more and more dry land appeared and an enlarged Loch Spynie was the last remaining area of open water. The present loch is but a minor pool in that area, much of which has been drained artificially. All these islands were subject to marine erosion and many, if not most, were built of glacial deposits. Their erosion, especially on their north sides, produced a great deal of the material that is now incorporated in the western part of the link. With each successive fall of sea-level, the amount of open water in the archipelago concontracted, and so also did the degree of erosion. Fig. 48 shows the shorelines at the several stages. Ogilvie points out that the lines separating different zones do not always follow contours, partly because of varying exposure. This implies that greater storms eroded to a lower level in exposed places, and also piled up shingle to higher levels elsewhere. Some features are explained as a result of halts in the general fall of sea-level.

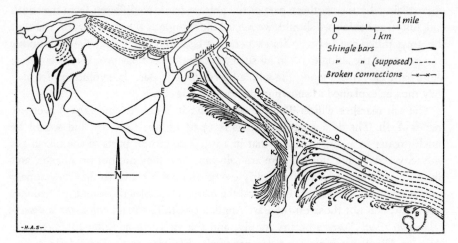

Fig. 48. Shingle-bars and shorelines – present and past – from Binn Hill westward
(after A. G. Ogilvie)

The low cliffs between Burghead and Branderburgh are for the most part old cliffs, now out of reach of the waves. There is locally a path or rough road at their foot. They contain some caves, and are also associated with a rock bench. They are cut in rocks of Permian age. This small outlier implies that these rocks once covered a much wider area. The cliffs are largely overgrown with vegetation (see p. 209).

We shall follow Ogilvie closely in describing the eastern part of the link. In recent years the seaward, shingle, part of the area has been afforested so that it is now almost impossible to follow many features which, when Ogilvie wrote, were clearly visible. In this part of the coast there are four main features, the hummocky drift of the interior, the terraced delta of the Spey, the lagoon of the former Loch Spynie, and the shingle bars on the coast. The shingle begins at Portgordon where there is a change in trend of the high ground at the coast. Ogilvie argues that this convexity 'caused the north-east storm-waves and the currents of the flowing tide to build offsets so as to maintain the previous direction, and so to smooth out the re-entrant'. But in order to produce the present conformation the former promontory of Binn Hill, about two miles west of the mouth of the Spey, had to be cut back to form the conspicuous cliff inside the modern ridges. It is a hill of Old Red Sandstone swathed with glacial deposits and prolonged westwards by them. It is probable that the truncation of the hill took place at several levels of the sea; there seem to be traces of erosion at 70, 50 and 40 feet (26, 15 and 12 m). Probably the main attack was in '25-ft' times, and even later. Whatever the precise history it was only after the cliff had been formed that conditions favoured the accumulation of the shingle ridges which cover a space about half-a-mile wide. The ridges are often remarkably parallel. After passing the cliff (see Fig. 48) the ridges maintain their direction to the point marked G, beyond which their trend is solely in response to the marine forces working on them.

The oldest hooked bars (BB) probably formed the support for a long bar CCC', reaching beyond the *present* bed of the Lossie. Beyond C' this bar developed a series of claw hooks which left as outlet for the Spynie lake only the gap D. Then the main bar, prolonged by D', must have blocked this outlet, and the lake may well have drained only through the gap at E, north of Kineddar. This was followed by the steady growth of straight bars nearly parallel to CD', but one, probably that marked G, extended, partly as a double bar, through the whole stretch to D'. Finally the two ridges (HH) reached the headland at H'H'H' . . .

After this stage a breach was opened in the barrier at J, and a rearrangement of the bars took place. A bar KK' was built, to be followed by the fan east of KK'. The bar CC, east of the mouth, was broken and its material re-arranged into some recurves. Something similar happened to the bars F, G and H. With the fall in sea-level the shore line became adjusted to the new conditions. The youngest shingle bars are those labelled Q, and Q'R in a sandspit which has once again forced the mouth of the Lossie to its former outlet near D. The eastern tie is broken by the mouth of the Spey, a short stretch of coast that probably changes as much as, even if not more than, any other section of the coast of these islands (see Plate 14). Grove (1955) has summarized the nature of those changes. He also points out that the ridges through which the river reaches the sea are probably not all of the same age; the minor ones were formed at a somewhat higher stand of the sea. There is a tradition that the river once reached the sea about three miles (4.8 km) farther west, owing to deflection by the shingle. There is no positive evidence of this. That it once had its mouth about a mile west of Kingston where there are several recurves is most likely. At the end of the eighteenth century ships of 350 tons could use the mouth, and vessels were built at Garmouth until about 1875. But the shifting mouth was troublesome, and defence works were necessary to protect the small settlement at Tugnet. In the Moray floods of 1829 a breach 400 yards (366 m) wide was made in the shingle. In 1835 the mouth was in much the same position as in 1955. In 1844 houses on the north-west part of Kingston were washed away, and the river was later diverted by piles. This was an unsuccessful remedy; the cut soon filled with gravel. Another unsuccessful attempt was made in 1857, but a new outlet was made in 1860. Change, however, continued; a new mouth might be successful for a time, but the shingle was continually moving to form a spit across the mouth. It is thought that a great deal of the shingle is brought down by the river itself since it flows in its lower course in braided channels over a gravel fan fringed by raised beaches and deltas.

PORTGORDON TO FRASERBURGH

At Portgordon the old cliffs behind the shingle and sand of Spey Bay coincide with the present sea-cliff which continues as far as Quarry Head, west of Rose-hearty. For the first two or three miles the cliffs are cut in the Old Red Sandstone

8

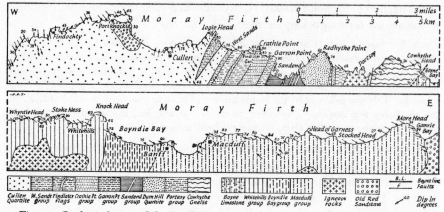

Fig. 49. Geological map of the Banffshire coast (after H. H. Read: Geological Survey)

which, at the coast, is relatively horizontal. The unconformable junction with the metamorphic rocks can be examined at Buckie, on the west side of which a knob of slate projects. But the older rocks soon rise up to form the cliffs east of Portessie. The coast from, approximately, Portessie to Troup Head affords a most interesting and continuous section in Dalradian rocks of the Highlands. They are exposed on the coast in belts striking in a general south-south-westerly direction. Originally they were clays, sands and limestones, but as a result of metamorphism they have been converted into slates, phyllites, schists, gneisses and quartzites. Moreover, they have been closely folded, and these minor folds have been combined with earth-movements on a much larger scale. The coastal area is plateau-like, and the cliffs range from about 100 feet (30 m) to more than 300 feet (91 m) at Troup Head. Deep valleys intersect the coast and, in the eastern part, give great beauty and diversity to the landscape. The cliffs, as in so many other parts of Scotland, are by no means wholly the production of present-day processes. They are often faced with a rock platform associated with one or other of the lower raised beaches. On the western part of the cliffs, continuous with those of the Old Red Sandstone between Portgordon and Portessie, there is an almost unbroken line of urban development; it is only on the east side of Cullen that a marked break occurs.

On a small-scale map the coast looks fairly straight; in fact it is much indented, and to a considerable degree the sinuosities are related to the rocks, but it would be dangerous to push this suggestion too far. That many of the minor indentations represent the interaction of marine erosion on varying rock types will be explained in the following pages. At this point it will be convenient to introduce the rock types as they appear on the coast from west to east. In this and in many other ways we shall follow the detailed work of H. H. Read (1923) (see Fig. 49).

TABLE OF ROCK GROUPS

KEITH DIVISION

(a) Cullen quartzite	$\begin{cases} a_1 \text{ Findochty beds} \\ a_2 \text{ Logie Head beds} \end{cases}$	Granulitic quartzites with subordinate dark garnetiferous mica-schists.
(b) West Sand group		Garnetiferous mica-schists with a thin calcareous band.
(c) Findlater Crags		Muscovite–biotite flags with scarce quartzite.
(d) Crathie Point group		Calcareous biotite flags.
(e) Garron Point group		Actinolite schists, silvery mica-schists.
(f) Sandend group		Limestones, black schist, staurolite schist and mica-schist.
(g) Durn Hill group		Quartzite and quartzose gneiss.
(h) Portsoy group		$\begin{cases} \text{Mica-schist, graphite-schist, tremolitic limestone, lime-} \\ \text{stone, quartzite and transitions between types.} \end{cases}$
(j) Cowhythe Gneiss		

The Boyne Line – a line of discontinuity, along which j and k, j and l, j and m, and h and m are brought in contact.

BANFF DIVISION

(k) Boyne Limestone		
(l) Whitehills group		$\begin{cases} \text{Rapid alternations of pebbly grits, pebbly limestones,} \\ \text{flags, phyllites, limestones.} \end{cases}$
(m) Boyndie Bay group		Andalusite schist and pebbly grit.
(n) Macduff group		Slate, greywacke, and pebbly grit.

Between Portknockie and Cullen Bay there is an unconformable mass of Old Red Sandstone, but it rests on the cliffs, and only reaches the sea for a very short space at Portknockie. Dykes and sills of igneous rocks also outcrop on the coast, most conspicuously at Portsoy. Boulder clays and other superficial deposits rest on the cliff tops in many places.

The Cullen quartzites are often folded into sharp anticlines, good examples of which can be seen a little west of Findochty, at the headland in that place, at The Lammies, and west of Portknockie. The rocks throughout the outcrop dip at angles between 25° and 45°, the higher dips being the more common. Brecciated zones correspond with the boatyard at Findochty and a cove just west of Tronach Head. The harbours at both Findochty and Portknockie are cut in breached anticlines. Tronach Head is a good view-point. Westwards the view to Findochty shows numerous stacks, doubtless belonging to the rock bench, but many of them submerged in part at present high water. Immediately to the east the bedding indicates minor faults so that two or more large masses of rock are now separated from the cliff by deep and narrow ravines. The upper parts of the cliffs hereabouts are often covered with boulder clay.

Cullen Bay is a most interesting locality from the physiographical point of view. There is a continuous exposure of the rock platform, which in the western part of the bay is overlain by a mass of Old Red Sandstone. Near Scar Nose, at the western end, there is an arch, called the Whales Mouth, where the unconformity between the metamorphics and the Old Red is clearly visible. The Bow Fiddle arch, at the same headland, is cut out of thick bedded quartzites and then mica-schists. In the bay itself the rock bench is covered by the golf course, on which there are one small and two large stacks of the Old Red. Farther east, two quartzite dykes run seawards and give rise to three prominent stacks known as the Three Kings of Cullen. The old cliff is a fine feature all round the bay, and the lower parts of the town of Cullen are partly built on its face, but more particularly on the rock platform in front of it.

The Logie Head group contains several varieties of rock, which are often thinly bedded and flaggy. Throughout the sections the dips, to the north-west, are high, varying from about 60° to 70°. On the east side of Logie Head a narrow band of the West Sand group forms the coast for about 500 yards (457 m); its eastern boundary is a fault, and for another 500 yards or so the Logie Head group re-appears to give way to another narrow (*c.* 150 yards: 137 m) outcrop of the West Sand group.

The Findlater group shows high dips, often of 80°, to the south-east, although in the many folds north-westerly dips can be seen. At the castle there is a band of more resistant quartzite. The castle is not on the cliff top, where there are remains of an entrenched camp. There has been a certain amount of alteration in the cliffs, both in making an excavation into which to build the castle, and in cutting through the isthmus for defence purposes. The Crathie Point group commonly show west-ward dips of about 80° and the craggy forms assumed by these rocks are locally conspicuous. The Garron Point group also shows high dips to south-east and north-west.

East of Garron Point the coast falls back into Sandend Bay where the main rocks are black schists and limestones. In the limestone there is a fine pot-hole of Old Red Sandstone age. The same high south-easterly rock dips prevail. The bay itself may reflect the less resistant nature of the rocks and is backed by sand dunes. The succeeding Durn Hill quartzite is displaced north-westward by a fault which runs south-eastwards from Reidhaven and this, combined with the massive nature of the quartzite, which forms craggy cliffs, probably accounts for the shape of Sandend Bay. In the Durn Hill quartzite there are several steep-sided and narrow indentations which coincide with joint and bedding planes. Redhythe Point is cut in well-bedded and much jointed quartzite showing westerly dips of 50° to 60°.

The re-entrant between Redhythe Point and East Head, in which is situated the town of Portsoy, is complicated by the intrusion of igneous rocks. The

Portsoy group proper form craggy cliffs on both sides of the town, but the town itself and the old and new harbours are built in the older igneous rocks. The form of Links Bay, into which the Burn of Durn flows, may be associated with that stream, and the fact that the bay corresponds with the junction of the igneous rocks and the succeeding Cowhythe Gneiss. The dips in this rock are similar to those in the other groups except in Strathmarchin Bay, where they are about 20°. The gneiss forms fine craggy cliffs which stand out on East Head, King's Head, and Cowhythe Head. Immediately to the east of these headlands, the Boyne Line reaches the coast on the western margin of the cove known as Old Hythe. The Boyne Limestone itself gives rise to the rock ledges at and near the Craig of Boyne.

For the next three or four miles (5 or 6 km) the coast is formed of the Whitehills group, a group consisting of rapidly changing beds with the usual fairly high dips. There are three main headlands in the stretch, Whyntie Head, Stake Ness, and Knock Head. The cliffs are less high, especially near Stake Ness. The two western headlands owe at least part of their form to dykes of amphibolite which reach the coast at those places. At Knock Head layers of pebbly grit are significant, and may be compared with those at Macduff. In Boyndie Bay the dips, mainly to the east, are still high. There is a good beach. At one time erratics were more common on the foreshore; many were used in the building of Banff harbour. Those that remain are still known as the Tumblers. In the west of the bay some erratics of black diorite are, or were, locally called the Boyndie Heathen. The Burn of Boyndie flows into the bay, and the coast is relatively low. The mouth of the burn is deflected slightly to the west. Near Banff pebbly grits cause some irregularities, and the junction of the Boyndie and Macduff group is at Banff itself. Meavie Point, in the Macduff group, is made relatively prominent by pebbly grit beds and dykes. Banff Bay, and its shingle beach and bar, corresponds with the mouth of the Deveron. The Macduff group reaches the coast over a distance of about seven miles (11 km); in addition there is the detached portion which forms Troup Head (see Plate 13). Throughout the group the beds are much folded, and the high dips are maintained. It is here, especially, that the pebby grit beds have a pronounced effect on the pattern of the coast. They occur throughout the stretch from Banff to Gamrie Bay; the main rocks in the section are varicoloured slates and slaty flags. Perhaps the best place to see the effect of the hard bands is just east of Macduff where there are several wall-like promontories including Dillyminnen, Stendreach and Drooping Craig; between them are small coves cut in steeply dipping blue slates and crags. Melrose beach, in Old Haven, is attractive, and is one of the constituent parts of Old Haven, a locality rich in rock plants. Garness is another headland where grit bands largely control its form. The outline of the coast from Stocked Head to More Head is occasionally broken by similar hard bands; the fossil cliffs are conspicuous in this area.

At More Head the coast recedes into Gamrie Bay in which the two interesting settlements of Gardenstown and Crovie are situated. The metamorphics are replaced by the Old Red Sandstone, and the two villages rest on the steep slopes and also on a low ledge a few feet above high-water mark. There are several steep and deep ravines, each with a small stream, leading down into the bay which, at any rate in part, is fault-controlled. The cliffs are bright red in colour and there is a marked shore platform in front of them. The metamorphic rocks reappear at Crovie to form the high and prominent mass of Troup Head, the rocks of which are generally more gritty than in the main outcrop of the Macduff group. The fault reaches the sea on the east side of Troup Head a little to the south of the deep inlet near Northfield. There then follows a short but picturesque stretch of Old Red Sandstone. The Lion's Head is a bluff of greenish conglomerate pierced by a cave which passes in to the great chasm known as Hell's Lum. The deep-cut valley of the Tore of Troup reaches the sea in Cullykhan Bay, to the east of which is Pennan Bay where the village occupies a position similar to that of Gardenstown. Pennan Head is in the Old Red conglomerate which weathers into little aiguilles. To the east the metamorphic rocks return to the shore and form, for a short distance, one of the finest lines of cliffs in eastern Scotland. Strahangles Point is magnificent; its form depends much upon the effect of marine and subaerial erosion along joint planes. Aberdour Bay contains a shingle bar and sand beach and former cliffs of the '25-ft' raised beach. It lies between Strahangles and Quarry Head, beyond which the cliff level falls rapidly to give place to the low coast which extends as far as Peterhead. Between Strahangles Point and Kinnairds Head there is an almost continuous section of knotted and andalusite schists. Rosehearty stands on this platform, and there are some low sand dunes along the main road. Although this stretch is interesting, it is not particularly attractive. There are sand beaches at Phingask shore, and Rosehearty, but otherwise a low rock platform, which shows various degrees of folding, forms the coast. Kinnairds Head is formed of tough schists. Fraserburgh Bay carries a fine beach backed by dunes, and is enclosed between Kinnairds Head and Cairnbulg Point (with Inverallochy), points formed by the outcrop of metamorphic rocks. Whitelinks Bay is, on a smaller scale, somewhat similar, but there are also expanses of rocky platform. The dunes in Fraserburgh Bay occupy almost all the space between the beach and the main road. They reach almost 50 feet (15 m) in height. Near Kirkton, the relief is irregular – small hollows, ridges, steep slopes, a little vegetation and some small stable areas. Much sand is moved in high winds. The prevailing winds are westerly, but those from between north and east have the greatest effect, and they blow especially in winter and early spring, when there are also stormy conditions at sea. Thus there is a good deal of change along the seaward side of the dunes. Fortunately the sandy beach offers considerable supplies of sand to replenish that blown from the dunes. There has been very little alteration in the high-water mark since 1869 when the area was first mapped by the Ordnance Survey, but there are small variations from time to time. The dunes and links widen to the east, and the Water of Philorth has also been deflected in that direction. The western part of the sands was carefully mapped in 1970 by R. Crofts and N. Rose (Dept. of Geography, Aberdeen University).

FRASERBURGH TO ABERDEEN

One of the most interesting sections of the coast of Aberdeenshire is that between St Coombs (Inzie Head) and Rattray Head. It has been investigated in considerable detail, and we are fortunate because there are historical records of some value which enable us to reconstruct the evolution of this part of the coast. Inzie Head and Rattray Head are low outcrops of schistose rocks, and near Rattray there are also extensive spreads of boulders, washed out from the local boulder clays. Between the two heads there is now a wide sandy beach backed by extensive dunes, behind which is the Loch of Strathbeg which is connected to the sea by a partly artificial channel at its north-eastern end.

Walton (1956) has studied the area, and has shown that a former opening, about four miles (6.4 km) wide, extended inland to about the position of the present main road, A952, which in places follows at the foot of an old cliff line, and perhaps even to the point where the road crosses the small burn about half-a-mile north-west of Netherton. This extension occurred at the time when the Yoldia Sea existed in the Baltic area. This indentation can be traced by means of ancient cliffs cut in glacial deposits, and it contained (see Fig. 50, main map) some islands of boulder clay. Probably at this time the waves began to form spits; there is some indication of one on the northern side of the bay (Fig. 50(a)), and a deposit of sand of this age seems to have formed under the shelter of one of the islands, all of which are now covered by grey dunes. At a later stage there was a fall in sea-level, so that in Littorina times the sea was about 15 feet (4.6 m) higher, relative to the land, than at present. Once again this can be traced clearly by means of old cliff lines which are conspicuous. For a time the Yoldia sea bay continued to exist, although smaller in area, and the former boulder clay islands had, as a result of the fall in sea-level, been incorporated in the mainland. But sooner or later the waves began to build a major spit which steadily grew southwards. It begins at the Castle hill of Lonmay, and there it is composed of several parallel ridges of shingle, with some recurved ends, which terminate a little southwards of the present artificial drainage channel from the loch. From this point onwards there is one main ridge, although broken in one place. Near its distal end there are more recurves, and both they and the main ridge are turned somewhat inwards, i.e., to the west. This bar does not quite reach the fossil cliff at Old Rattray where there is more shingle at the foot of the cliff. Walton is confident that all this shingle has travelled south from the boulder clay near Inzie Head. It is not, however, clear that the lagoon behind the bar was at that stage ever completely enclosed; it is more than likely that a tidal channel existed at its southern end. It may also be assumed that at this

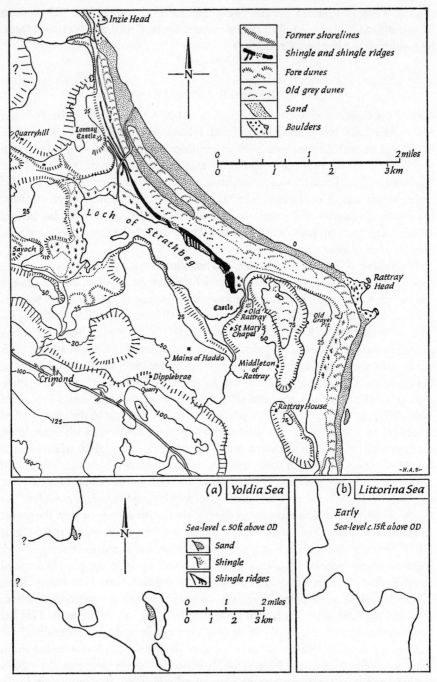

Fig. 50. Coastal features around the Loch of Strathbeg; and (a and b) suggested evolution of the coast in the Rattray area (after K. Walton)

stage sand dunes were forming on the shingle, and that marsh deposits accumulated in the lagoon.

Since the Littorina stage is now represented by fossil cliffs, and the bar is well above present sea-level, there has clearly been a negative movement of sea-level; but it is probable that the change of level was not sufficient completely to drain the lagoon, and that at high water the sea came into it so that somewhat later it became a small harbour. It was probably never easy of access, but the fact that there are Castle hills at both Lonmay and Rattray imply some importance. If the castles were, as was probably the case, partly defensive, and built of wood, they were presumably to withstand Viking raids. Rattray became a village of local importance, and the harbour, somewhat sheltered by the boulder clay promontory, gradually increased in significance. A chapel was built at Rattray in the thirteenth century, probably by William Comyn, and was connected with the Abbey of Deer. In 1324 the Lordship of Rattray and the harbour went to Sir Andrew Douglas, but by the mid-fifteenth century they were broken up, and in 1459 the Earl of Errol received from James II the lands to the west of the small burn east of Haddo, whereas those on the other side of the burn became the property of the Keiths of Broadland. There is no need to follow the ownership of the lands in any detail. There were various disputes in the sixteenth century, and there is every reason to believe that the harbour, called Starna Kippie, at Rattray was flourishing. The trade was mainly in white fish, and there was considerable contact with Holland, but details of its trade are lacking. By this time Rattray had been made a Royal Burgh, and in 1564 the burgesses were given the right to set up a market cross, to hold a weekly market, and two fairs a year. But the title availed little, and in 1696 the settlement contained only seventeen adult persons over sixteen years of age. But the harbour was still in use, not only by Rattray but also by inhabitants of near-by villages. The cause of the decline was the usual one, the choking of the harbour mouth with sand. Even in 1654 Sir Robert Gordon of Straloch had commented on the parlous state of the harbour, but this may have been but a temporary blockage. Hepburn in 1721 wrote:

Crimond is divided from Lonmay by the river Rattray, at the mouth of which river on the south side is situate the village of Rattray, famous for codfish which the inhabitants take in great plenty . . . There are many sea calves in the mouth of the river and this is the reason why there are no salmon there.

On the other hand Alexander Reith, in 1732, says that the harbour was then choked up, suggesting that Hepburn's account may have been written some time before it was published. The *Old Statistical Account* says that about 1700 the lake was smaller than it now is, but was still used by small vessels. This entrance, however, was closed about the year 1720 by an easterly storm. This seems to have been sudden, and a small ship is said to have been trapped and its cargo of slates was used in roofing the Mains of Haddo.

Map evidence of this and early changes is lacking. Roy's map of 1747–55 shows the bar along which runs a road which crossed the old outlet in the south. It also shows a new channel at the north end which is now filled so as to form the modern fresh-water loch. With the closing of the harbour the Royal Burgh of Rattray soon ceased to exist, but a few fishermen operated from the open beach. The level of water in the loch rose, partly because seeping had stopped, presumably because the sand and shingle bar had become more or less water-tight as a result of deposition of silt and sand and the growth of vegetation in the loch. This caused some flooding of the western shore of the loch. Some attempts were made to drain it by a canal, parts of which are still visible on the northern part of the area. In the First World War the level was raised so that the loch could be used as a sea-plane base.

The movement of beach material along this part of the coast is now apparently directed to the north, since a modern bar, in front of the older ones, is growing in that direction. There is no doubt that in Littorina times the drift was to the south. There is, perhaps, a rough parallel between what has happened here and the old shingle bar on which Stonar (near Sandwich, Kent) was built, and the newer northward deflection of the Stour.

South of Rattray Head similar features are found, but details have not been worked out. Rather more than a mile north of Scotstown Head a small burn, draining into a loch, is dammed back by the beach. Scotstown and Kirkton Heads are generally similar to Rattray Head, and as far as Craig Ewan and the mouth of the Ugie the beaches are backed by lines of dunes (see Plate 12).

Sandy beaches extend as far as the mouth of the Ugie. The granite mass, named after Peterhead, reaches the coast near St Fergus. Scotstown Head and Kirkton Head are composed of morainic debris, and it is at Peterhead itself that the granite outcrops. The harbours to north and south of the road joining Keith Inch to the mainland are largely enclosed by skerries, and the main jetties enclosing Peterhead Bay deprive its shores of effective wave action. From Peterhead to Cruden Bay the granite is generally uniform in character. The master joints run usually north and south, and at right angles to them are the slides, as the other set of joints is called by the quarrymen. Along the coast there are many stacks and caves. Some of the caves and inlets have been cut along dykes of dolerite which waste more easily than the granite. These narrow inlets are locally called yawns, a name which applies as far south as Aberdeen. At Boddam a long narrow gully indicates the position of a dyke, and a somewhat similar feature is found a few miles farther south at Dunbuy where a cave penetrates some distance inland along a dyke. There are fine natural arches at Dunloss and The Bow, and at the Bullers of Buchan the coastal scenery is bold, there being stacks, an arch and a blow-hole in a good range of cliffs. Cruden Bay is wholly within the granite area. Its inner edge is largely dune-covered, and the dunes hide a terrace, a former beach at about 50 feet (15 m) above present sea-level. The lower beach, at about 25 feet (8 m), is usually distinct, but is often hidden under blown sand. Port Errol, and its harbour,

are at the north end of the bay, underneath the Ward Hill. Today the Cruden
Water enters the sea at this point, but before 1798 the Water found a mouth to the
north of Ward Hill, a course now marked by a line of woods. The stream which
runs south from Longhaven, reaches the sea by the old mouth of the Cruden
Water which was diverted to its present course to keep the sand from silting the
harbour. The present harbour dates from about 1875 (Smith, 1950–2).

At the Skares the granite gives way to the schists; and there is a cliffed coast
extending to Forvie Sands. Close to the granite the dip of the schists is about 20°
to the north-east, and the rock is really a clay-slate. A little farther away from the
granite the cliffs are cut in pebbly quartzites and grey schists, and the dip lessens.
Towards The Veshels the rocks become almost entirely knotted and andalusite
schists, and maintain this character as far as Slains where, at the Old Castle, isoclinal folds are visible in the cliffs in which caves are cut. In Broad Haven the dip
is east at about 45°; the dip varies somewhat in direction. At St Catharine's Dub
the promontory is made of pebbly quartzites with partings of knotted schists, and
quarrying has altered the natural appearance of the ground. A little to the north the
Old Castle of Slains stands on a small peninsula made of foliated schist with bands
of pebbly quartzite. On the rock platform there are fine grained reddish deposits
which may be of wind-blown origin from inland moraines.

The peninsula . . . is therefore a composite feature. Two stacks, tied to one another and to the
mainland by glacial deposits, are separated by a narrow channel from another stack which also
has a capping of drift. The largest stack, closest to the mainland, exhibits a cliff cut in drift
behind the fishermen's houses, while the smaller tied stack bears the remnants of the Castle
and the single cottage. The cliff, on the northern side . . . of the peninsula, has a small section
of raised beach conforming in height with that of the outer end of the peninsula. The whole is
thus an assemblage of capped and plugged ancient elements which have been only partly
re-excavated by later marine action. (Walton, 1959)

Along all this part of the coast glaciation plays an important part in the scenery.
The schists are covered with fine-grained deposits which, as Walton remarks,
give the cliff profile a 'subtly different appearance' from those covered in true
boulder clay. The schist coast is generally smoother than that of the granite to the
north, but in both rocks yawns and inlets are cut along lines of weakness.

Another interesting part of the east coast of Scotland is that which contains the
sands of Forvie and the estuary of the River Ythan. As long ago as 1865 Jamieson's
work on the raised beaches made the name of the Ythan well known to all concerned
with fluctuations of sea-level. The sands of Forvie have, for centuries, been the
subject of many articles, but it is only in the last few years that they have been
investigated scientifically. It is still too early to suggest that they are fully understood (see Plate 10).

The sands occupy the area between the road from Collieston to Newburgh and
the sea. Foveran Links on the south side of the Ythan mouth are quite separate
from them and, in fact, form the first unit in the line of dunes which extends from
the Ythan to Aberdeen. The northern part of the Forvie sands, rather more than

half of their total area, rests on metamorphic rocks, the remainder, except for the southern tip, cap raised beaches, but metamorphic rocks occur at a lower level. Their total area is approximately 1,780 acres (720 ha). But it is important to appreciate that between the sands and their rocky basement there is commonly a layer of glacial drift. The evidence of the quartz and boulders makes it clear that the first ice-sheet to invade this part of the coast came from the west and north-west. The second, on the other hand, reached Forvie from the south-west and south and brought with it the red Strathmore boulder clay which can be readily examined between Hackley Head and Collieston. The third ice-sheet did not reach the area now covered by the sands. Kirk (1958) states clearly that the Strathmore ice 'brought with it either from Strathmore or the bed of the North Sea, considerable quantities of red drift derived from Old Red Sandstone formations and this was dumped outside the lower Ythan in the form of boulder clay and *thick beds of sand* and gravel, thereby filling in the irregularities of an ancient coast line' (the italics are mine, J. A. S.). In this way the river was dammed and a lake was formed. This, however, disappeared with the melting of the ice and when this was complete the Late-glacial sea invaded the estuary. Kirk is of the opinion that this change took place when sea-level was, relative to the land, about 75 feet (23 m) higher than at present. This is demonstrated by the occurrence of raised beaches and terraces in the southern part of the present sands. As the sea-level continued to fall, beaches and terraces were left at about 50 and 30 feet (15 and 9 m) in the Ythan valley. As the ice melted and sea-level fell, the rivers cut down their beds and extended seawards. It is suggested by Kirk that some of the earliest wind-blown sands date from this time. A later transgression followed, during which clays were deposited and the sea rose to about 30 feet (9 m), corresponding to the '25-ft' beach of Jamieson. This was a warmer and damper period, and forests throve along the Lower Ythan. This Atlantic period (= Littorina) is that in which we find the earliest traces of man in this area. He lived on the terraces and beaches which were gradually exposed with the fall of sea-level to the present. There are also many remains of the Early Bronze Age on the Lower Ythan. Kirk (1958) points out that the Atlantic transgression led to the deposition of great quantities of silt and sand on the coast

partly as a result of rivers reducing their loads in sympathy with heightened base level, and partly in consequence of marine erosion of the sands and silts of the Strathmore glacial series such as are well exposed at Collieston, e.g., in the fossil cliffs of the '25-foot' beach. In the subsequent withdrawal of the sea during the Bronze Age and later, large quantities of this material came to be exposed along the emergent beaches and in drier periods appears to have been blown inland by south-easterly winds.

In this way dunes began to form along the coast, not only at Forvie, but all along the coast to Aberdeen. There seems no doubt that at Forvie great volumes of sand in this way reached the higher beaches which now underlie the present surface of the sand.

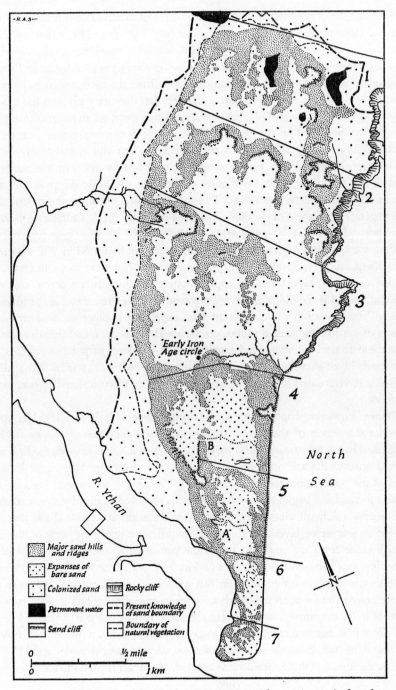

Fig. 51. The Sands of Forvie showing positions of seven 'waves' of sand
(after S. Y. Landsberg, 1955)

At the present time there are seven main waves of sand, and it is usually assumed that the most northerly wave is the oldest (see Fig. 51). The waves numbered 1, 2 and 3 are now more or less immobile, although there are small movements resulting from local erosion. All the other waves are as yet uncolonized. They are broader north to south and less long east–west than the earlier waves. They still appear to be moving, and it has been estimated that they may advance from one to fifty yards (46 m) a year. This, however, does not mean an advance of the whole dune by that amount but 'mainly by a shallow deposit of a few inches' (Landsberg, 1955). The figure also shows that a ridge joins waves 1 and 2, and nearly reaches the eastern end of 3, that a pronounced V-shaped hollow exists on the north and west of wave 5, and that the orientation of the three southerly ridges is somewhat different from those farther north. The most detailed work on the morphology and vegetation of the Forvie sands has been done by S. Y. Landsberg, but, unfortunately, is only available in her Ph.D. thesis, and I am most grateful to her for lending me a copy. (A good summary and maps will be found in *The Vegetation of Scotland*, ed. J. H. Burnett (1964), Ch. 4.) Her general view seems to be that the successive waves have travelled approximately north-eastward across this area, and that wave 1 is the oldest and wave 7 the newest. Reference to Fig. 51 gives her view of the evolution of the present landscape. She appears to hold that large masses of sand first gathered at the southern end of the area, and that in course of time these accumulations moved northwards in the form of parabolic dunes, and expanded, in an east–west direction, as they travelled northwards. We shall see later that she has dated the movement of the dunes on historical and archaeological evidence.

Before, however, considering the historical aspect, it is worth while to reconsider the origin of the dunes. It seems that she supposes each wave to have originated in the narrow southern tip of the peninsula. Once started, the first wave moved northwards and attained certain positions at certain dates, until it finally reached the position numbered 1 on Fig. 51. Presumably at a somewhat later date wave 2 formed and began to move northwards followed in turn by 3, 4, 5, 6 and 7. This seems a difficult view to maintain even if she allows, as she does, that sand existed as a layer or layers on the wider middle and northern parts of the area. The distance from the southern tip of the peninsula to the northern end of the dunes is about three-and-a-half miles (5.6 km). There is certainly no reason why a dune should not move this distance, but what is more difficult to explain is why the first dune should continue to advance when one or more had formed to windward of it. These, surely, would shelter the first formed dune to a marked degree and allow it to become colonized by vegetation. It is partly a question of how far the first dune had advanced before the second one was formed and started to chase it. The three oldest dunes are now vegetated, and apart from quite local and small changes do not appear to be moving.

Moreover, if each dune originated in the narrow southern tip of the peninsula it is not easy to see why it expanded so much east and west unless, as it clearly must have done, it made use of the sands, associated with former sea-levels, lying on the main mass of the area. But if this is so, why is it necessary to assume that each dune began in the far south and moved northwards to its present position? There is no reason to object to a northerly movement of the various dunes, but the evolution suggested by Landsberg seems to be almost too precise. We have seen on pp. 236 and 237 that there is every reason to suppose that considerable amounts of sand existed over much, or all, of the area well before the Christian era. Winds blowing over the area would easily have blown this into the form of dunes and, as Kirk (1958) remarked, much may have been blown inwards by south-easterly winds. If we therefore suppose that the several lines of dunes have originated on the broader parts of the Forvie peninsula we have perhaps an easier explanation of their present topography. There is no need to assume a train of dunes travelling northwards and expanding sideways at the same time, an explanation that presents the difficulty that once a new dune had formed on the windward side of its pre-decessor it would cut off the sand supply from it. The oldest dune appears to have reached approximately its present position about 1680. It is suggested that the rate was far quicker after about AD 1400, although by that time it was a much bigger dune than in earlier years. We shall note the historical evidence of the overwhelming of certain buildings below, but there is no need to assume that burial was always the direct result of a northward travelling dune. Local storms and winds might easily cause sufficient movement of sand for that purpose, and the ridge on the eastern side of the area covered by lines 1 to 3 on Fig. 51 suggests a movement from the west. There is no need to query the dates of inundation given by Landsberg; the main query is that of the mechanism by means of which inundation took place.

She has pointed out that the orientation of the dunes is somewhat anomalous in that they do not seem to conform to the prevalent winds of today. If, however, we assume that the older lines of dunes have originated more or less *in situ* and have not travelled all the way up from the south, and also that (cf. Kirk) south-easterly winds may have played a part in their formation, there seems less difficulty in accounting for their shapes. This in no way contravenes Bagnold's hypothesis quoted by Landsberg but seems to accord with her view that 'In the light of available information, it thus appears that local changes in orientation are likely to be caused by the effects of local topography on the wind régime.'

The archaeological and historical interest of Forvie is considerable, and what is known of it throws some light on the physiographical evolution of the area. Reference to Fig. 52 will show the general nature of the sands in their central section. From west to east there are ten zones recognizable: I the alluvium of the Ythan; II a terrace rising from 6 to 10 feet (1.8 to 2.7 m) above high-water mark; III a terrace at 30–40 feet (9–12 m); IV the westerly dune ridge containing layers of humus and cut into by blow-outs; V a raised beach at about 55 feet (17 m); VI bare, hard, sand without beach material; VII a much-eroded line of dunes

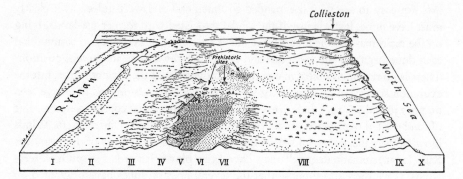

Fig. 52. Prehistoric sites at the Sands of Forvie (after W. Kirk)

which dies out northwards, but to the south runs into one of the east–west dunes; VIII a damp depression with rushes; IX the young eastern dunes; and X the present beach along which, in slacks in the dunes, raised beach material is found at 10 feet (3 m) above it.

Farther south, another beach at 20 to 25 feet (6 to 8 m) is found. In zone V there are Early Iron Age sites. These are not necessarily at the same height at which they were when in use, since they may, and almost certainly have, sunk lower as a result of blowing sand. However, the sites and their associated material indicate that a small community lived here at that time and cultivated the light sandy soils and pastured sheep, pigs, and oxen, and also fished. Eventually moving sand obliterated the area until it was re-exposed in recent years. However, the evidence indicates clearly that human occupation goes back for about 2,000 years. Landsberg's map (Fig. 51) suggests that the sand had covered this site by about AD 100, but the probability is that the site was deserted well before this. I find it difficult to follow Landsberg's reasoning, viz.

Thus, some years, probably centuries, before o B.C. sand could not have extended as far as [line 6] on fig. [51], site A., yet by o B.C. ± 100 it had reached site B and was locally covered by a sandy heath vegetation. It had, however, not spread further than [5] since in this locality a single circle of similar type . . . is found on a clay, not a sand substrate.

Why, in fact, is it necessary to assume all the sand travelled always in the same direction?

Farther north, near Rockend and elsewhere there are medieval remains. The chapel at Forvie, which is built on clay, was dedicated to St Fidamnan who died in AD 704. The date of the building of the chapel is uncertain. It is first mentioned in the records of the thirteenth century of the Chartulary of Arbroath, and something of its history can later be traced in the ecclesiastical records of Aberdeen and St Andrews and in Papal Grants (Kirk, 1958). In 1573 the parish was united with Slains. It is not known to what extent, if any, the parish had suffered from blown sand up to this time and, as Kirk remarks, 'The amalgamation of the parishes could just as well be interpreted as a post-Reformation administrative regrouping as an indication of the decline of Forvie.' However, in 1680 it was certainly sand covered, and by 1732 the name was almost forgotten. There is good reason to think that sand movements have been spasmodic rather than continuous. The storm of 1413 is well authenticated, and there were certainly stormy periods at the end of the fourteenth and early in the fifteenth centuries. It is worthy of remark that the gales in 1950 and 1951 set great masses of sand in motion in the southern and central parts of the area, and that at that time the parabolic form of some of the dunes was accentuated. It is not surprising to find that at Forvie, as in some other comparable places in Britain, a legend has grown up about the inundation by blown sand. At Forvie it is related that an heiress of the estate was dispossessed by a wicked uncle, and put to sea in a boat. She was rescued by Northumbrian fishermen and saw her curse fulfilled that 'nought be found in Forvie's glebes but thistle, and sand'. Boece also tells us that in 1526 there was much sand movement on the coast, and that a chapel built in about 1012 at Cruden Bay was 'as oftimes occurris in thay parts ouircassin be violent blast of sandis' (see Plate 11).

Air photographs have shown up the former pattern of plough rigs and furrows, and soil investigations have shown that there had been negligible intermixture of sand with the peaty clay before the sand finally overwhelmed the cultivated area. Much of this area is more than 150 feet high (46 m), and forms a part of the coastal plateau. Kirk suggests that this was an important factor in restraining the sand from covering it. The plateau, partly covered with boulder clay and raised beach sediments, runs out to sea to form a line of cliffs in metamorphic rocks which, like those to the north, present an intricate outline of stacks, gullies, and caves, by no means wholly of modern origin. The old church of Forvie, which stands close to the cliffs, has been exposed for some time and was partly excavated at the end of last century.

A glance at the map will emphasize what has already been said, namely, that there are several different types of habitat in the Forvie sands. Since much of the area is a plateau more than 100 feet high (30 m), and much of the rest consists of raised beaches and terraces, it is in some ways not a typical coastal area. Nevertheless, the dunes, separated by lows and flats, carry an interesting vegetation cover. Landsberg distinguishes the following types:

(1) Since there is nothing at Forvie to correspond with small hummocky dunes which gradually coalesce into a foredune the pioneer community occurs locally on the crests and lee slopes of moving dunes. The main plants include *Ammophila*, *Festuca rubra*, *Senecio jacobaea* and *Cirsium arvense*.

(2) There are large areas of grassland covered almost completely by *Carex arenaria* and *Ammophila*. These are not grazed.

(3) *Carex* grassland and dune pastures occur in habitats similar to (2) and thus in well-drained sand. The grassland is lightly grazed; the pasture is heavily grazed.

(4) *Grey dunes* and grass lichen heath occur in small areas and are distinct from the great parabolic dunes.

(5) *The Empetrum–Lichen* community is always associated with well-drained conditions, and belongs more to the dunes than the plains. It is for the most part not more than 3 to 9 inches high (8 to 23 cm), and the abundance of bare sand indicates immaturity. Erosion may take place readily and is only checked when the water table is reached.

(6) *Dune heath* develops from *Empetrum* heath where erosion of the sand does not reach the water table. The dominant plants are *Empetrum* and *Ammophila*, and they attain 12–15 inches (30–8 cm) in height. It is limited to stable phases in dunes and elsewhere.

(7) In the *Dune Callunetum*, *Calluna* is dominant, although in places *Empetrum* may be a co-dominant. It is extensive, and grasses are rare. 'In the Callunetum the dune plain is colonized; in *Empetrum*-lichen heath and dune heath it is the actual dunes . . . which are colonized.'

The occurrence of plants such as *Juncus* spp., *Nardus*, *Erica tetralix*, and *Salix repens* in general indicate wet or damp areas, and in winter the water table rises and forms extensive winter lochs.

From the Sands of Forvie in the north to Girdle Ness in the south, Aberdeen bay is 13 miles (21 km) wide, and forms one of the longest stretches of lowland beach and dune coastlines in Highland Scotland. The bay has a wide radius of curvature and faces east-south-east. Three major rivers, the Dee, Don and Ythan, drain the extensive hinterland of this distinctive coastline. Although the estuary of the Dee is now occupied by docks, wharves, sea-walls and piers, it once had the same estuarine form as the Don and the Ythan. Each estuary has a broad tidal basin which is preceded by a gorge section. Broad sandspits seal these tidal pools and the rivers break through to the sea through constricted channels. (See Plate 9.)

The coastal section between the Dee and Don has changed beyond recognition. Within living memory the wide beaches and vigorous dunes have been replaced by golf links, an esplanade and a sea-wall. The remnants of the beach are retained by an extensive system of groynes.

Between the Don and the Ythan however, the coastline retains a natural appearance. The beaches are wide (average 490 ft: 150 m) and broken only by shifting tidal pools and the meanderings of the small streams which drain the dune belt. The dunes are high (often more than 72 feet: 20 m OD) but broken into groups, by wind erosion. Series of conical sandhills are more common than parallel lines of dunes: the undercut coastal edge is more common than lines of embryo dunes. In a few areas, as for example Balmedie (seven-and-a-half miles: 12 km, north of Aberdeen) erosion and deflation have created tracts of sandhills and hollows which are completely bare of protective grasses. A similar situation pertains wherever vehicular access can be made to the dune and beach zone. Elsewhere blow-outs cut deeply into the dune ridge systems. The directions of individual blow-outs vary, but the dominant trend of sand movement is generally south-east to north-west.

Landwards of the dune belt, which varies in width from 110 yards (100 m) to over 710 yards (650 m) (average 220 yards: 200 m), there is a low marshy slack zone, which is frequently underlain by peat. These areas flood in winter and have a distinctive flora which includes some marsh-loving species. Where the slack is absent, smaller redepositional sandhills pile against the more prominent parts of the Post-glacial cliffline which is cut in the soft boulder clay and outwash sands and gravels of the surface deposits of this lowland coastal plain. Bedrock is at a considerable depth beneath the unconsolidated deposits of this part of Aberdeenshire, and rock outcrops are almost unknown along the length of this low, soft coastline. Several small meltwater channels breach this degraded Post-glacial cliff which appears to have been formed when sea-level was approximately 23 ft (7 m) above its present level. There is thus considerable variation to be found in the junction between the sand dunes and the interior – raised shoreline plains, degraded Post-glacial cliffs and meltwater stream valleys are all found within a short distance from the beaches.

Agricultural use, rifle ranges, golf courses and recreational zones utilize the dune and coastal environments, and plans exist for expanding tourist and amenity facilities.

The origin of the dunes can be seen as part of the long-term transfer of vast quantities of siliceous sand from the shallow off-shore zone to the coastline. Considerable fluvial and meltwater deposition must have taken place in Aberdeen Bay in Post-glacial times. The beaches and dunes probably accumulated in association with a rising sea-level which pushed the sediment bank landwards. Observations today suggest a diminution in beach nourishment and a tendency for the beach and dune system to readjust to this reduced supply by assuming a morphology characterized by evidence of erosion and retreat.

The Don reaches the sea at the north end of Aberdeen, the Dee near the south end of the city. The valleys are unlike one another; that of the Dee is more open, whereas the Don valley is narrower and the river winds much more in its lower parts. The coast at Aberdeen is now protected and in its southern part stabilized by breakwaters. But between the two rivers there is a broad sandy beach which is a great amenity to the city. As in almost all parts of the east coast, there are remains of raised beaches. That at about 25 feet (8 m) lies immediately behind the coastal road and extends northwards beyond the Don. In the eighteenth and early nineteenth centuries when much reclamation took place, both rivers entered the sea through estuaries choked with sandbanks. It is possible that the Don was deflected southwards by a spit, so that its mouth was at or near Footdee where a spit now slightly deflects the Dee, although harbour works effectively mask the work of natural processes (Walton, 1963). However, where the Don made its present mouth a creek seems to have existed roughly along the line of the old course. Now the outlet is diverted somewhat to the north, and there is erosion on the north bank of the Don. The Dee has also altered its course. There is no doubt that at one time it reached the sea in the Bay of Nigg. Its former course can easily be traced between Torry and the hill of Tullos. Borings show clearly that this course is floored by water-borne detritus.

ABERDEEN TO GARRON POINT

From the mouth of the Dee to Stonehaven the crystalline rocks of the Highlands form a line of cliffs of considerable variety and beauty. In the same direction, north to south, the degree of metamorphism decreases. Near Nigg Bay all the original sedimentary features of the rocks have been almost obliterated, and the rocks are completely crystalline. It is fortunate that many aspects of the cliffs have been studied by Walton (1959) who has shown conclusively that only to a relatively small degree are the present features to be explained by present conditions. We shall have occasion to say more about coastal platforms later on, but it should be borne in mind that all along the east coast of Scotland one or more are present in numerous localities, and that the lower one which is so often most conspicuous is certainly being trimmed by the waves today, but its full history is long and complicated. As Walton remarks, 'The most conclusive evidence against the assumption that yawns, stacks and platforms are modern derives from the fact that they are both plugged and covered with boulder-clay which the late-glacial and post-glacial seas have only partly removed'.

The range of cliffs in metamorphic rocks between Aberdeen and Stonehaven is interrupted at Cove by a small mass of granite, and throughout there are veins and dykes which usually have yielded more readily to marine erosion, and it is often along them that yawns, commonly trending either east–west or north-east to south-west, have been formed. It is partly for this reason that the stretch of coast is intricate in minor, but picturesque, detail. The contrast between the less metamorphosed rocks in the south and the far more crystalline ones farther north, as well as the effect of the Cove and Aberdeen granites, are also controlling factors. Moreover the cliffs are capped by boulder clay and other glacial deposits of three separate advances of the ice. The earliest of these advances laid down a clay, the colour of which varied with the nature of the rocks, so that between Aberdeen and Stonehaven it is usually grey, but at, e.g., Catterline it is deep red-brown. Somewhat later ice moving in from the sea deflected the land ice and left on the coast a stiff black and shelly clay which can be examined south of Bervie. The second glaciation left a red clay seen to advantage at and south of Muchalls. This was the Strathmore ice, and is associated with a number of overflow channels, some of which reach the cliffs. This ice reached as far as Peterhead, and even beyond. The third glaciation was more limited and is found north and south of Muchalls where it left a grey-coloured moraine, sometimes reddened by mixing with the Strathmore ice. It has left moraines on Girdle Ness and on the inside of Greg Ness. Only occasionally do glacial deposits form the cliff. In Nigg Bay the sequence from below upwards is: (1) weathered rock; (2) bedded sands and gravels (= ancient beach); (3) greyish till; (4) reddish till (= Strathmore ice); (5) laminated brick clays and silts; (6) beach gravels and sand of the '100-ft' sea. This section has been exposed by wave action. As a rule on this part of the coast the drift caps the

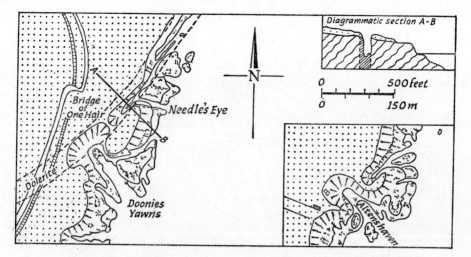

Fig. 53. Cliff features along the north-east coast of Scotland (after K. Walton)

cliff and gives it a bevelled appearance. Since the drift–solid junction near Stone-haven and elsewhere is sharp it suggests that the drift is resting on a platform cut by marine action and now about 50 to 70 feet (15 to 21 m) (at Stonehaven) above sea-level. The clay slope may be gullied, and occasionally ancient sea-stacks rise through it, a fact which clearly implies that the ice (the Strathmore advance) was not at all powerful. Drift which originally covered lower platforms has usually been destroyed. There is often a well-marked platform or notch at about 30 feet (9 m).

Walton has analysed some features of this coast in detail and we cannot do better than follow his explanations. About 500 yards (457 m) south of Greg Ness is an intricate example of erosion. Fig. 53 shows that a dolerite dyke runs approxi-mately north-east to south-west through the gneiss, and that a great yawn has been formed along the seaward contact of dolerite and gneiss. This contains a smaller yawn, and the lowest trench runs some way seawards. It is possible to relate the two yawns to high and low water, but it is far more probable that they are much older features. The sea cutting inwards along lines of weakness running in a north-west and south-east direction has produced a series of stacks and clefts such as the Needle's Eye. The figure shows that the mainland is connected with the boulder clay on this stack by a narrow ridge, the so-called Bridge of One Hair. The ridge separates Doonies Yawns from the Needle's Eye and presumably fills a part of the yawn not yet cut away, and so proves it is earlier than the boulder clay, in this case of Strathmore age.

Several other yawns in this neighbourhood tell the same tale. At Altenshaven, a mile farther south, the main yawn trends north-west and south-east, and there is also one running in from the north-east, and so in this example the boulder-clay

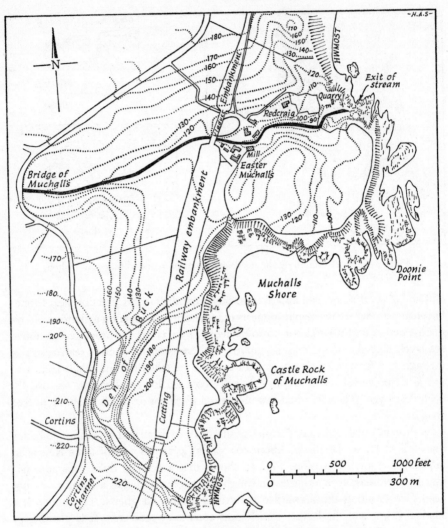

Fig. 54. Muchalls Shore (after S. Simpson and G. K. Townshend)

bridge is wider. Similar features occur in several other places. At Muchalls the glacial overflow channel named the Den of Buck reaches the coast. It is a broad and dry channel but is crossed, in an artificially maintained channel, by a small stream, called the Muchalls burn (see Fig. 54). After crossing the glacial overflow the burn runs into a steep gorge which, about 30 feet down (9 m), gives place to a vertical tunnel. After another 40 feet (12 m) it reaches the shore by means of a narrow cleft. That it once reached the sea about 200 feet (61 m) east of the present tunnel is clearly indicated by an old channel in which there were several small falls. This change of course is of glacial origin. The Pre-glacial course was cut well below present sea-level in what is now called Muchalls Shore. North and south of this pebbly beach there are cliffs of Dalradian Schist, but the shore itself is backed by a steep boulder-clay slope rising to about 130 feet (37 m). In other words, the Pre-glacial valley has been filled with more than 130 feet of boulder clay. When the ice finally disappeared melt water ceased to flow in the Den of Buck, and the Muchalls burn, in its turn, followed the lower course of the Den, i.e. some distance to the north of its own Pre-glacial course. The burn was now on solid rock and on top of the cliffs which at that time were much lower than now since the sea-level stood higher relative to the land. Then, as sea-level fell, the burn made a series of water-falls, and this, together with headward cutting, produced the present gorge. The tunnel, however, is less easily explained. Simpson and Townshend (1945–51) suggest three possible origins – blow-hole, pot-hole and erosion along joints. They regard the third as the most probable. Around the tunnel the rocks are broken and so easily removed.

This downward movement (on account of the sinking of the water-table) of the water would certainly be strongest where the rock was affected by shatter . . . The tunnel may then be satisfactorily accounted for if it can be suggested why the water should have followed a particu-ar course . . . We suggest that the erosion of the sea-cave may have opened up a fracture of some size reaching right underneath the stream . . . This would strongly have favoured the downward movement through joint cracks . . . [but] mud and silt carried by the stream would have tended to block any cracks and crevices . . . If, however, the waves beating into the cave were to compress air into all the cracks in the rocks, it is possible to picture the sediment in the cracks being continuously disturbed. Thus a continuous movement of particles in the major crack might ultimately abrade a definite channel along a preferred line. (Simpson and Townshend, 1945–51)

South of the Burn of Muchalls the coast is much indented, and faced by the rock bench in front of the fossil cliffs. There are also several prominent stacks. The beach at Muchalls shore is the largest for several miles. The headland of the Castle Rock of Muchalls is on the south side of the bay. The fossil cliff and rock bench run on to the south, and the cliff is cut back in such a way as to form several small headlands separated by narrow inlets. A little to the north of the Limpet burn there are four prominent stacks in schistose grits which strike to the north-east. Perthumie Bay contains a broad rock platform, the northern part of which is cut mainly in pelitic schists in which mullions are developed in several places. They are formed of closely-folded grits. The middle and southern part of the bay is in psammitic schists. The dry valley named the Den of Cowie reaches the coast about a quarter-of-a-mile north of Garron Point. It is probably a glacial over-flow, and cuts the cliff edge about 100 feet (30 m) above sea-level. On Skatie shore there are two prominent grit ridges, about 20 feet high (6 m), which cross the platform. Garron Point, where the Highland Boundary Fault runs out to sea, is made of spilites and forms a prominent headland. In Craigeven Bay, across the mouth of which the fault runs, the schist cliffs, north of the fault, are capped by reddish boulder clay; the rocks on the south edge of the fault are in the black shales and chert series, and Ruthery Head is, like Garron Point, in the spilites in this series (see Plate 8).

GARRON POINT TO ARBROATH

The coast from Garron Point to Stonehaven is in the Old Red Sandstone rocks. Near Slug Head the Downtonian beds make a line of pinnacles which runs seawards. Several small faults cut the platform and usually make small notches at its seaward edge; Cowie Harbour is on a tear fault. Nearer Stonehaven there is a good deal of sand and shingle. Nevertheless the beaches have suffered a continuous loss of material. The drift carries material to the south, and it is not fully replaced from the north. Every now and again large amounts of material are dumped on the Stonehaven beaches, and across the mouth of the River Cowie. This may lead to a change in the course of the river near high-water mark, since its velocity is decreased, it is unable to carry as much suspended material as usual to the sea. A gradual rise in the bed of the river then takes place. Should a heavy flood ensue, the Cowie breaks through the shingle, and this may in turn cause severe scouring of the bed and banks of the river. A sudden lowering of the bed by as much as four feet (1.2 m) has been known to occur in a few hours following a single rainstorm. (This paragraph is based on information given to me by the Department of Health for Scotland (1952).)

Downie Point is formed of conglomerate which is cut by large joints. Strathlethan Bay owes its shape largely to erosion of softer tuffs and sandstones between resistant conglomerates to north and south. Dunnottar Castle stands on a headland of conglomerate which strikes at right angles to the coast; the beds are more or less vertical (see Plate 6). Fossil cliffs and geos and the raised beach platforms are present on the headland and also farther south. The '100-ft' beach is conspicuous in this area. Carter Craig is a small island of augite-andesite. The bays south of Dunnottar are nearly all in less resistant strata; in Tremuda Bay the high cliffs show terracing largely controlled by basalt flows. Basalts in the Old Red rocks have considerable effect on the coast scenery as far south as Catterline. At Crawton there are several interesting features. The inlet named Trollochy has nearly vertical sides. There are some closely spaced east–west faults at its western end, and there is little doubt that it owes its origin and form to differential erosion along the faults. A small Post-glacial stream flows into it on the north side. Its former course, now plugged with boulder clay, is about 70 yards (64 m) to the west and lies along a prominent joint; a blow-hole is sometimes active at high spring tides. Less than 200 yards (183 m) south of this inlet the coast turns westward, but is broken by a series of small inlets and promontories, all facing south. The promontories are mainly basaltic, the inlets are cut in the sandstones, although there are one or two exceptions. There is a storm beach, facing south, at the north-west corner of

Crawton Bay; the boulders on the beach are derived from the Old Red conglomerates which, on the west side of the bay, form a vertical wall about 100 feet high (30 m). A little farther south there are two large stacks.

South of Crawton the same type of coast extends to Inverbervie Bay, but lavas play less part in the cliffs. At Catterline and Braidon there are small bays; that at Braidon coincides with a dip fault. The minor indentations on this stretch of coast are the work of waves acting along joint planes and on the somewhat less resistant parts of the sandstones. The cliff is continuous, but its fossil nature is not always apparent since the present waves reach the cliffs and give the impression, locally, that they are wholly of modern origin. That this is not so is easily seen if the cliffs are followed and the numerous traces of fossil cliffs show that those which are now reached by the sea are but rejuvenated forms. The number of off-shore rocks and reefs as at Forley Craig and the Slainges point to the same conclusion. The cliff tops are for the most part farm land and for several miles grass fields reach the cliff edge and give a most pleasant setting to the coast. The cliffs average about 100 feet in height (30 m), and the surface rises steadily inland. There is higher ground at Craig David on the north side of Bervie Bay.

South of Inverbervie the cliffs retreat inland, and a broad platform lies in front of them. In Inverbervie Bay there is a good shingle beach, and the former line of railway between that place and Montrose ran on this platform for some distance. To the south of Gourdon the platform is broader. It is, in one sense, all part of the '25-ft' beach, and it is now being cut into by the present sea. It forms a rough and jagged foreshore. Extensive pebble ridges at its rear separate those parts still over-washed by the waves from the fields or grassy sward which extends back to the cliffs. About two miles south of Johnshaven there are several settlements named Mathers. Sir Charles Lyell noted that a former settlement of this name was built on an old shingle ridge and sheltered by a ledge of limestone. Unfortunately this was quarried for lime so thoughtlessly that in 1795 it was finally broken through by the sea and the whole of the hamlet was destroyed. Along all this part of the coast there are traces of one or more raised beaches, and in the promontory between Tangleha' and Rock Hall there are three, at approximately 100, 40 and 25 feet (30, 12 and 8 m).

At Milton Ness, immediately south of Tangleha', the coast turns westwards, and at St Cyrus the cliff line runs well inland so that between that place and the North Esk there is an expanse of dune and also some salt marsh about a quarter-of-a-mile wide. At, and for a short distance west of, Milton Ness the cliffs are fronted by a rock platform on which there are several stacks, which were formed at an earlier period. One at least of them is almost surrounded by the sea at high water. The whole line of cliffs is formed of basic lavas of Devonian age. In the northern part of the area there is a mixture of red and green andesites; farther south the lavas are largely basaltic, but in places almost covered in scree and blown sand. They are locally much faulted, and wet or dry channels often indicate lines of fault.

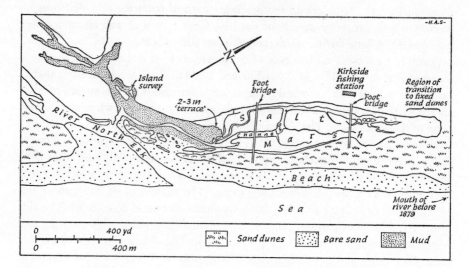

Fig. 55. St Cyrus salt marsh, 1951 (after C. H. Gimingham, 1953)

Much of the area north of the North Esk is now a National Nature Reserve and rests on the '25-ft' beach. On the seaward side there is a fine sand beach, and behind it a line of dunes which reaches a maximum height of about 50 feet (15 m). Inside the dunes is a plain which (see Fig. 55) is mainly dune pasture in the north and salt marsh in the south. The salt marsh runs as a long tongue northwards from the river and represents a former lower end of the river when it was deflected more to the north than it is today. The change in the mouth of the river took place in 1879 when a flood broke through the dunes and beach somewhere near its present mouth. South of the river there are extensive links and a wide beach which continue as far south as Montrose and the mouth of the South Esk.

The vegetation of the dunes is interesting and has been worked out in some detail (Robinson and Gimingham, 1951; Gimingham, 1953). Five succession stages are recognized: (1) the foreshore stage in which *Agropyron junceiforme*, *Atriplex glabriuscula*, *Cakile maritima* and *Salsola kali* are the most frequent species; (2) the foredunes. As on so much of our east coast *Agropyron* dunes represent the earliest stage, and soon give place to *Ammophila* dunes. Other plants at this stage include *A. glabriuscula*, *C. maritima*, *Festuca rubra*, *Honkenya peploides*, *Cirsium arvense* and *Rumex crispus*; (3) mobile and young fixed dunes. *Ammophila* is now driven out, and among a considerable number of plants *Cerastium tetrandrum*, *Senecio jacobaea*, *Cirsium arvense*, *Bryum pendulum* and *Ceratodon purpureus* are the most common; (4) on the older fixed dunes the community is closed. *Ononis repens* is dominant and *Ammophila arenaria* is still abundant; (5) dune pasture. *Carex arenaria* is now usually dominant, but other grasses often form a dense turf. The total number of species in this stage is large. However, the general nature of

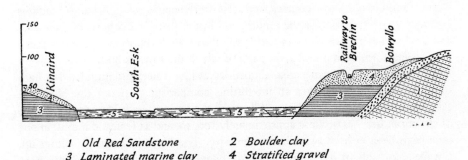

1 Old Red Sandstone   2 Boulder clay
3 Laminated marine clay   4 Stratified gravel
5 Carse clay

Fig. 56. Section of the valley of the South Esk (after J. C. Howden)

the dunes and dune pasture, and the plants associated with them are closely comparable to other similar localities on the east coast of Britain.

Partly because of the way in which the salt marsh came into being there is no very clear zonation of plants in it. The upper part shows a greater proportion of silt, the lower part is composed of bare sandy mud, on which a few species of algae flourish. At a rather higher level *Puccinellia maritima* is dominant, and with it are *Aster tripolium*, *Triglochin maritima*, and *Atriplex glabriuscula*. *Puccinellia* is dominant over most of the marsh except near the edges. There is a good deal of *Festuca rubra* in the lower part of the basin where it is associated with *Cochlearia officinalis*, *Plantago maritima*, *Aster tripolium* and *Glaux maritima*. In some of the wetter parts *Agrostis stolonifera* replaces *Festuca rubra*. Where the marsh abuts on the dunes a rather more definite zonation is noticed and is dependent largely on tidal range. *Festuca rubra* and *Armeria maritima* are found in places where there is at least one submergence a month; at somewhat lower levels *Plantago maritima* is more abundant.

The cliff vegetation at St Cyrus has been investigated more fully than in almost any other coastal locality. It will not be described here, but it may be said that the vegetational characteristics of talus slopes, rock faces, rock ledges, sandy slopes, exposed cliffs (only in the north of the area) and cliff summits are distinguished.

South of the North Esk there is no salt marsh and the coastal strip, nearly a mile wide, is low and composed of beach deposits. The backbone of this peninsula (see Fig. 55) is a shingle bar of rounded boulders, mainly of quartz-porphyry and other rocks identical with those of the conglomerate of the cliffs and of the boulder clay. Most of the boulders appear to be derived directly from the boulder clay. This ridge divides the town of Montrose into two parts, the sandy part near the sea and the clayey part on the land side. It is thought that all three rest on glacial marine clay (Fig. 56). Near the river mouth unstratified red boulder clay rests on Old Red Sandstone *in situ*. A laminated red clay makes a terrace about 40 feet (12 m) above present sea-level, and stratified sands and gravels reach to about 100 feet (30 m).

Peat is found in old water-courses, and rests on the marine clay. A bar of boulders (see above) stretches across the mouth, and inside there is a deposit of carse clay and sand which may reach to 15 feet (5 m) above sea-level. Finally, there is much blown sand forming the Links, on parts of which the town is built.

The Basin is an extensive area, almost dry at low water, and inundated at high tides. Flett (1915–24) makes an interesting comparison between the Montrose Basin and Scapa Flow in Orkney. All round the basin there is abundant evidence of elevation. Beaches occur at approximately 100, 50, and 25 feet, and it must at one time have been at least twice its present size. The basin is gradually silting up. On the south side of the basin and of the channel (the mouth of the South Esk) connecting it with the sea, the land rises sharply to 100 feet and more.

If the 40-foot contour is followed between Milton Ness and Scurdie Ness some idea of the former outline of the coast can be obtained. With the fall of sea-level acting in association with longshore processes the bay shallowed and, apart from the present basin, eventually became dry land. The waves piled up a major shingle ridge which grew southwards to make the peninsula on which Montrose now stands. This ridge may have been of the nature of a bay-bar reaching from about St Cyrus to Montrose. Unfortunately so little is known about it that we are driven to speculation. The North Esk is deflected to the north and before the change of course in 1879 the northerly deflection was considerably greater. How this fits in with the southerly growth at Montrose is uncertain. The ecology of St Cyrus has been carefully investigated; there is great need for a careful physiographical explanation of the whole complex; the St Cyrus area should not be separated from that between the mouths of the two Esks. There is a *prima facie* similarity between this stretch and that between Don and Dee at Aberdeen.

The coastal cliffs from Montrose harbour to about the middle of Lunan Bay are made of lavas with the exception of a fragment of upper Old Red Sandstone which outcrops at Boddin Point and rests on the lavas. North of Boddin Point the cliffs are but moderately high, and are by no means outstanding. The rock platform is present, and at Fishtown of Usan, where it is beautifully displayed, a dyke has weathered so as to leave a conspicuous canal running through the platform. At Skae, just north of Boddin Point a porphyrite dyke is intruded near a fault. Dyke and fault breccia form a great wall of rock rising up to about 100 feet from the beach and project across a small bay flanked by steep cliffs. The dyke is about 16 feet wide (5 m), and has been tunnelled at its base where it is some 30 feet wide (9 m). There is a corresponding, but less noticeable feature on the Lunan Bay side of Boddin Point. Lunan Bay is beautiful as well as interesting. There are fine exposures of raised beaches, and the Red Castle stands on the edge of the old cliff in the rear of the bay. In front there is a line of sand dunes which, with the beach, have deflected Lunan Water somewhat to the south.

The sea formerly ran much farther up the valley of the Lunan Water. The 50-foot contour crosses the stream just above Balmullie Mill and the 100-foot contour crosses it near Hatton Mill. Rice (1960) has studied the deposits in the valley and notes that at and east of Bandoch (rather more than a mile west of Inverkeilor) outwash gravels of glacial origin rest on beds of sand and clay which occupy all the lower part of the Lunan valley and also the hollow on its south side at Anniston. The upper surface of these gravels is about 70 feet (21 m), and they are the equivalent of the laminated brick clays associated with the '100-ft' sea on the east coast of Scotland. The cliff of the '25-ft' beach is cut in these gravels where they face the sea. South of the Red Castle the cliff runs behind the dunes and encloses a long green strath; about half-a-mile from Ethie Haven they close in to the present coast. The whole length of Lunan Bay is fronted by a fine beach; the rock bench is present in the northern part from about Braehead to Boddin Point, and in the south-east at Ethie Haven. The nature of the coast on the south side of Lunan Bay and as far as Red Head is shown in Fig. 57. The effects of faulting are apparent, and some of the rugged cliff scenery is the result of wave action in hollowing out relatively soft amygdaloidal lavas and conglomerates, and ignoring the compact lavas. Just south of Rock Skelly there is a conglomerate with large boulders of volcanic rock resting on lavas. At the shore the waves cut away the softer conglomerates, and are exposing hard stacks of lava which were buried under them. A little farther south a good deal of faulting may be seen in the cliffs, especially between Rumness and Maw Skelly where a band of conglomerate about 15 feet thick (5 m) is, as Hickling (1908, 1912) says, 'tossed about in quite a remarkable fashion'. On the other hand the faults have little or no effect on the cliff form. As far as Auchmithie and Castlesea Bay the massive conglomerates and sandstones make fine, but broken, cliffs. Lud Castle stands on a small promontory, on the south side of which there is a valley in Arbroath Sandstone filled by Upper Old Red. Some 150 yards (137 m) inland the Gaylet Pot is a fine blow-hole, reached by a cave one hundred yards (91 m) or more in length. Carlingheugh Bay, bounded to north and south by faults, corresponds with the outcrops of the relatively soft Upper Old Red Sandstone which here consists largely of irregular alternations of conglomerates and soft false-bedded sandstones. In the bay there is a fine display of horizontal ledges, called The Floors. They form a broad and interrupted platform by no means all at one level, some 300 yards wide (274 m), and correspond to the lower platform that is so commonly developed on the east coast. A small promontory on the north of the bay is cut into by the Dark Cave which follows a line of fault.

Between Carlingheugh Bay and Arbroath the cliffs are easily accessible, and present some remarkable erosion forms. They are cut in the Upper Old Red Sandstone, a large slice of which is faulted down to form the coastline. Throughout this line of cliffs the influence of jointing on marine erosion is beautifully displayed. The Deil's Heid is a conspicuous stack, but dates back to a time when sea-level was higher relative to the land. The joint along which it is cut off continues southwards through a spur in the cliff where a similar stack is being formed, The Three

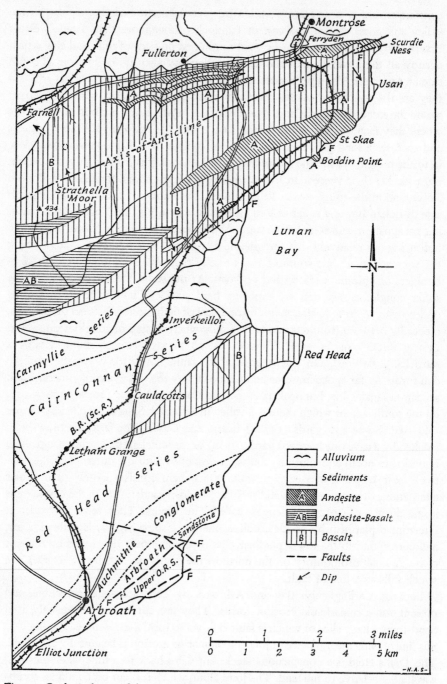

Fig. 57. Geological map of the coast of eastern Forfarshire (after D. A. Robson); sedimentary boundaries from a map prepared by Prof. G. Hickling

Storey House. Dickmont's Den is a small geo cut along a line of jointing. There is a blow-hole near its entrance. There is another blow-hole, the Mermaid's Kirk, connected with a passage that runs parallel to the general line of the cliffs. The junction, unconformable, between Upper and Lower Old Red rocks can be seen at Whiting Ness. All the way from here to Carlingheugh Bay the cliffs are cut by numerous gullies, some of which perhaps began as caves, but all, as noted above, are associated with lines of weakness.

THE FIRTH OF TAY AND THE COAST OF FIFESHIRE

At Arbroath the character of the coastline changes. The former cliff, now behind a wide platform, is low and inconspicuous. At Arbroath itself the '25-ft' beach is well developed. The rock bench fringes the shore at Carnoustie and for three or four miles towards Arbroath. It is similar to the coast south of Inverbervie; the broad rock platform is backed by a shingle ridge which gives place to low dunes and a grass-covered belt reaching back to the old cliff. The natural appearance, however, is marred by the railway line which runs close to the sea. For a mile or two south-west of Arbroath the rock platform is absent, and at East Haven where the platform is notably broad the small harbour corresponds with a gap in it.

Buddon Ness is a large triangular raised beach foreland. There is a good deal of sand on it and a little shingle on parts of the beach. There are some lines of dunes, but since the Ness is used by the Ministry of Defence it has suffered a certain amount of alteration. Near the lighthouses the dunes are relatively untouched. The Ness is flat and presents interesting problems, the chief of which is its stability. There are several maps of the Ness covering the last 200 years, and they show extremely little change and suggest that the present feature is based on some more stable foundation which may well be an ancient beach or rock platform. It may possibly be compared with Chanonry Point at the mouth of Inverness Firth. The Ness is surrounded by a sand beach and there is no surface indication, along the beach, of underlying rock. Its triangular shape is nevertheless somewhat puzzling if we assume a solid foundation. If it is, or was, primarily an accumulation form, it is interesting to speculate why it grew in its present position. A full investigation would be rewarding.

At Barry there are deposits of marine brick-clay which are probably to be associated with the '100-ft' sea. West of Elliott Junction (about two miles south-west of Arbroath) and for some distance towards Carnoustie there are beaches at about 70 and 50 feet (21 and 15 m), but the most continuous is the lowest ('25-ft') beach which fringes the coast as far as Dundee (Rice, 1962). The coastline from Monifieth westwards is built up, and in Dundee Law we meet the first of the volcanic hills which are more characteristic south of the Tay.

Along the foreshore at Dundee there is a wide stretch of reclaimed land. Above Dundee the lowland extends into the Carse of Gowrie and also up the Earn valley.

The Carse of Gowrie represents a former wide open bay on the north side of the Tay. At Errol there is, or rather was, a well-known pit section showing 8 to 10 feet (2.5 to 3 m) of yellow-brown clay and silt, the true carse clay, resting with a sharp boundary on 5 to 7 feet (1.5–2 m) of fine blue clay, below which is a red clay 4 feet (1.2 m) or so thick. The two lower deposits are characterized by an Arctic fauna and correspond with the '100-ft' beach (Davidson, 1932). The south side of the Firth is steep and formed of the Old Red Sandstone and associated lavas. Near Wormit the andesitic lavas are cut by a buried glacial valley. The lavas run on to Tayport (Sissons et al., 1965) (see Plate 5).

Between the Tay and the Eden there is a wide expanse of blown sand known as Tents Muir. The Muir is almost entirely made of sand; the proportion of shingle and shell is extremely low. Today much of it is afforested, and to obtain a better picture of the area the One Inch Ordnance Survey Map of 1854 is worthy of study. It shows, by means of contours, that the dunes are aligned parallel to the coast in a regular succession of ridges. There are also less regular features roughly parallel to the Tay and Eden. Since 1854 there has been considerable accretion, and doubtless this was also the case before that date, but earlier maps are not helpful in this respect since they are far too inaccurate. This is true even of Airlie's map of 1775. Evidence given to the Royal Commission on Coast Erosion (Final Report, 1911) showed that between Leven and Tay 534 acres (216 ha) had been gained and 23 acres (9.3 ha) lost between 1853–5 and 1893–5. Since these figures were given by the Ordnance Survey they may be regarded as accurate. The gain was on the seaward side, and particularly on the northern part of the seaward front. This process has continued up to the present time, and recent growth is usually estimated with reference to the line of concrete blocks erected as defences in 1940. It is assumed that these were placed somewhat above the high-water mark of ordinary tides of that time. Some of them are now as much as two-thirds of a mile within high-water mark, and some are completely sand-covered. The rate of accretion has, of course, been inconstant, and in recent years has undoubtedly been affected by the growth of the forest.

There has also been some accretion at the south end of Tents Muir. There is a small spit about a mile east of Shelly Point which has grown in a south-westerly direction. Its growth has been rapid in recent years. On the south side of the Eden there has also been accretion. The West Sands at St Andrews, alongside Pilmour Links, have extended both seawards and, to a lesser extent, northwards. On the other hand, there has been a little erosion on the west side of the Links. These changes appear to be intimately connected with the changes in the mouth of the Eden. Until about 1930 the mouth was being pushed to the north, and in so doing it caused some local erosion at the south end of Tents Muir. It has been estimated (Grove, 1953) that some 75 acres (30 ha) of low dunes were lost. About 1940 the river changed its course, and broke through the northward-growing spit, and certainly by 1944 air photographs show the mouth about 1,000 yards (914 m) north of Out Head. The current edition of the One Inch Map (1952–4, published 1957) shows that the mouth has moved once again to the north.

Tents Muir cannot logically be separated from the various offshore banks on both sides of the Tay estuary. Abertay Sands now join Tents Muir sands, but Admiralty Charts up to 1947 show a narrow channel between them. Fig. 58 shows that the eastward extension of Abertay Sands is indicated by the three-fathom (5.5 m) contour, and also that this contour has itself changed somewhat. On the north side of the Tay the submarine contours east of Buddon Ness indicate a similar change, so that the entrance to the Tay (1950) is now almost due east and west.

How has all this sand come to this part of the coast? This is not an easy question, and the various suggestions put forward hardly afford a complete answer. Grove believes that there are three possible main sources: (1) that carried down by the Tay and Eden; (2) the off-shore sediments; and (3) that derived from local coastal erosion. There can be little doubt that in the past the rivers brought down large quantities of sediment. Allen (1947) states that nowadays the weight of material in *suspension* in the upper part of the Tay estuary is 1:5,000 parts of water; in the lower part of the estuary the figure is 1:20,000. These figures do not include the bed-load, which averages slightly more in the estuary than on the sands of Abertay. Just how much the Tay carries down now, or has carried in the past, is uncertain. In the past, during and after the melting of the ice, the amount transported by the river must have been much greater, but now a great deal remains in lochs along the course of the river, although the disappearance, for one cause or another, of the inland cover of vegetation should have increased the supply in the last two or three centuries (Grove, 1953).

The second possible source is in the abundant off-shore sediments, but Grove thinks that this is a less likely source now than in the past. Today there probably exists a state of equilibrium between the deposits and sea-level, and he suggests that if material comes from off-shore, it probably does so from a more distant source. It is a matter which is difficult to prove. Thirdly, some of the material may be produced by coastal erosion. On Tents Muir itself there has been but local erosion in the south-eastern part, and since the erosion has been not only local but recent, it can scarcely be reckoned as a source of supply of any significance. More material may have come from St Andrews. It is known (see p. 260) that there has been a good deal of erosion there in the past, and some still goes on in the boulder-clay cliffs.

These sources apply only at the present day. Most of Tents Muir rests on the '25-ft' platform. Presumably it was covered with marine sediments, and some of the present sand may have originated in this way. Is it necessary to assume all the sand in the dunes is associated with present conditions? However, since the area is so rapidly increasing not only the flats, but also the dunes, are quite modern. Dunes which presumably existed where the Forestry Plantations now are, may well have been built in part of glacially-derived sands. Any dunes on the seaward

9

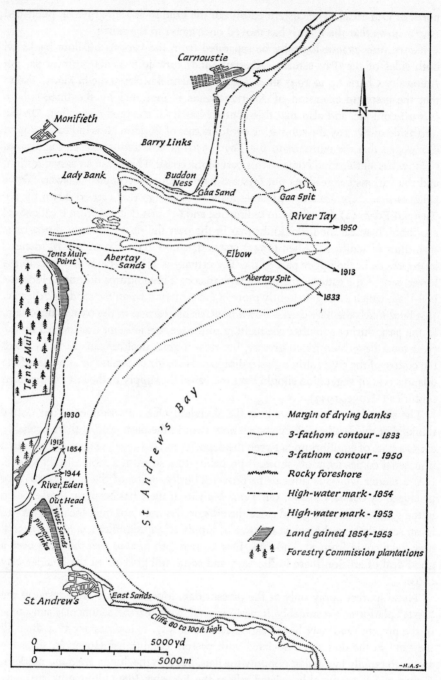

Fig. 58. Tay mouth and St Andrew's Bay

side of the plantations have formed since about 1905 or a little later. It is undoubtedly wave action that is the main agent in the transport of sand to the beach
today. The tidal currents may help to a limited extent, but even at springs they
seldom exceed a rate of half-a-knot. The general direction of movement of the
currents is northerly, and so they may to a small extent aid the work of the waves.
The prevailing winds at Tents Muir are west and south-west except perhaps in the
spring months. Winds exceeding 30 m.p.h. (48 km) are nearly always from the
quarter between south and west. The coast runs almost north and south, and it
must be presumed that it is onshore winds which set up beach-drifting. At Leuchars
east and south-east winds are more frequent than those from north and north-east,
and the growth of the beach must largely be ascribed to these winds, and also
possibly a greater tendency for northerly directed beach-drifting. With the growth
of the forest the effects of westerly winds on the open part in front of the forest
must become increasingly less significant. The vegetation of Tents Muir is not
dissimilar from that of other comparable coastal areas on the eastern side of
Scotland; there is, however, no precise record of what the tree-covered area was
like before afforestation took place. The growth of the natural vegetation covers
roughly a period of 60 years. The formation of the foredune alongside the Ice
House took place between 1905 and 1915. A series of dunes and slacks has
developed since then and a well-marked succession of vegetation can be traced.
There have, however, been certain interferences with the natural growth; sea
buckthorn, for example, was planted by the Forestry Commission in certain places
in 1925, and has spread considerably since then. In the southern part of the area
there are now four main dune ridges with intervening slacks.

On the seaward side there is first a normal development of small dunes which
often begin to accumulate around masses of seaweed. *Agropyron junceiforme* is the
first species to appear in the south, but near Tents Muir Point *Elymus arenarius*
is usually the pioneer, and grows luxuriantly. Sands well out from the main dune
complex are also being colonized. *Ammophila* follows, especially in the south, and
is dominant in all the unstable dunes. On the tide line *Cakile maritima, Honkenya
peploides* and *Salsola kali* are common. In the north-east corner of the area, flood
tides sometimes inundate hollows behind the coastal dunes where the first,
embryo, slacks occur. They are usually colonized by *Glaux maritima* and *Festuca
rubra*. However, there are nearly always some slacks completely protected from
sea-water by the growth of the dunes. The bare sand is then, or even before full
protection, carpeted with *Armeria maritima, Juncus balticus, Aster tripolium*, and
other species.

Behind the slacks is another range of dunes which has been described as semi-
stable. The dunes may reach about 15 feet in height (5 m), and it is these that surround the line of concrete blocks put up in the war. They show the usual features,
including blow-outs. The commoner plants include *Festuca rubra, Agrostis tenuis,
Sedum acre, Galium verum, Tortula ruraliformis* and *Peltigera canina*. On the more

stable dunes *Ammophila* gradually gives way to other plants, mosses, and lichens so that a more or less closed community is formed. Behind these dunes is a line of slacks which are drier than those mentioned above. There may be a little local flooding associated with the Powie burn, and the water table is close to the surface. *Salix repens* with *Juncus balticus*, *J. Gerardii*, and *Triglochin maritima* are common, and the less common species include *Corallorhiza trifida*, *Centaurium pulchellum* and *Teesdalia nudicaulis*.

Behind this line of slacks is a line of fixed dunes, more marked in the southern part of the reserve. The vegetation is dominated by mosses and the soil is more acid than in the frontal dunes. The associated slack, especially south of the Ice House, is well defined and contains an abundance of birch scrub. The soils are increasingly acid, a fact which shows itself in the widespread *Calluna* and *Empetrum* heath, together with various mosses. The grey dunes behind make a conspicuous ridge all along this area. The colour is produced by the species of *Cladonia* which dominates. In 1963 there was one Juniper bush. The dunes are separated from the Forestry Commission plantation by an area of dune heath and marshy slacks. The drier parts carry *Calluna vulgaris*, *Empetrum nigrum*, *Erica tetralix* and *E. cinerea*. The wetter areas are dominated by *Filipendula ulmaria*, and *Salix aurita* is spreading. Pines, spreading by seeds, are growing rapidly in parts of the Reserve.

South of the Eden estuary the coast of Fife, apart from certain igneous outcrops and a large number of vents of former minor volcanoes, is formed wholly of rocks of Carboniferous age. The rocks belong mainly to the Calciferous Sandstone Series. All around the coast there is usually a broad platform which, in a general way, may represent the '25-ft' level, but it undoubtedly has a longer and more complicated history. Glacial deposits locally form short lines of cliff, and much of the scenery at and immediately inland from the coast is of glacial origin (see also p. 208). St Andrews is built on platforms associated with the lower raised beaches, and the position of its castle and cathedral relative to the present coast allow some estimate of erosion to be made. It is regarded as unlikely that any considerable space would have been left between the beach and castle when it was built at the beginning of the thirteenth century. It is, however, argued that there was at that time a strip of the '25-ft' platform on the outer side of the castle, and of sufficient width so that there might be 'a religious house and room ... for the playing of bowls'. The rocky foreshore is notably wide to the north of the town, and is backed by a low cliff. The castle is now protected. On the south side of the town, at the East Sands, a good deal of St Nicholas' farm has been destroyed by the sea, and walls of concrete now front rocky cliffs near the harbour. The Royal Commission on Coast Erosion (1911) reported that

The erosion [on the East Sands] has been going on for a long time, and within the last 10 years the coastline appears to have receded about 20 or 30 feet [6–9 m]. The erosion principally affects the property belonging to the University of St. Andrews, and it appears to have been accelerated by the removal of materials under the authority of the Corporation – who claim the foreshore – from the beach. The representative of the Corporation who gave evidence admitted that the removal greatly assisted the erosion.

A peregrination of the coast of Fife will seldom reveal spectacular scenery, but nevertheless will afford a great deal of interest in detail. Sandy beaches are few and rather scattered; most of the foreshore is a rocky platform, which contains many interesting structures. The rocks are frequently folded and faulted, and cut by a number of igneous intrusions. The platform thus resembles a full scale geological map. Even so it is by no means easy to follow the structures since they are locally overgrown with algae, obscured by boulders and occasionally by sand, and the structures themselves are often complicated. Various parts of the coast have been described in considerable detail. Just west of Kinkell Ness, about $1\frac{1}{2}$ miles east of St Andrews, Balsillie (1915–24) has identified five necks on the foreshore (see Plate 7). Some are small and more or less circular; others are elongate. Fig. 59 shows their general nature, and their relation to the sandstones which are much folded and also faulted in a complicated fashion. On the east side of Kinkell Ness there is a classic section. The most striking feature is the vent called The Rock and Spindle. This stands near to the low cliffs. Its greatest breadth is about 280 yards (256 m), and a projection runs out from the main mass. The walls of the vent are much disrupted. The material in the vent is for the most part a greenish granular tuff mixed with blocks of basalt. 'The whole aspect and structure of the rock vividly suggests to the observer the intermittent energy of one active volcanic crater, from which fine dust and lapilli were ejected, while occasional more vigorous explosions threw larger blocks of basalt over the inner and outer slopes of the cone' (Geikie, 1902, Memoir). The Rock represents an intrusion of molten material in the chimney of the cone; the Spindle is a mass of basalt, probably pushed into a cavity, so that its columns, as a result of cooling, now converge to the centre like the spokes of a wheel. The word rock is used in the Scottish sense of distaff. In the next half-mile or so there are several more vents, and the strata about 1,700 feet (518 m) to the east are bent into a perfect dome which, being cut through at the same level as the rest of the platform, is displayed to perfection. The vents are, in fact, stacks, often much worn down, on the raised beach platform.

To describe the whole coast in detail is unnecessary; it will suffice to call attention to certain particular features. At Buddo Rock there are a number of caves, all associated with the '25-ft' level. At Craig Hartle the skerry is submerged at high water, and is comparable to the sandstone reefs which form Cambo Brigs on the east side of Cambo Sands. Just north of Fife Ness there are several dangerous reefs. They are all formed of sandstone and are known as Tullybothy Craigs (= Balcomie Brigs) and, farther seaward, the Carr Brigs. Between Fife Ness and Crail the rock platform is somewhat narrower than that north and west of the Ness. At Crail Point the ruins of the priory stand at the edge of the beach, and it is clear that since it was built a considerable amount of erosion has occurred at this place. This degree of erosion is also apparent farther west. At Pittenweem and Anstruther

262

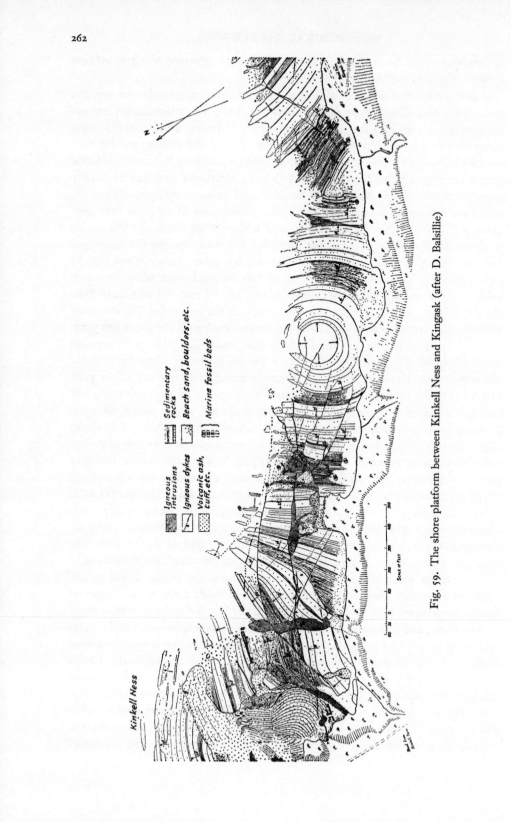

Kinkell Ness

Igneous
intrusions

Igneous dykes

Volcanic ash,
tuff, etc.

Sedimentary
rocks

Beach sand, boulders, etc.

Marine fossil beds

SCALE OF FEET

Fig. 59. The shore platform between Kinkell Ness and Kingask (after D. Balsillie)

Geikie suggested (1902) that much of the lowest beach has been cut away in places in consequence of which the cliffs are higher. This, if taken at its face value, also indicates the long evolution of the rock platform. At Billow Ness (Anstruther Wester) the strata are more easily eroded since they consist of shelly and thin sandstones. The small bay at Marsfield is cut in boulder clay. There are several isolated rocks or islets between Fife Ness and Pittenweem – the Claverance Rocks at Crail, the East and West Wolf of the Cuttyskelly off Cellardyke, the Breakboats at Pittenweem. They are all made of sandstone. Pittenweem Harbour and the gap in which the town is situated are associated with, and more than likely controlled by, a fault which runs nearly north and south across the beach. Another interesting feature of this coast, especially at Anstruther and Cellardyke, is the abundance of erratic blocks on the foreshore. They doubtless have a restraining effect on the erosion. Some may be as much as 10 or 12 feet long (3 or 3.5 m). At St Monance, on the west of which town there has been considerable loss by erosion, igneous dykes are apt to form chasms and not crags. This is by no means an unusual effect and may possibly be associated with the jointing of the igneous rock.

The foreshore between St Monance and Elie has been described by Cumming and others (1931–8). Here again there are several necks which cut through sandstones and limestones. But a feature of major importance is the Ardross fault which runs parallel with the coast for about two miles. It largely determines the structure of the platform (see Fig. 60). On the shore at Newark, where the remains of the castle indicate considerable erosion, there is a neck made classic by Geikie's description. It is mainly an agglomerate of the local strata embedded in green tuff. There are also some larger blocks and veins of basalt in the neck. The relation of the other necks to the fault is clear from the figure. Elie Ness is a mass of agglomerate and in the harbour the small island, now artificially tied to the mainland, is similar. The inlet between this islet and Elie Ness probably owes its origin to the line of the Ardross Fault which runs into it. The headland at Earlsferry is basaltic, and the West and East Vows are detached parts of it. The next headland to the west, Kincraig Point, is the most conspicuous as well as the most interesting and picturesque on the Fifeshire coast. It is a nearly vertical precipice, about 200 feet high (61 m) and extends a half-a-mile or more along the coast. Masses of agglomerate and basalt have been isolated, and there are several caves and chasms. The view from Ruddon's Point, also agglomeratic, is equally striking, since it shows three marked raised beach terraces on the western side of Kincraig. The coast to the west falls back into Largo Bay with its fine sands to east and west of the central area, where the rocky platform is conspicuous. The setting of Largo Bay is the more attractive because of the ancient volcano of Largo Law rising up a mile or two inland. At Lundin Links there are numerous erratic blocks, mainly doleritic. Most of them are probably too large to be moved under present conditions, and they are, in a sense, a defence to the coast. It has been suggested that their arrangement indicates that they may have originated as a moraine-like

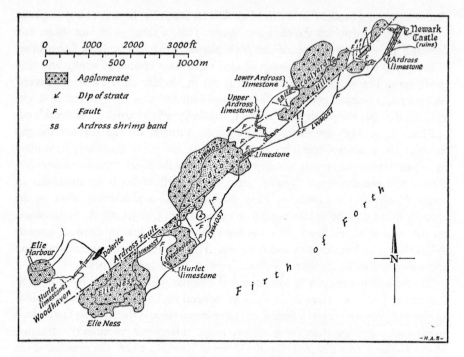

Fig. 60. The volcanic rocks along the Ardross Fault. Elie Ness and Ardross, Wadeslea and Coalyard Hill, are regarded as forming two necks which have been dislocated and displaced by the faulting (after G. A. Cumming)

feature on ice, and that they were stranded where they are now found. A little to the east of Largo large numbers of erratics can be seen, many of which stand on low pedestals formed of tuff. The height of the pedestals represents the amount of down-cutting of the platform since the blocks were dumped in their present positions. The blocks are usually far more resistant than the rock on which they rest. It is relevant, perhaps, to remark at this point that all along this coast cliffs cut in tuff are much more uniform in structure and resistance than are those in sandstones and shales. On the other hand, the tuff matrix yields more readily to erosion than the dykes and sills of basalt which cut it, and they consequently project as knobs and crags upon the beach platform, or as buttresses in the cliffs.

At Leven, Methil, Buckhaven, East Wemyss and Coaltown of Wemyss the coast is either built up, or marred by mining. The barren and productive Coal Measures both reach the coast hereabouts and a rock bench is usually present. There is also a good deal of shingle resting on the bench or forming a beach. The small interval between West Wemyss and Blair Point is relatively unspoiled. There is a striking contrast between the industrialized coast and its harbours and the picturesque fishing towns and harbours east of Elie. Kirkcaldy occupies about three miles of foreshore. From that town to Kinghorn Carboniferous sediments,

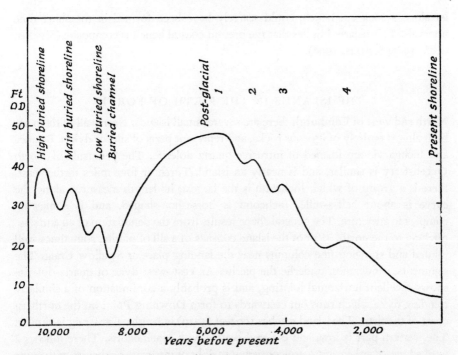

Fig. 61. Sea-level changes relative to the land during the last 10,000 years at a point immediately east of the Menteith Moraine

lavas and sills reach the coast; their strike is almost north–south. Long Craig, the West and East Vows, Craigfoot and the Chest are small islands, all parts of sills. Along the shore there are low cliffs, and near Seafield Tower limestones and sandstones form ridges traversing the shore platform. To the north of Kinghorn the lavas and sills run southwards out to sea and form minor projections and at Kinghorn there is a large mass of basalt which continues to within a mile of Burntisland where there is a wide beach to the east of the town. The Black Rocks in the sands are basaltic and run roughly parallel to the intrusives in the Calciferous Series at Burntisland. To the west in Silversands (Whitesands) Bay the coast is fairly steep; the Common Rocks are teschenitic.

In the Forth tidal water in high raised beach times reached well beyond Stirling. The details of the coast above the bridges will not be considered in this book. Ancient beaches, some raised and some buried, have been intensively studied by Sissons and his colleagues. The relation of the beaches to the fluctuations of the ice have been carefully investigated, but how the beaches relate to the rock platform on the more open coast is uncertain. Fig. 61 shows Sissons's views on the fluctuations of sea-level in south-east Scotland during the last 10,000 years. It must be remembered that many fluctuations, often of much greater range, preceded these. Each was associated with a distinct shoreline. In resistant rocks even small notches, let alone a rock cut bench, could scarcely have been formed in the short time intervals. This clearly implies that the present coastal bench is composite (Sissons, *et al.*, 1965; Sissons, 1966).

THE ISLANDS IN THE FIRTH OF FORTH

North and west of Edinburgh there are several small islands. Nearly all of them are built almost entirely of igneous rocks, and represent parts of sills or dykes. Beamer and Inchgarvie are formed of intrusive quartz dolerite. The peninsula of North Queensferry is similar, and is nearly an island. Three or four miles farther east, there is a group of which Inchcolm is the largest; its length measured along the curve is about half-a-mile. Inchcolm is horseshoe-shaped, and is somewhat complex in structure. The general shape results from the denudation of an anticline pitching to the north. Most of the island consists of a sill of picrite, sometimes well jointed and showing fine columns near the landing place at Swallow Craig. The sediments, sandstones, underlie the picrite. An east–west dyke of quartz-dolerite shows excellent horizontal jointing, and is probably a continuation of a similarly trending dyke which runs out eastwards to form Downing Point on the northern coast of the firth. The island reaches 100 feet (30 m) in height. Car Craig is similar. The western part is low, and covered largely with beach debris. There follows a marked westward-facing scarp consisting mainly of intrusive teschenite with some thin interbedded sediments. There are a few minor scarps to the east of the main one, but the total area of land above high water is small, a few hundred square feet. Some interesting investigations of movement of bed material in the Firth of Forth between Rosyth dockyard and Oxcars deep, near Inchcolm island, were made by Smith and Parsons (1968). The experiment was made with scandium-46 for long-term measurements and with lanthanum-140 for short-term measurements. It was found in 1961 that material from Oxcars deep moved upstream to the area already dredged at Rosyth. In 1964 these findings were confirmed by a fresh trial. In 1965 and 1966 other investigations were made in a channel 8–10 fathoms deep (14.5–18 m) about half-a-mile south of Oxcars Light. The 1965 experiment was exploratory and short term, but was sufficient to support a down-stream movement of material. The fuller investigation in 1966 showed that movement from this site is predominantly down-stream and suggested that 'the ebb channel ... provides an alternative deposition site which is as accessible as Oxcars deep ... The long term effects of spoil deposition will have to be watched in case the regime in the area is altered.'

Inchmickery, the Cow and Calves, and Oxcars are all intrusives; so also are Long Craig and Haystack. Most of these are rocks awash at high water. Cramond is larger, but is of the same type. The Buchans are intrusives and awash at high water. Petrologically there are slight differences between these various islands, but they are irrelevant from the present point of view. It is noteworthy that several head-

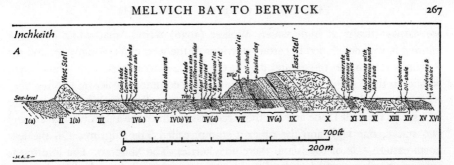

Fig. 62. Section across north shore of Inchkeith. i(a), Cawcans sediments; ii, Cawcans in-trusive band; i(b), Cawcans sediments; iii, Cawcans lavas; iv(a), Kinghorn sediments (oldest beds); v, Kinghorn lavas; iv(b), Kinghorn sediments (beds above v); vi, Kinghorn intrusive band; iv(c), Kinghorn sediments (beds above vi); iv(d), Kinghorn sediments (central beds); vii, Main intrusive band; iv(e), Kinghorn sediments (youngest beds); [viii not shown on this section] ix, Upper intrusive band; x, Lighthouse lavas; xi, Kirkaldy sediments; xii, Kirkaldy lavas; xiii, Skerries lavas; xiv, Skerries sediments; xv, Skerries intrusive band; xvi, Outermost eastern basalt [small letters in line above braces indicate sub-divisions] (after L. M. Davies)

lands on the north shore are similar: Hawkcraig Point, Braefoot Point, and the less marked headland on the west of Dalgety Bay are all formed of olivine dolerites. The bay is in the Calciferous rocks. Craigdimas, a low-lying skerry, differs only in that the rock of which it is formed is extrusive rather than intrusive.

Somewhat farther east is the larger island of Inchkeith (Davies, 1936). It is nearly a mile long in a north-north-west, south-south-east direction and is formed almost entirely of intrusive and extrusive rocks, the outcrops of which are parallel with the long axis of the island. It reaches to the 100-foot contour, and traces of the '100-ft' beach have been recognized on it. The rocks dip to the east-north-east, and as the section (Fig. 62) shows, present a scarp facing westwards. The long coasts on either side are relatively simple, but where the rocks run out at right angles to the sea, the coast is indented, especially in the north. The East and West Stells are the main headlands.

The small islands, including the Bass Rock, are described later in association with the numerous vents on the southern shore of the firth.

The Isle of May lies in the east part of the Firth of Forth about eight miles (13 km) south-east of Anstruther Wester, and administratively the island is a part of that parish (see Plate 2). The island is about one-and-a-third miles long (2,140 m) in a north-west and south-east direction, and about a quarter-of-a-mile wide (402 m). It consists of one sill of dolerite which dips at a low angle to the north-east. This means that the western side of the island, which reaches 150 feet (46 m), is steep and the eastern side passes gradually below sea-level. Several faults cross the island with a roughly east and west trend. Although they do not appear to have caused much displacement, they have guided erosion since they have caused a good deal of fracture so that the waves have cut out gullies or geos, the most noteworthy of which is a little south of the lighthouse. The northern part of the island is low and divided into separate parts by gullies which are narrow straits at high water. Walker (1936) writes 'that while columnar jointing is well developed in certain portions of the west cliffs, notably around the Pilgrim's Haven, yet the greater part of the façade is so seamed with joint-planes, parallel to the cooling surface of the sill, that the columnar structure is often very effectively masked'. Rock stacks and arches are usually associated with master joints, e.g. the Mill Door and another arch just north of the Pilgrim's Haven, and two stacks, one high and the other a stump, called The Pilgrims. On the low north-eastern side of the island there are skerries, The Middens. The jointing is here more or less parallel to the dip. There are raised beaches on the island, corresponding to the '50-ft' level of the mainland, and some traces of the '25-ft' and '100-ft' levels.

The Bass Rock, perhaps the best-known landmark in the Firth of Forth, is a plug of phonolitic trachyte and is comparable to North Berwick Law and Traprain Law on the mainland (see Plate 4). It reaches a height of 420 feet (128 m); its sides are precipitous except on the south-east where there are three great steps or terraces separated by steep rises. The lowest terrace sinks towards the two landing places. The rock rises from water 60 feet deep (18 m), and the steep sides, although not in any sense columnar, show a vertical marking. The rock is penetrated by a tunnel, 100 feet high (30 m) at its eastern end, but soon dropping to about 30 feet (9 m). There is a small beach of gravel near its western end. The whiteness of the rock is misleading. The true colour is dusky to black. The friable white crust is hard and solid and adheres firmly to the rock. It is produced by the excrement of birds; the natural colour of the rock is seen near sea-level where the waves continually wash the rock.

Sheep were grazed on the rock, the upper surface of which is grass-covered. Balfour recognized eleven species. A considerable number of flowering plants grow on the rock. *Beta maritima* and *Lavatera arborea* (tree mallow) attract most attention. There was once a small cultivated garden on the rock; although a few traces of other plants remain, the abundance of *Urtica dioica* effectively marks the position of the garden. The pink campion and the common sorrel are found especially where the dung of sea birds is mixed with the soil. The number of phanerogams on the Bass is less than half of those on Ailsa Craig. This may be partly owing to different climatic conditions, the very steep sides and absence of any shore except in the south-east, and also because of the small size of the rock. Great numbers of birds breed on the island, including the Solan Goose (Gannet). The island has a long history, and St Baldred seems to have been its first inhabitant (died 606). For many years a fortress existed on the rock; it was also used as a prison. (This brief account is based on *The Bass Rock* (Edinburgh, 1848), edited by Thomas M'Crie, which contains a very discursive chapter on the geology by Hugh Miller; an account of the botany by Prof. J. H. Balfour, and of the zoology by Prof. J. Fleming: the history is by M'Crie, and a long account of the Martyrs of the Bass is written by the Revd. James Anderson.)

EDINBURGH TO BERWICK

The Firth of Forth is situated in comparatively low country. As has been seen the Fife coast, *sensu stricto*, is low and the shoreline itself somewhat dull from the scenic point of view, but the country behind is pleasant, and ancient volcanoes such as Largo Law and Kellie Law give it a character of its own. This is even more noticeable on the southern shore. There the hills near the coast are almost entirely formed of igneous rocks. Some are built of lava flows, others are plugs of ancient volcanoes. The Pentlands, about 12 miles (19 km) inland from Edinburgh, have a core of Silurian sediments, which sink below conglomerates of Old Red Sandstone age which, in turn, are swathed by lavas. The Castle Rock at Edinburgh is a column of basalt. Calton Hill is a detached part of Arthur's Seat, itself the remnant of three volcanoes. The Braid Hills consist of Old Red Sandstone lavas, and farther east are North Berwick and Traprain Laws, which stand up abruptly from the coastal plain; they are built of phonolithic trachyte. The Garleton Hills, just north of Haddington, are lavas and show terraced features, largely as a result of glacial action working on more or less horizontal flows. The evidences of former volcanic activity give a striking background to the coasts of the firth as seen from the sea. They are, however, only the more notable features. The detailed study of the south shore reveals the enormous importance of igneous rocks in the minor structures of the coast. For the most part the south shore of the firth is low. It is a little difficult to give an eastern limit to the firth, but the change in direction of the coast at Tantallon Castle is a convenient point. At that place the cliffs are higher and more prominent than any farther west. The upper limit for coastal purposes is equally arbitrary, and the railway bridge will be used as a definite even if artificial line. On the other hand, it corresponds in part, for practical reasons, with a dolerite sill that crosses the Forth from Hound Point, through the island of Inch Garvie to the Ferry Hills in Fife. Hound Point itself is rather more than a mile downstream from the bridge, but projects largely because it is more resistant than the Calciferous Sandstone Series into which it is intruded. Cramond island is also part of a sill. East from the village of Cramond the coast is built up, protected, and largely artificial for some miles through Edinburgh, Leith, Musselburgh, Prestonpans, Cockenzie and Port Seton. All the way from Dalmeny to Prestonpans the '100-ft' beach is well developed. In Edinburgh it extends to the Hospital, Botanic Gardens and Abbeyhill, and to Whitecraigs in Musselburgh. The junction between it and the lower beaches, particularly the '25-ft' beach, is marked by a fairly distinct feature, usually close to the present sea, except at Leith Links and Inveresk. Interesting details of minor changes of coast, as a result of erosion, near Leith and Musselburgh are given by Stevenson (1811–16). He also refers to Morison's Haven (see below), and to erosion on the north coast of the Firth at Elie, Anstruther and Crail. Apart from the hills of igneous rock just inland (see above) the coast is low lying and formed in the same Calciferous Series. Reclamations have been made, and harbour works at Leith and elsewhere have made some use of the off-lying craigs and rocks of sandstone. Locally there are spreads of pebbles and small boulders as at the mouth of the Esk at Musselburgh. Between that place and Prestonpans there was formerly a small haven, called Morison's or Acheson's Haven. It was a safe refuge in the eighteenth century, and was not finally disused until after the First World War. Then the area was reclaimed. As a harbour it dates back at least to 1526. Graham (1961–2) notes that the only surviving plans of the harbour are those on the Ordnance Survey Maps, that in 1913–14 showing it in its final stage. There was also at one time a small village there, and is marked on Adair's map of 1682. It has now disappeared.

From Musselburgh to Cockenzie and Port Seton the coast is largely built-up. Preston Links foreshore has been extended between 80 and 100 yards (73 and 91 m) since 1892 largely as a result of dumping. The generating station is partly on this reclaimed land, and some open space has been provided to the west of the station. At Seton Shore there has been some accretion as a result of planting lyme grass on the sand strip between the road and high-water mark. Towards Long-niddry the coastal strip widens and the dunes may be as much as 250 yards wide (229 m) between the road and the beach. These dunes suffered serious erosion in the 1953 storm, mainly as a result of wind, not sea, action. Since then they have been thatched and planted. The ridge of rock, the Long Craigs, although broken by a gap about 500 yards wide (457 m), is a protection to the sands since it forms two long arms at either end of the bay. Ferny Ness is surrounded by a rock platform, but is fully open to westerly winds. In the 1953 storm the road in Gosford Bay was almost undercut. It is worth noting that this erosion was aided by storm-water gullies draining from the road and also by the scour around the defence blocks which were erected in the war. The shelter belt of trees was planted in the late eighteenth century. Between Gosford Bay and Aberlady Bay is the rounded headland of Craigielaw Point. On the western side of the headland the rock platform encloses a small bay and carries a small islet, Gremcraig island. The platform consists of limestones and shales of Calciferous Sandstone age and is continuous to Aberlady Point. In the platform an anticline and a syncline trending north-west and south-east, three small faults and several minor reversals of dip can be traced. The sea has etched out the shales so that scarp and dip features are developed.

Aberlady Bay is a Local Nature Reserve and has been carefully described by Usher (1967). The reserve includes Craigielaw Point and extends to Hummell Rocks in Gullane Bay. Aberlady Bay (= Gullane Sands) shows several interesting features. On the mudflats there is usually no vegetation, but in a small re-entrant south of the Peffer burn *Zostera* spp. is abundant. Where the mud gives way to sand *Salicornia* spp. is plentiful. There is a considerable area of salt marsh on the south-west facing point of the bay. *Puccinellia* and *Glaux* are the main colonizers of the sand; locally *Honkenya peploides* is abundant. On slightly higher ground *Festuca rubra*, *Armeria maritima*, *Plantago maritima*, *Aster tripolium*, *Spergularia media, Triglochin maritima,* and *Carex* spp. come in. Since 1965 a few plants of *Limonium vulgare* have appeared.

The dunes are in the northern part of the bay. They have prograded noticeably since 1954. Embryo dunes often begin around clumps of drifted sea-weed. *Agro pyron junceiforme* is the pioneer, and is soon followed by *Ammophila arenaria* and *Elymus arenarius.* In the older dunes other grasses, e.g. *Dactylis glomerata,* and herbaceous species come in. The seaward dunes contain just over 10 per cent of shell material and a pH of 8.85. In the older dunes these figures drop to 2.4 per cent of shell in the upper layers, and 5.7 per cent in the lower; the pH figures were 6.85 and 7.80 in July 1967. Only one of the dune slacks is moist throughout the year. In it are found *Polygonum amphibium, Mentha aquatica, Equisetum fluviatile, E. palustre* and *Menyanthes trifoliata.*

The grassland, which forms most of the links, is divided into two parts. The greater part is calcareous; in the north there is some grassland lying on the tesche-nite rocks which outcrop at Gullane Point. The grassland rises up above the 100-foot contour at Gullane Links.

The Management Plan contains full lists of plants, bryophytes and fungi, and also of birds, mammals and amphibians, coleoptera and spiders found on the Reserve.

Gullane Bay, between Hummell Rocks and Black Rocks, is backed by dunes. It is one of many places on the coast characterized by sea buckthorn. The area seems to have been inundated with sand in the seventeenth century, probably because marram had been pulled in too great quantities for house thatching. From the late eighteenth century golf was played. The anti-tank blocks and various exercises carried out in the area preparatory to the landings in Normandy led to serious erosion. This, and erosion caused by visitors, has been checked by planting, and new foredunes have been created. (See East Lothian County Council, 1970, for a great deal of information about the coast from Preston Links to Thornton Loch.)

The rock platform is present at Black Rocks and, although locally sand covered, is continuous to Eyebroughty Scar, an outcrop of trachyte with several sharp crags striking east-north-east (Fig. 63). The scar is within low-water mark, and there is a wide expanse of sand between it and Weak Law rocks, mainly built of trachytic lavas, ash and agglomerate. They are mostly covered at high water. Near Weak Law there are also tabular masses of felspathic sandstone lying on the felstone which forms the rock platform. Partain Craig, and the Yellow Man, both former vents, follow to the east. The off-lying rocks, Fidra (= North and South Dog), Lamb and Craigleith, are all volcanic (basalts and dolerites). The Hummell Ridges and associated skerries to the west of North Berwick Bay are masses of volcanic ash, and the small headland, The Sisters, separating this bay from Milsey Bay is faulted, and consists of intrusives and ashes. The Leithies are similar, intrusive basalts and ash, and many large blocks of basalt. Leckmoram Ness is a neck. Basalts and agglomerates make the rock platform, including Podlie Craig,

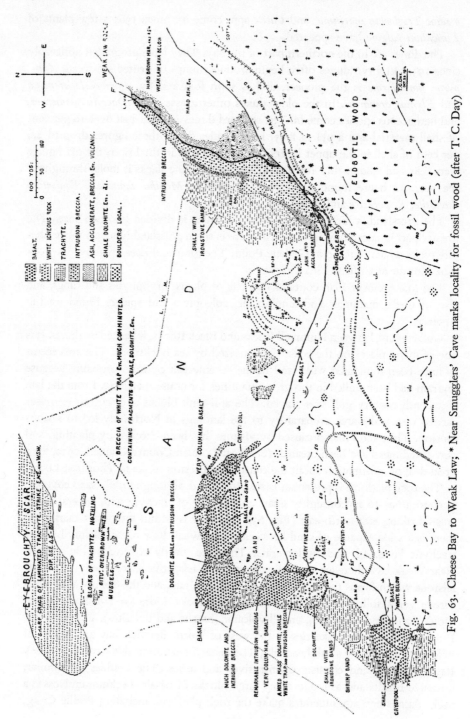

Fig. 63. Cheese Bay to Weak Law; * Near Smugglers' Cave marks locality for fossil wood (after T. C. Day)

as far as Gin Head. Hereabouts the coast becomes bolder, and there are high cliffs of ash and breccia, sometimes cut by dykes, at Gin Head and Tantallon where the ruins of the Castle lend distinction to what is naturally a fine piece of coast. The bay between Gin Head and Tantallon corresponds with a fault. The cliffs at the castle reach about 100 feet (30 m), but show no clear stratification. Oxroad Bay, to the south of the castle, is also associated with a fault, and the cliffs on its east side consist of ash beds, thin limestones and shales. Near the Gegan there is a local unconformity between the cliff rocks and those forming the shore platform.

The Gegan itself is a mass of stratified ash. The projecting reef running out to St Baldred's Boat is on the landward side sedimentary, sandstones and marls, but the outer part consists of two vents named the Cars. St Baldred's Boat is a lump of basalt. The broad platform of Scoughall Rocks is cut in both sedimentary and igneous rocks, and farther south gives place to Peffer and Ravensheugh Sands. St Baldred's Cradle and Whitberry Point are carved out of a mass of intrusive basalt. (The numerous outcrops of vents and related features on the south coast of the Firth of Forth were described in a series of papers by T. Cuthbert Day in *Trans. Edin. Geol. Soc.* in the 1920's.)

The effects of erosion on all this stretch of shore depend, as elsewhere, much upon the resistance and nature of the rocks. Since the sedimentaries are often folded and dip at appreciable angles, the harder bands stand out as low ridges. It is almost impossible adequately to describe the appearance of the igneous rocks. Commonly intrusive basalts form minor eminences and since the ashes are often bedded, they may resemble sedimentary rocks. Because the vents are irregularly distributed, the outline of the coast is also irregular. Locally there is a good deal of blown sand, and the contrasts of the dark, irregular rocks, the bright sand and its partial cover of vegetation make the coast most attractive. In Tyne Sands and Belhaven Bay there is a wide expanse of sand, and Sandy Hirst and West Barns links are, in effect, small barrier beaches. The skerries between Belhaven Bay and Dunbar are parts of a long east–west dyke. The ash cliffs on the western side of Dunbar are thinly bedded, and the layers on the beach yield unequally to wave erosion. Dunbar Castle stands on a mass of basalt intruded into Upper Old Red Sandstone rocks. The Dove Rock, a conspicuous feature just west of the castle, is a basalt plug, and a little farther seawards is a neck filled with basic tuff. The Old Harbour and adjacent part of the town rest on another neck, and between Victoria Harbour and Broad Haven there is a columnar sill (Graham, 1966–7).

Apart from local necks, the coast east from Dunbar as far as Torness Point is low lying. The rock platform, the inner part of which is often sand covered, is remarkably well developed, and is cut in the Carboniferous Series. The strata are, broadly, arranged so that the coast cuts across a widely open syncline. The minor headlands are nearly all in limestone, whereas the small re-entrants are in the shales and coal-bearing beds. The platform continues eastwards to and beyond Thornton Loch. There were formerly about a dozen cottages here together with a maltings, smithy and kippering house. These were allowed to go derelict after 1930; the farm was divided into small holdings. The dunes along the raised beach were planted in 1963. Although the general nature of the coast remains similar, the occurrence of Old Red Sandstone between Cove (Graham, 1963) and Siccar Point gives a much warmer colour to the local cliffs and also to the sand and small dunes in some of the bays.

The cliffs rise in height towards Bilsdean. Since shale usually underlies sandstone, erosion is rapid and the process is much helped by springs and rain which, working along joint and other planes of weakness, lead to considerable falls of cliff. A conspicuous stack, called the Standalone, used to rise from the platform some 500 or 600 yards (457 or 549 m) north-west of Bilsdean Foot; it was destroyed in a storm in 1906. Near the Dunglass Burn the bedding in the cliff is nearly horizontal and there are long reefs on the beach.

Differential erosion is well displayed in Cove Harbour. On the eastern promontory, where the dip reaches 80°, the sandstones project; the bay is mainly cut in shaley beds, and on the west side is another sandstone headland, followed by another small bay.

At Siccar Point is the unconformity made famous by Hutton. On p. 458 of the first volume of his *Theory of the Earth* we read:

At Siccar Point we found a beautiful picture of this junction (between the Silurian and Carboniferous) washed bare by the sea. The sandstone strata are partly washed away and partly remaining upon the ends of the vertical schistus; and in many places points of the schistus are seen standing up through among the sandstone, the greatest part of which is worn away.

Eastwards from Siccar Point, red and white sandstones and marls form the cliffs for a mile or more, but near Redheugh give place to the Silurian rocks which, with the igneous mass at St Abb's Head, form one of the finest lines of cliff in these islands. Both cliffs and the land rise considerably in height towards Fast Castle which stands on a promontory of highly-inclined beds. The castle marks an angle in the coast; to the west it turns and runs for a time parallel with the strike of the rocks so that the structure is not so clearly seen. But at and east of the castle the folding in the cliffs is extraordinary. It is impossible to describe it in detail. Fig. 64 after Geikie (1863) shows its general nature. At the Souter and Brander the folds are spectacular; other noteworthy features occur at the Muckle Pits, where the cliffs reach 500 feet (152 m), the highest point on the east coast of Scotland, the Little Pits, the double fold on the skerry named Biter's Craig, the arch at Thrummie Carr, and the rapid and sharp folds and waterfall near Heathery Carr (see Plate 3). The Silurian rocks run inland at Pettico Wick and give place to a mass of felsite which forms St Abb's Head (see Plate 1). On the landward side the felsite is bordered by a marked valley of glacial origin. The headland is fronted by magnificent cliffs, rugged and precipitous. There are numerous clefts, gullies, and coves, and many stacks, reefs, and skerries just off-shore.

To the south-east the felsite gives place to rocks of Lower Old Red Sandstone

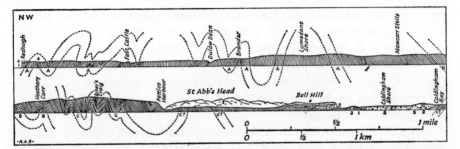

Fig. 64. Coast section of Berwickshire from Redheugh to Coldingham Bay. 1 (ABC), Lower Silurian; 2, Lower Old Red Sandstone ash; 3, Intrusive igneous rocks; 4, Upper Old Red Sandstone; f, Faults (after Archibald Geikie)

age. From Horsecastle Bay to Coldingham Bay (Shiells and Dearman, 1963–4), apart from a short space at Halterem's Loup where the Silurian outcrops, the cliffs are largely cut in ashes. In and near Horsecastle Bay there are several caves and the cliffs are rugged. The Silurian reappears between Coldingham Bay and Callercove, although there are also knobs of felstone, one of which forms the stack known as Linkim Kip. The strata are somewhat less folded in this locality, but frequently pierced by igneous dykes which help to diversify the coast. From Callercove felspathic rocks form cliffs 75 to 100 feet (23 to 30 m) high almost as far as Eyemouth. It is sometimes difficult to distinguish between lava and ashes.

The bay at Eyemouth (MacKenzie, 1954–6) corresponds with a fault, and the Silurian reappears on its south side. Between Eyemouth and Burnmouth (McLennan, 1892–6) there are numerous folds visible in the cliffs, and other features include Scout Cave, hollowed out in the lower part of a dyke. At Gull Rock the cliffs are more or less vertical and the beds in them much folded. The rock itself shows twisted strata. At the Breeches rock is another dyke, the lower part of which has been worn into an arch. Another deep ravine meets the coast at Burnmouth, and this too is partly associated with a fault which brings down the Carboniferous rocks to the coast. These rocks extend beyond Berwick. The strata in these beds are almost horizontal, and they make a line of steep cliffs, nearly uniform in height. The Lamberton Limestone runs along the cliffs for about two miles, and is conspicuous in Marshall Meadows Bay, just south of the Border. The limestone is often underlain by shales and under-cutting occurs. It reaches sea-level at St John's Haven. To the south the red and false bedded sandstones offer excellent examples of differential weathering.

Eyemouth harbour presents several points of interest. Whenever sand is brought into it, a bar begins to form since the ebb currents are not strong enough to remove it. Sand may come in with each flood tide, and particularly when there is severe wave action and stronger currents at springs. Inside the harbour wave action rapidly decreases and the sand, brought in in suspension or in other ways, is deposited to form a bank. In storms the bank may be built rapidly, as in October

1955, or it may take some months to form. However, once it is formed, it stays until it is removed by dredging or flushing by a spate in the river. The matter was investigated at the Hydraulics Research Station (Bull. No. 48, 1958) and of the several schemes tested to prevent the formation of the bank two appeared to be successful, sluicing or the building of a new entrance to the harbour. Incidentally, this was the first model of this kind in which radio-active material was used to show movement of the sand.

Along all this part of the coast the shore platform is well developed and the cliffs are fossil, but in some places, for example St Abb's Head, they are washed by the present waves.

Chapter VIII

The Orkney Islands

The structure and pattern of the Orkney Islands are in marked contrast with those of Shetland. Apart from a small area of igneous and schistose rocks near Stromness and one or two other places, and occasional lavas and dykes, the islands are almost entirely built of Old Red Sandstone rocks. The Upper Old Red and associated volcanics are restricted to Hoy; the other islands and skerries, about 90 in all, are formed of the Middle beds of the Old Red, the Stromness, Rousay, and Eday beds. Apart from Swona and the Pentland skerries, the islands make a fairly compact group (see Fig. 65), and it has been pointed out that a fall of sea-level of 120 feet (37 m) would convert them into one island and one of 60 feet (18 m) would make three islands. There are well-marked depressions between Mainland and Rousay and Shapinsay, and also between Westray, Eday and Stronsay and Rousay and Shapinsay. These depressions trend approximately north-west and south-east, and are parallel to valleys such as the Oykell and Shin on the mainland of Scotland. It is held that these deeps were once former valleys, part of an old river system, that drained south-eastwards from a watershed which formerly extended along a course from the Minch to the Faroes, when that area stood much higher as a result of the accumulation of Mesozoic sediments and Tertiary lavas, the remains of which can be seen today in the Inner Hebrides and the Faroe Islands.

On the eastern side of the islands the form of the sea-bed is very different from that to the west of the archipelago. A continuous and steep bank can be traced; in the south its top is about 20 fathoms (37 m), and its foot 30 fathoms (55 m); in the north the base falls to about 35 fathoms (64 m). It follows a remarkably straight course to the north-north-east, and there is little doubt that it is a continuation of the steep eastern coast of Caithness. It is probably an integral part of the Great Glen Fault. Its northern continuation may be the Walls fault in Shetland. The scarp is regarded as an old land feature, and presumably corresponds to a time when Orkney formed an extension of the Caithness plain. The Pentland Firth is deeper than any of the inter-island channels, and may also have been followed by one of the south-easterly flowing rivers.

Most of the islands are relatively low except for Hoy. Several parts of north-eastern Mainland exceed the 700-foot contour (213 m), and on Rousay Blotchnie Field reaches 821 feet (250 m). Fitty Hill in Westray is 521 feet (159 m), and in Eday and South Ronaldsay various places rise above 300 feet (91 m). But Hoy

Fig. 65. Outline map of the Orkney Islands (based on Ordnance Survey; Crown Copyright
Reserved)

presents quite a different picture. Several peaks are more than 1,000 feet (305 m),
and Ward Hill reaches 1,565 feet (477 m), the highest point in Orkney. This
difference may be partly explained by the fact that most of the island is formed of
rocks which do not occur elsewhere in the group. These Upper Old Red rocks
have a volcanic zone at their base, and consist of about 3,000 feet (915 m) of red
and yellow sandstones which give rise to hills of steep rounded outlines. Ward
Hill lies between two steep valleys which reach the sea at Rackwick near which

place they join to make a wide valley facing south-west. Glacial deposits partially close the north-eastern ends of these glens. It is unlikely that the present small streams cut the two valleys, and they have no connection with the present drainage system. It is suggested (Wilson et al., 1935) 'that they are pre-Glacial in age and were widened and deepened during the Glacial period ... these streams were tributaries of a great river that flowed down through what is now the Pentland Firth, and that the valleys were cut when the Upper Old Red Sandstone extended over the west Mainland'.

Scapa Flow is also a Pre-glacial feature. Its floor is noticeably flat and is, in fact, a basin partly filled with sediment. In form it has much in common with the firths of Stronsay and Westray. The deepest part, Bring Deeps, may be an old glacial lake not yet filled with sediment (Flett, 1915–24).

The larger islands of Orkney, especially Mainland and South Ronaldsay, are traversed by a number of faults. We are not here concerned with their effect on structure except along the coast. When a fault cuts the coast it is common to find that erosion has produced a geo or some form of indentation. Garthna geo and Clay geo are two of many examples. On the other hand, Scapa Flow probably owes its general form to faulting; the North Scapa fault follows its northern shore, the East Scapa fault follows its eastern shore in Mainland, and it is probable that several faults, approximately parallel with the North Scapa fault, traverse it. Undoubtedly much of the detail of the cliffed coasts of Orkney depends directly upon marine erosion working on joints, bedding planes, faults and also along dykes of igneous rock. The rock in contact with such dykes is often removed fairly easily; the Hole o'Row on the south side of the Bay of Skaill is a good example. Sometimes dykes may form wall-like features; the camptonite dyke at Banks in Birsay is well-known. On the other hand, dykes may be eroded more easily than the rock through which they cut, and then they form hollows.

Raised beaches of Late-glacial or Post-glacial age seem to be rare in Orkney. The Memoir (Wilson et al., 1935) records two gravel patches at Muckle Head and Selwick, in north-western Hoy 'which are difficult to account for except on the assumption that they are connected with one of the Raised Beach periods'. They are 20 to 40 feet (6 to 12 m) above sea-level and rest on a well-marked notch or bench cut in solid rock. When the Memoir was written the old views prevailed concerning the '100-ft', '50-ft' and '25-ft' levels and the higher 'Pre-glacial' beach in the western isles. It was thought unlikely that the occurrences in Hoy belonged to any of the later beaches, but that they might be contemporaneous with the rock-notch in the western isles. The problem is still an open one. These remains should, however, be considered in relation to the wave-cut platform which can be followed in many places in solid rock along the northern shore of Scapa Flow. It is most improbable that this is a modern feature. It is often several yards wide and stands at the foot of a low and dead cliff. So far as I am aware it has not been mentioned elsewhere. It is also relevant to call attention to what appear to be dead cliffs in one or two places in South Ronaldsay, e.g. Wind Wick. The problem of the age and formation of cliffs is as important in Orkney as elsewhere. The fine range of cliffs, reaching 1,100 feet (335 m) at St John's Head in Hoy, plunge directly down to deep water. The Old Man of Hoy implies a certain amount of recession, and as late as the early part of last century this famous stack had an arch on its landward side. This is now destroyed, but the part where it was attached to the main stack is shown by a thickening of the column. The stack rests unconformably on a weathered surface of lavas and is 450 feet high (137 m). At the base its diameter is about 30 yards (27 m), and the lavas form a small platform, a minor projection, at about high-water mark. Elsewhere in this line of cliffs weathering has etched out vertical chasms along joint planes; at Bre Brough a pillar has nearly been separated from the cliffs. But how far the whole cliff line has receded is unknown.

The west coast of Hoy is best seen from the sea (see Plate 22). The highest cliffs are in the north-west. They are almost vertical and consist of massive red sandstone. At Rora Head, about one mile south of the Old Man of Hoy, the coast turns eastward, and for nearly a mile a lava is intercalated between the Eday sandstone below and the Hoy sandstone above. Rack Wick is the only low part of this coast. There is a good beach and boulder clay reaches sea-level. There is also some blown sand. The cliffs south of Rack Wick are magnificent. Their height varies between 300 and 600 feet (91 and 183 m). There are numerous geos and there is usually a narrow platform or bench at their base. Near Little Rack Wick the cliffs are of red sandstone which is undercut so that they rest on pillars. There are many coves and some long and narrow geos. The cliffs remain almost vertical for some distance; near to Tor Ness, where the coast turns to the east, the colours of the cliffs incline to yellow and white. Several burns reach the sea between Rack Wick and Tor Ness; they cut deep valleys and finally reach the sea over a fall. The cliffs are washed by the sea today, but the bench and fossil geos, such as Geo of the Lane and Lyrie Geo, indicate that the cliffs have had a long history.

In western Mainland the effects of erosion are beautifully demonstrated in several places. There are several geos and stacks, and the cliff is often riddled with caves, several of which extend some way inland. Usually these features are associated with jointing and sometimes with faults or with dykes of igneous rocks. Along much of this west coast of Mainland the flagstones which form the cliffs have a seaward dip, and there is no doubt that erosion is active. Locally there may be rock-cut benches and occasionally the height to which wave action is active is demonstrated by the nature of the cliff top, for example, at Yescanaby, where a small mass of granite-schist underlies the Yescanaby sandstone. The coast as seen on a small-scale map is fairly straight from Breck Ness to Brough Head, but in detail it is much indented (see Plate 24). The Bay of Skaill lies in the continuation of the low ground of the Loch of Stenness, Loch of Clumly and the Loch of Skaill and, in that sense, may be compared with Birsay Bay which is similarly associated with the line of low ground of the Loch of Boardhouse, the Loch of Banks, and the Loch of Harray. In both bays there are boulder beaches and extensive dunes. The Bay of Skaill also contains Skara Brae. That this bay has been flooded, at least in part, in comparatively recent times is proved by the occurrence of a submerged forest which, under certain conditions of wind and tide, can be seen near the north shore. Skara Brae is in the south-east corner of the bay, and is a great sand dune rising to about 35 feet (11 m) from the beach. Fifteen feet (5 m) of this height is the work of man. The huts on this ancient settlement were first exposed in a storm in 1850, and when found intact the huts gave the impression of a rapid evacuation since pots and implements were left just where they would have been used. There are also traces that some of the inhabitants returned. Gordon Childe (1951) concludes that the village was buried in sand, probably during a hurricane from the north-west. If this coincided with a high tide, conditions must have been similar to those in 1850. But sand continued to be a menace in later times. Childe concludes that

the peculiarities of our village and in particular its 'subterranean' character are the final outcome of a series of progressive modifications in plan all dictated by the ever-present necessity of maintaining the access to dwellings impeded by sand drift and of excluding damp and wind from the interiors . . . The village was not originally subterranean; it began in an agglomeration of free-standing huts which became embedded by successive steps in heaped-up refuse. The enemy that eventually overwhelmed the site was . . . sand . . .

Skara Brae culture is earlier than that of the Brochs, and the village probably dates from just before the Christian era.

How far inland it was built in the first place is unknown. It is based on rock; sand has covered it. It is now protected artificially from marine erosion, but what is known of it does not suggest that erosion has been notably severe at this place in the last two thousand years. North of this bay the cliffs are formed of grey flags, and are cut by geos and caves as farther south, but are not particularly impressive. Mar Wick is a smaller but not dissimilar feature from Skaill Bay; the erosion of the rock platform has led to the formation of a bank of slabs and boulders impounding a shallow lagoon, below high-water mark, on the inner edge of the inlet, but outside the beach proper. The cliffs at Marwick Head are fine; they are cut in grey and ochreous flags, are often almost vertical, and near the Kitchener Memorial are more than 200 feet high (61 m). It would be interesting to know if any measurements of their recession had been made since the memorial was erected. The Brough of Birsay is separated at high water from the Point of Buckquoy. Eastwards therefrom the coast scenery is interesting. Whitaloo and Ramly geos are associated with faults, and near Costa Head the cliffs are magnificent and about 400 feet high (22 m). They are fronted by a narrow rock ledge, and this and the Standard, a prominent stack, indicate the degree of erosion. The cliffs are vertical. A little to the east, near Haafs Helia, the beds are disturbed and smashed, and two faults run northwards to the sea. Farther to the south-east, the cliffs fall in height, and the coast is somewhat more sheltered. The sands of Evie, held in the re-entrant formed by Aiker Ness, are notably calcareous.

A traverse along the west and north-west coasts of Mainland shows that erosion of the cliffs is proceeding actively. On the other hand, in several places there are rock benches; for example, at Broad Shore, Harra Ebb, and in Birsay Bay. There is also a bench along the outer part of Eynhallow Sound. Since raised beaches are rare in Orkney, it is to be expected that benches such as characterize so much of the mainland of Scotland should be scarce or absent. It is, however, difficult to believe that Brough Head has been separated from Mainland entirely under present conditions. If we consider any part of the coast of Orkney in relation to the rock bench in Scapa Flow and the rock platforms near the Point of Ayre (in south-east Mainland), the fossil cliffs in South Ronaldsay, and many other comparable features we must admit that, despite appearances, all the cliffs are inherited from a past time and that modern erosion is modifying rather than creating them (see Plate 23). This is consistent with the shore features on the north coast of Sutherland, and if, as seems likely, there is a tilt downwards in a northerly direction the more limited developments of the bench in Orkney and especially in Shetland are in accord.

Rousay is unlike many of the other islands. It has a relatively unbroken outline, and only in a few places do the cliffs reach any height, 70 feet (21 m) at Scabra Head, and 200 feet (61 m) at Bring Head where they overhang. A number of dykes intersect the coastline. A noticeable feature is the marked terracing of parts of the island produced by gently dipping strata. The west facing coast of Westray presents a fine range of vertical cliffs cut in grey flags, and the power of wave action is well demonstrated at Aiker Ness where (cf. The Grind of Navir, p. 289) there is a storm beach of large slabs 40–50 feet (12–15 m) above sea-level.

To describe, even in outline, all the cliff coast of Orkney is not possible in the space available. But something must be said of the cliffs facing eastwards. In the Deerness peninsula of Mainland, cliffs are prominent, and there is also a rock platform around much of it. This feature should be studied in much greater detail. It seems to be directly comparable with the rock platform which occurs along so much of the east coast of Scotland. The Brough of Deerness and the Gloup a little farther south are excellent features, but immediately north of the Brough the cliffs do not give the impression of actively receding. On the other hand the east coast of Shapinsay is locally completely undermined by the sea; there are numerous caves, and the cliffs are supported on legs. In South Ronaldsay there is also some fine cliff scenery, and in several places there is a well-developed shore platform which would be difficult to explain as having been formed wholly under present-day conditions. The cliffs at Wind Wick and several other places are certainly fossil cliffs. The cliffs on the south-facing coast are good. Details of the smaller islands must be omitted, but the fine Gloup in Swona should be mentioned and the platform and the north-east and south-west features in Copinsay.

Ayres and oyces are perhaps even more conspicuous in Orkney than they are in Shetland. This may be because some of them join together what were formerly large islands, e.g. Hoy to South Walls and Deerness to Mainland. These features have never been fully analysed, and beyond ascribing them primarily to wave action there is little more to be said until a careful examination of them has been made. The Survey Memoir, 1935, makes some interesting comments on some of them. That between Hoy and South Walls is made of shingle and was breached until the road was made about the turn of the century. The ayre is now curved to the north, and that it is probably still in movement is shown by the fact that the line of telegraph poles when first put in were arranged along the northern side of the beach; in 1935 they were on the southern side. It was estimated that the advance of the ayre to the north had been about 20 feet (6 m) by the late 1930s. It is difficult to agree, in any literal sense, with the view in the Memoir that there 'is no doubt [that the northward movement is] due to the pressure of the Pentland Firth tide sweeping up Aith Hope being greater than that of the tide in Long Hope'. If the effects of wave and storm action superimposed on the tides is also allowed for the explanation is probably correct. The view is also expressed in the Memoir that 'Ayres of this type most probably represent the last stage in submergence before a peninsula becomes an island, but as long as the isthmus is complete, there is a tendency for material to accumulate on both edges. This is caused partly by the action of the tides and partly by the formation of sand dunes'. This is an interesting view and seems to imply that, if submergence continues, the ayres, at any rate of this type, will sooner or later disappear. This may be so, but it would be interesting to investigate the full significance of this in terms of rate of submergence as determined by tidal observations, and also in terms of age relative to submerged or partly submerged peats. The rate of supply of material of one sort or another to an ayre is also a relevant consideration.

The Memoir distinguishes three slightly different types of oyces. The first includes sheets of water 'held up at high-tide and not completely drained away at low-tide'. The Peerie water on the western edge of Kirkwall is an instance. The area behind an ayre may be dry at low water, as in the Ouse of Aikerness and the Ayre of Fribo in Westray. Thirdly, the ayre may block the entrance to a bay and form a fresh-water loch; Ross Loch in Stronsay and Echna Loch in Burray are given as examples. This hardly seems a comprehensive view; how, for example, does Vasa Loch in Shapinsay fit in to it? The Memoir itself (p. 113) says that 'it [i.e. the ayre] juts out into the sea to form the headland of Vasa Point, and impounds the waters of Vasa Loch'. Others again, like Lea Shun in Stronsay, may be fresh-water areas held up behind blown sand. The interesting comment is made on this and similar oyces that 'Excepting in those cases where the primary embankment was a storm beach, it is most probable that in the first stages the floors of the lochs were above sea-level'. It is also stated in the Memoir (p. 10) that usually the eastern end of an ayre is higher than the western end, and that it is the eastern end that is commonly joined to the land. The outlet is at the west. In the exceptional cases the outlet is to the south. The Ouse of Firth at Finstown is to the north, perhaps the only example. These points all need detailed examination in terms of the direction of local drift of beach material. A careful study of the One Inch or Two and a Half Inch Ordnance Maps, as well as a general acquaintance on the ground, suggests many interesting local problems. Not only must direction of drift be considered, but also depth of water and the nature and gradients of the surface on which the ayres rest. It is also noted in the Memoir, and is indeed obvious when travelling in the islands, that ayres and oyces are more common on south and east coasts. We may readily agree with the view that this is because the west coasts are often, not always, higher and cliffed with few shallow bays. It is, however, unlikely that eddy currents produced in bays at high water are responsible for these features. There is great scope for research into these topics in all the islands, and perhaps particularly in Sanday where some of the ayres are more than a mile long. Their relation to areas of blown sand is also worthy of greater investigation.

Apart from ayres, there are several examples of cobble and boulder beaches in the islands. Those in the bays of Skaill and Birsay are known to most visitors, but one of the finest of these beaches is on the island of Stronsay where, along the south-western part of the island, there is a cobble beach which impounds Straenia Water and continues for about two miles to the north. It is formed almost entirely of well-rounded boulders, which, near Straenia Water, are up to three or four feet in diameter (*c.* 1 m). It is generally about 60 yards wide (55 m) and grass-covered. This, too, needs investigation. It faces Stronsay Sound and is consequently somewhat exposed but by no means to the same extent as a beach on the western side of Mainland. It might be worthwhile to see how far it is inherited from past conditions.

Stevenson (1811–16) remarks that the Start Point of Sanday was in his time (and still is), an island at every flood tide but, in the memory of some older people in 1811, there was a continuous tract of firm ground connecting it to the main island.

FAIR ISLE

This small island of Old Red Sandstone is situated about half way between Orkney and Shetland. It is 3½ miles long (*c.* 5.5 km) from north to south and 1½ miles wide (*c.* 2.4 km). Ward Hill is 712 feet high (217 m). It is a fertile island. It is surrounded by cliffs which, on its northern and western sides, are remarkably fine and in one place reach 600 feet (183 m). There are many caverns, geos, stacks and arches. Sheep Craig, consisting of massive red sandstone, rises sheer from the sea on the eastern side. The stacks of Skroo off Ward Hill are noteworthy. It is a well-known bird-migration station, and the island has one mammal peculiar to it, the field mouse – *Apodemus f. fridariensis* (Waterston, 1946) (see Plate 27).

Chapter IX

The Shetland Islands

The Shetland Islands are composed largely of rocks similar to the metamorphic and igneous rocks on the mainland of Scotland. In places, particularly in the west, there is also a cover of sedimentary and volcanic rocks of Old Red Sandstone age. The structure is complicated and, especially in Mainland, is characterized by north–south trends and dislocations. In the north of Mainland the trend turns towards the north-east. There are three main islands, Mainland, which is almost divided into two at Mavis Grind, Yell, and Unst. Fetlar, Whalsay and Bressay are much smaller, and Foula and Fair Isle are separated from the main group by several miles of sea. There is also an infinity of lesser islands and skerries; the largest of these include Papa Stour, West Burra, Muckle Roe and Noss. No place is more than a mile or two from the sea, so that almost all may be regarded as coast (Fig. 66).

A first glance at a geological map suggests a close correspondence between many of the long inlets, called voes, and the geological structure. In fact this is often more apparent than real. Major faulting has certainly played an important role in Shetland. One fault follows approximately the east coast of North Maven, passes inside Muckle Roe, continues south through Aith Voe and passes out to sea a little east of Roe Ness. This is the Walls Boundary Fault. The Nesting fault may be traced on the east side of Fitful Head, and then it probably follows, off-shore, the west coast of Mainland, running to the west of St Ninian's Isle, and then follows Clift Sound, and cuts across Mainland along a line that just includes the extreme inner parts of Cat Firth and Dury Voe. In Swining Voe it seems to split; the north-west branch *may* continue along Yell Sound; the east branch, named Blue Mull Sound Fault, traverses Yell from Hamna Voe to Hascosay Sound and thence more or less along the coast and Blue Mull Sound. Unst is almost bisected by a fault running from, approximately, Belmont to Burra Firth. There are also many lesser faults; in Walls they frequently trend approximately west-south-west and east-north-east. In Yell there is a major dislocation between Hamna Voe and Whale Firth. Flinn (1961) thinks that the Great Glen Fault may run far to the north, and suggests a line (see also pp. 82, 209) to the west of Fair Isle and then continuing as the Walls Boundary Fault, and possibly much farther north, a continuation that may be indicated by the contours of the sea bed between 61° and 62° north latitude.

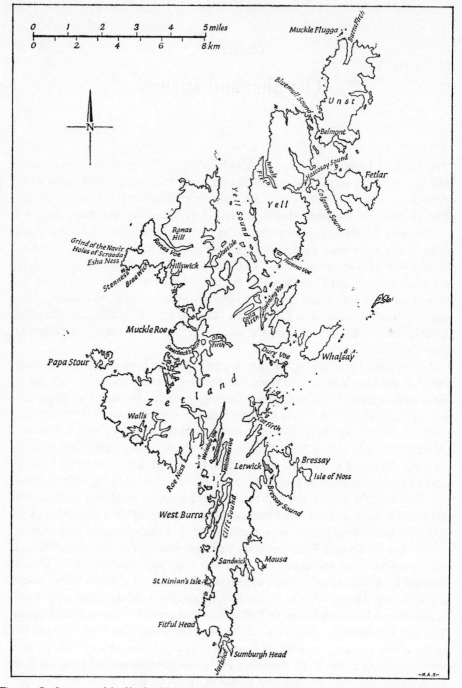

Fig. 66. Outline map of the Shetland Islands (based on Ordnance Survey; Crown Copyright Reserved)

The coastal scenery of Shetland varies greatly from place to place, and much of it is magnificent. The islands represent a part of the sea-floor which, for a reason unknown, stood higher than the areas roundabout. When they emerged above the waves is uncertain, but there is no doubt that in recent times Shetland has been an area of submergence. The numerous drowned valleys indicate this, and that submergence has continued until recent times, and is probably still in process, is shown by the fact that peat, still largely uneroded, has been taken below water level. (The precise significance of *un*eroded peat below sea-level is not clear. In many places in Britain and elsewhere peat in the 'submerged forests' is often largely uneroded. This may imply rapid submergence or rapid covering by some other deposit; it is puzzling.) Raised beaches are absent, but certain minor features on the coast need explanation. At Sandwick and in Brei Wick (Lerwick), to take two instances, there are traces of low-level platforms which appear to be ordinary wave-cut platforms. If this is not the case it is difficult to find an adequate alternative explanation. If these were, e.g. on the east coast of Scotland, they might well be taken as a low part of the rock platform which follows so much of that coast. Their exact significance in Shetland is yet to be decided.

In both Shetland and Orkney (and cf. the Isles of Scilly), there are numerous instances of tombolos, of bay head and mid-bay beaches and of other accumulation forms that hardly come into one of these categories, but nevertheless cut off and enclose an area of water. Flinn (1964) discusses them, and writes:

The most common type of bar or spit is the tombolo, of which there are forty, together with five double tombolos . . . Bay-head barriers and mid-bay spits are quite common in the voes, there being fifteen of the former and twenty-one of the latter. Six bay-mouth spits occur, partially cutting off shallow embayments of the shoreline, but never at the mouths of voes. There are ten looped barriers. Finally there are forty-three bars which completely cut off the voe or bay behind them from the sea. Seven of these are of bay-head type and seven are of mid-bay type, all occurring in voes. Most of the remaining twenty-nine are bay-mouth bars completely cutting off shallow embayments of the shoreline from the sea, but several are at or near the mouths of voes . . .

Flinn shows that a number of these features, which usually consist mainly of pebbles, but occasionally of sand, are associated with lines of tear faults, or in areas where there is glacial drift available. He follows C. A. M. King in associating them with shallow coasts – in Shetland caused by drowning. But no analysis is given of the different features. Why, for example, in Dales Voe are there three between Fora Ness and Mainland? The northernmost has almost certainly been formed by waves from the north-east; the southern ones by waves from the south-west. But why do tidal entrances cut the northern and middle bars at their south-eastern ends? The beautiful tombolo joining St Ninian's Isle to Mainland is probably constructed from both sides (see Plate 25); the same origin almost certainly applies to that connecting Gluss Isle to Mainland. I have seen a large number of these features, and there is abundant scope for research on the ways in which they are formed. What is their age? If the area is still sinking it is not easy to think of them as of any particular age, nor is it easy to see how they keep pace with the submergence. Bores might throw light on this point. How many have been formed by direct wave action across a re-entrant, or by beach drifting? What effects do major storms have on those that are relatively exposed, e.g. St Ninian's tombolo? How does the material of which they are formed travel along the voes – and some tombolos are a considerable distance from the mouth of a voe (Zenkovich, 1967). We may agree that all these features are associated with shallow water; we do not, however, know in what limits of depth they may be formed. Not only a full study of those in Shetland, but a comparative study of Shetland, Orkney, the Isles of Scilly and possibly the Farnes would, in this respect, be most valuable.

The general appearance of the coastline will depend, here as elsewhere, largely on the nature and disposition of the rocks adjacent to it. The west side of Shetland is more exposed than the east, but in Unst, Fetlar and other places, the north and east coasts face long fetches of open water. Flinn's map (1964) is most informative and deserves careful study. Cliff formation, however, is a subject that needs much fuller investigation than it has so far had. Flinn remarks on the absence of any modern wave-cut platforms at the foot of cliffs. He shows on lines of soundings that the cliffs usually descend below present sea-level to considerable depths. He believes that the cliffs are in active retreat in exposed places, but fully appreciates the difficulty of calculating the amount. He gives four estimates –

If it is assumed that the slopes on the eroded sides of the hills were originally about as steep as the preserved sides, a minimum estimate of the distance retreated may be obtained. For the 900-foot [274 m] Fitful Head this distance is about a mile, as it is for the 600-foot [183 m] Hill of Clibberswick. For the 600-foot Noss Hill the distance is about three-quarters of a mile, and for the 300-foot [91 m] Sumburgh Head about half-a-mile.

These and other cliffs are very steep, even perpendicular, and, according to Flinn, always in active retreat so that 'there is little or no sign of sub-aerial erosion on them'.

Flinn has done valuable work in taking lines of soundings at several places, and in analysing Admiralty Charts and data available at the Hydrographic Office. He has established that the sea-floor around the archipelago is stepped or terraced; he notes frequent occurrences 'of nearly horizontal surfaces at forty-five, twenty-five and thirteen fathoms' (82, 46, and 24 m) and regards these features as indicating erosion surfaces before submergence took place. He suggests that they may have been erosion surfaces formed by the sea in front of retreating cliffs, but much modified, before drowning, by sub-aerial erosion and glaciation. He is, however, emphatic (p. 331) that 'there is no wave-cut platform at present sea-level or at any depth below sea-level, despite the fact that the present cliffs must have retreated half-a-mile or more to have attained their present form'. The cliffs plunge below water; their gradient may be 1 in 1 or even more, whereas below water it becomes

1 in 10 and, as indicated above, may be interrupted by terrace-like features. This, as Flinn himself suggests, implies that sea-level has risen from about the level of the lowest platform at a more or less uniform rate which would drown valleys and rise upon the hill sides.

A steady rise of sea-level of this amount up to the present time is scarcely in agreement with the views of, e.g. Jelgersma (1966). On the other hand, if we assume, for the sake of argument, that Shetland has been steadily sinking then the conception may be more or less valid. But has anyone yet interpreted the cliff form correctly? A vertical, or nearly vertical, cliff plunging below sea-level, and showing no trace of an erosion bench or terrace, does not necessarily imply any considerable retreat. The waves are reflected from such a surface. It is, of course, true that they will cause erosion as a result of the compression of air in rock crevices, and that cutting back will take place more readily along fault or joint planes, and that subaerial erosion may lead to cliff falls and slow retreat of the cliff top. I agree with Flinn in thinking that in many Shetland and other cliffs little erosion takes place below half-tide level, and also that the algal growth at that level seems little disturbed by storms. He argues that cliff retreat takes place as a result of falls from above caused by undercutting. But what is the evidence of undercutting? (Some cliffs, cf. Orkney, are much undercut by caves. In this way certain cliffs may well retreat as a result of collapse.) Does it just balance falls so that it is not apparent? With a stationary sea-level beach material collecting at the cliff foot supplies plenty of ammunition for the waves to attack the cliffs; when the cliffs plunge into deep water, this material disappears and only locally can be used for erosive purposes. A steep or vertical cliff seems by its very nature to imply erosion, and also long continued erosion. But is this always so? How much retreat is implied in the great vertical cliffs of Costa and Hoy in Orkney? The projecting of slopes and similar exercises can be misleading, and it is far from certain how the height of cliffs is related to their rate of retreat. The higher a cliff the more material has to be removed for each foot of landward retreat, but much depends on rock type, structure, subaerial weathering and other factors.

In making these points I am by no means contradicting Flinn's views, but I find some of them difficult to accept. We are still largely ignorant of the processes by which cliffs in resistant rocks retreat. Shetland is in an area where submergence may have been taking place for a long time, and so cannot directly be compared with other parts of northern Scotland. On the west coast of Shetland the effect of storms is profound. In Esha Ness, Northmaven, the Grind of the Navir is a ridge of great boulders and slabs of rocks, two or three or even more feet in long diameter, piled up on top of cliffs of resistant Old Red Sandstone lavas and tuffs. They may well be wholly the work of the sea at its present level; the main ridge is situated in a place where the waves can follow up a steep slope. It is true that, since the relative levels of land and sea have changed in recent time, the boulders may have a longer history, but since the rock is much jointed it hardly seems necessary to suppose that the grind is of earlier date.* Close to the grind is another interesting feature. The Holes (now only one!) of Scraada form a great blow hole. The sea tunnelled into the cliff along a line of weakness, and originally there were two holes, but the bridge between them has collapsed. What is even more interesting is that the holes have cut back into a small burn, along which there were some mills, and has diverted the stream so that it now drains into the holes instead of following its original course to the north-west.

This observation provokes the view that cliff retreat may take place by undermining. There is no doubt that it may be an important factor, but it does not eliminate the difficulty presented by the absence of platform or erosive beaches at the foot of so many Shetland cliffs.

In very different circumstances we have a rough measure of erosion in the sheltered bay in which the interesting remains of Jarlshof stand. The most obvious way in which this can be seen is in the loss of about half of the original courtyard of the Broch. 'The area enclosed by the remaining arc of the courtyard wall is 2,000 square feet [c. 186 sq. m], suggesting that the original space was approximately double.' This change has taken place in a period of about 2,000 years, but as Hamilton (1956) remarks, the whole complex was built about 20 feet (6 m) above sea-level and on blown sand. It is true that the site is sheltered, except from a southerly direction, but the change is relatively small. It is by no means clear if this change has been aided by recent submergence.

If we look at the archipelago as a whole it is probable that, apart from relatively minor coastal changes resulting from erosion and accretion, the general pattern of the islands has been produced by sub-aerial erosion acting upon an uplifted mass of ancient and much folded and faulted rocks. The long voes often bear a clear, but by no means always simple or readily explicable, relation to rock structure. Some seem to follow outcrops of relatively less resistant rock. In Walls and Northmaven there is no similar relationship and in Unst and Yell some of the larger inlets do not trend with the rock outcrops. It is possible that the separation of the islands one from another is often the result of faulting which has produced lines of weakness. The highest part of the islands, Ronas Hill, is a mass of granite.

The outlying island of Foula has for several years been visited by expeditions from Brathay, and Pirkis (1963) gave a careful analysis of its coast. Most of the island is Old Red Sandstone, but a narrow eastern strip consists of igneous and metamorphic rocks, separated from the Old Red by a fault. The figure (67) shows that Pirkis defined seven distinct parts of the coast. The igneous coast (1)

---

* S. Hibbert (*A Description of the Shetland Islands*, 2nd reprint 1931, p. 266) notes that in Stenness, in the winter of 1802, a tabular-shaped mass 8 ft 2 in × 7 in × 5 ft 1 in was dislodged from its bed, and removed a distance of from 80 to 90 feet. He also measured the space from which a block was carried away by the waves in 1818, and found it was 17½ ft × 7 ft × 2 ft 8 in. This mass was removed about 30 feet and broken into 13 or more pieces. A block measuring 9 ft 2 in × 6½ ft × 4 ft was pushed up a slope for a distance of 150 feet.

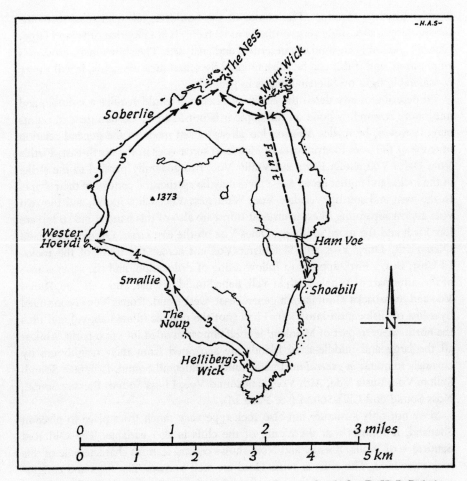

Fig. 67. Sketch map of Foula (based on Ordnance Survey; detail after D. H. B. Pirkis)

is, in detail, irregular and reflects the somewhat varied nature of the rocks, and is much indented by minor geos. The second section shows the influence of jointing well, and the third section is dominated by The Noup, a hill just inside the coast and reaching 700 feet (213 m). In the following section (4) the cliffs follow fairly closely the line of strike and show the influence of the dip. In section 5 the coast is high, usually exceeding 500 feet (152 m), and is spectacular. The northern part of these cliffs shows as a straight line on a map, but in detail they are more indented. Pirkis remarks that 'The cliffs, battered by north and north-westerly gales, must be receding very slowly for, despite their size (1,200–700 feet [366–213 m]), there is only a moderate accumulation of angular slabs at their base, locally known as the "hondins"'. The sixth section shows many caves, arches and stacks and also morainic deposits. The final section (7) is the lowest, and is the only one where a boulder beach is forming. The nature of the boulders, taken in relation to the surrounding rocks, suggests a southerly movement. It is a pity that other and larger islands have not been similarly described and analysed. The slow rate of erosion is emphasized, and if this can be substantiated by actual measurements, it will throw considerable light on Shetland coasts in general.

To describe in any detail the coasts of Shetland would require a volume, and until more research is done on this topic it is not feasible. Certain general points may, however, be made. Allusion has already been made to the general relation of some of the voes to structure. In Mainland some voes in the north-east, Firth's Voe, Dales Voe, Colla Firth and Vidlin Voe, run generally parallel to the strike of the rocks, and in that broad sense there may be a structural control of their shape. In the west and south, Weisdale Voe, Whiteness Voe, Clift Sound, and the two voes almost separating West from East Burra are also of this nature, and so too are Lax Firth and the more southerly Dales Voe on the east coast. On the other hand Olna Firth, Dury Voe, and Wadbister Voe cut across the trend of the rocks. In Unst, Burra Firth appears to follow a line of dislocation, and the same is true of the outer part of Whale Firth in Yell. Balta Sound in Unst, Mid Yell Voe, Ronas Voe and Swarbacks Minn have a general east–west trend. Ronas Voe is continued by a line of dislocation into Colla Firth (not the one mentioned above) and thus the north-western part of Mainland is itself almost divided into two parts. Almost all the large and middle-sized islands are separated from their neighbours by channels trending *in general* north and south – Bluemull Sound, Colgrave Sound, Sullom Voe, Busta Voe, Aith Voe, Sandsound Voe, Linga Sound, Bressay Sound, Noss Sound and Clift Sound (see Plate 26).

Since not only structure but also rock type vary much from place to place in Shetland, it follows that the nature of the cliffs is also variable. The Old Red Sandstone of Walls, Bressay and Noss shows coastal features characteristic of this rock elsewhere. The cliffs at Noss Head are well known. The flags and conglomerates that make most of the coast from Lerwick to Sumburgh Head are somewhat sheltered; the coast is lower and broken by many minor and several large inlets, some of which, including Red Stane and Channer Wick, reach westwards into the metamorphics. Mousa island, separated from Mainland by a channel trending west of north, is grassy. It reaches 150 feet (46 m), and there are some interesting beaches at its southern end. The best preserved broch in Scotland is on this island. The extreme south of the peninsula is Sumburgh Head in the Brindister Flags. The south-western end of the peninsula is more complicated. Gneiss forms the coast of Colsay island and Mainland to the south side of the Wick of Shunni. Fitful Head is in metamorphics and the west facing cliffs are imposing and reach to more than 900 feet (274 m).

The granite cliffs of south-eastern Walls, Muckle Roe, Brae Wick and Ronas Hill offer some magnificent scenery; the cliffs and stacks east of Brae Wick are very fine. The andesites and basalts at Esha Ness and neighbourhood, together with the islets and stacks off-shore, and the intricate coast along the Vullians of Ure and of Hamna Voe, give rugged and jagged outlines. The stacks named The Drongs, near Hillswick, are in metamorphic rocks.

Yell is built almost entirely of gneiss and is reminiscent of the gneissic parts of Scotland. The western coast, in Yell Sound, is somewhat sheltered. In the northern part of the west coast pegmatite veins play a part in the cliff scenery. The east coast is more indented, but Yell is the most uniform of the islands. Unst and Fetlar are much more complicated in structure. The several belts of rock in Unst run roughly in the same direction as the main axis of the island; in Fetlar the trend is to the north-west in the west and to the north-east in the east of the island. All the rocks of both islands are metamorphics which have been subjected to strong earth-movements which have produced nappes and schuppen. Neither the amount nor the direction of these movements are known in either island. However, during the movements the rocks were folded along axes trending north-north-east and south-south-west. The outcrops of the west coast suggest a fairly definite relation between rock type and cliff form, but detailed field work is required before any worth-while explanation can be given. The west coast is faced by steep cliffs in the northern half; in the southern part the cliff height is less. There are numerous geos and stacks. The northmost stacks, Tipta Skerry and Muckle Flugga, are gneissic. The north coast is much indented and cliffed and so too is the east coast as far as Harold's Wick. South of that bay there is more shelter and the cliffs are less prominent (see Plate 28).

Chapter X

The remote islands of Scotland

There are several small islands which are some distance from the mainland and cannot be considered with any of the larger groups. They are seldom visited, but their geological and physiographical significance is important. Several of them are perhaps best known to ornithologists.

ST KILDA

St Kilda is the largest of these islands and has an area of about 1,575 acres (607 ha). It is one of a group of four main islands with which are associated several smaller ones and many stacks and rocks. The fine peak of Conachair reaches 1,396 feet (426 m). The island is smooth, rolling and grassy. Except along the coast, exposures of bare rock are rare, and much of the surface is peat-covered (Fig. 68).

Dùn is separated from the main island by a channel about 100 yards (91 m) wide. The island itself is nearly a mile in length, but its width is generally less than a quarter-of-a-mile. On its south-western side there are fine cliffs, and the waves are in process of reducing the island to a series of stacks. Levenish may be a former extension of Dùn. Boreray and Soay are best defined as large stacks; there is no level ground on them but there are fresh-water springs. The other rocks and stacks are all noteworthy, especially Stac an Arnim, 627 feet (191 m), and Stac Lee, 544 feet (166 m). The former is the highest stack in the British Isles (see Plate 30).

The coasts of all the islands are cut into caves or small geos. Some may be as much as 100 feet long (30 m) and correspondingly deep. Sloping beaches are often found at their ends. As a rule the caves are cut along dykes or a sloping sill. If a dyke crosses a narrow promontory it is more than probable that it will be cut through to form an arch or tunnel. Two of these tunnels, that near Gob an Duin and that under Gob na h'Airde, are major features; the first is 150 feet long (46 m) and nearly 80 feet high (24 m). The second is more than 300 feet long (91 m) and over 100 feet high (30 m). Blow-holes are also frequent. The only low part of the coast on any of the islands is that around Loch Hirta or Village Bay. Conachair is a granophyre hill which descends steeply to sea-level. The west sides of St Kilda, Boreray and Soay are made of gabbro, dark in colour. The jointing of this rock is marked but somewhat irregular. Marine and subaerial erosion have pro-

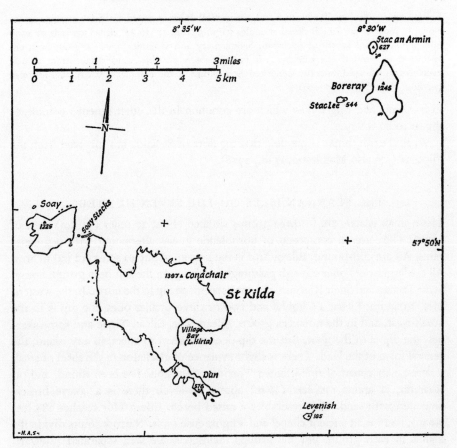

Fig. 68. Outline sketch of St Kilda (based on Ordnance Survey; Crown Copyright Reserved)

duced great bastions and castellated masses separated by deep chasms and screes. On the north side of Boreray the ridge is cut into spire-like forms. The north coast of Soay is a cliff about 1,000 feet high (305 m).

Although there is no direct evidence of glaciation certain features suggest that ice, possibly in the forms of floes, reached the islands. There are boulders of gneiss on Hirta beach; they may, however, be ballast. Soil samples from near the house in which the factor used to live show fragments of pink garnet and other minerals associated with gneiss boulders. No certain remains of any raised beach have been found.

All the islands are built of igneous rock, and represent all that is left of a major complex, perhaps six or seven miles (10 or 11 km) in diameter. If this is so, the centre was somewhere between St Kilda and Boreray. It was an igneous complex directly comparable to those of Skye, Rhum, Ardnamurchan, Mull and Arran.

The intrusions are sheet-like in form and, the vertical dykes excepted, intrusive margins can be shown in most cases to be inclined at angles varying from 15° to 80° either towards or away from the supposed centre of intrusion. Sedimentary and volcanic rocks have not been encountered on any of the islands ... It has not been possible ... to demonstrate arcuate structures in the field, and the suggested ring nature of the complex must remain an open question. (Cockburn (1933–6))

Cone sheets and ring dykes which are common in the other igneous complexes appear to be absent.

W. B. Turrill thinks it possible that the flora of St Kilda has survived from the Pliocene (see also Mathieson, *et al.*, 1928).

THE FLANNAN ISLES OR THE SEVEN HUNTERS

These small islands are in three groups situated about 20 miles (32 km) west of Lewis. They are all composed of hornblende gneiss alternating with pegmatite veins. All are cliff-bound. Eilean Mór is the largest, about 30 acres (12 ha) in area. All the larger ones offer enough pasturage for sheep. In the southern group, Soray, Sgeir Toman and Sgeir Righinn, the gneiss seems to dip to the north. In the western isles, Roareim, Eilean a'Ghobha and two or three smaller ones, the dip is to the north-west, and in the northern group, Eilean Mór, Eilean Tighe and some skerries, the dip is to the west. But the dip does not seem to affect, in any island, the general form of the land. There is slight evidence of glaciation in the form of small erratics; a specimen of fine-grained Torridon Sandstone has been found, and on Roareim, at about 150 feet (46 m) above sea-level, there is a coarse breccia cemented with sand. This resembles a raised beach. Eilean Mór reaches 288 feet (88 m); there is an ancient chapel and a lighthouse on it. Narrow straits divide the islands in each group, but the groups themselves are well separated (Stewart, 1933).

SULA SGEIR

This isolated rock, about half-a-mile long in a north-east to south-west direction, and everywhere less than 200 yards broad (183 m), is another fragment of hornblende gneiss with veins of pegmatite. The gneiss appears to dip to the north in the southern part of the island, and more to the west in the northern half. The island is nearly divided into two parts by a low col over which the waves wash in severe weather. Both parts reach rather more than 200 feet (61 m), and are surrounded by cliffs in which a number of geos and caves have been cut. There is also a tunnel cutting right through the island. There are some smaller and lower rocks near the main island; Bogha Corr is nearly a mile to the north-east and Gralisgeir about half-a-mile to the south. Both are very small (Stewart, 1933).

NORTH RONA

North Rona is a triangular-shaped island each side of which is, measured in a straight line, about a mile long. The highest point, 355 feet (108 m), is in the southeast corner. The northern arm, Fianuis, is relatively flat and about 100 feet high (30 m). The rocks of which the island is formed are assumed to be Lewisian in age, but they have not been exactly matched on the mainland. There are many veins of pegmatite and locally they are an important element in the scenery. One large vein makes the imposing cliff, facing north, of the central ridge of the island. The gneiss shows marked dips, and where they are steep they have a pronounced effect on the landscape. In such places weathering of the scarps and marine action in cutting geos and caves give a coast of great variety and interest. Marcasgeo separates the Sceapull peninsula, and there is a tunnel which divides Fianuis from the main part of the island. Since the dip of the rock is to the south-south-west, the scarps face a little east of north. This is marked on the west where the dip may reach 50°.

The power of the waves is well illustrated on the western side of the Fianuis peninsula. There is a storm beach on top of the cliffs and standing about 20 yards (18 m) back from their edge. It is about 600 yards long (549 m), 10 feet high (3 m) and 20 feet broad (9 m). The average height of the beach above sea-level is 70 feet (21 m). The rock which produces the boulders of the beach is easily broken by the waves on account of its fissility. There are off-lying rocks and skerries at each corner of the main island. St Ronan's Church is near the centre of the island (Stewart, 1932).

SULE SKERRY

This small islet is formed of gneiss which has a bedded appearance. It is for the most part low-lying; the maximum height is 40 feet (12 m). The dip of the division planes in the gneiss is 30° to 40°. The sea has cut geos along shatter belts. The islet is about 550 yards long (501 m) in a north–south direction, and rather less in its maximum east–west width. There is a lighthouse on it (Walker, 1931).

SULE STACK OR STACK SKERRY

This small islet, like Sule Skerry, stands on the 30-fathom (55 m) Skerry Bank. It is well known for its gannetry. It is another fragment of gneiss, well-rounded, and rises to 120–30 feet (37–40 m). It is about 60 acres (24 ha) in area. There is no vegetation, but a complete cover of guano. The main axis of the island is nearly north–south, and a narrow gully almost cuts the rock into two parts. The rock is resistant and weathers into rounded surfaces. There is a marked contrast between the steep western cliffs and the smoother and sloping ones on the east. The gully is the result of marine and atmospheric action and is associated with structures in the cliffs which resemble giant stairs. There is another gully connecting with a vertical shaft at the northern end. As in the other isolated islets, pegmatite veins are characteristic. It will be appreciated that the Atlantic swell which is constant, the carpet of guano and nearly 4,000 gannets make observations difficult and unpleasant! (Stewart, 1938).

All these islands stand on the continental shelf, and suggest a former extension of the land. Since they are relics they have an especial appeal. The lighthouse of the Flannans is occupied. The permanent occupants of St Kilda were evacuated in 1930; since then there have been official visitors who spend some periods on the main island from time to time.

ROCKALL

Rockall is 70 feet high (21 m) and 80–100 feet wide (24–7 m) at the base (sea-level) and rises almost vertically from the sea. It is about 200 miles (322 km) west-northwest of Barra Head. It is somewhat conical in form and is the only exposed part of the Rockall Bank, which is about 60 miles long (97 km) and 30 (48 km) wide and 65 to 100 fathoms (119 to 183 m) deep. The bank rises from depths of about 1,000 fathoms (1,829 m) and is more than 100 miles (161 km) west of the continental shelf. Two reefs, Helen's Reef and Haslewood Rock, are nearby. Rockall is made of aegirine-granite, probably of Tertiary age. Its sides are very steep and the east face is almost vertical. Air photographs show a system of joints, and the east face corresponds with those trending north. Other joints strike roughly eastwards and determine the moderately dipping main northern slope and the vertical faces on the north-east of the island. Soundings suggest that the east face extends for about two miles or more. It is thought that the east face may be a fault. There is no wave-cut platform. Sabine suggests that the reason it remains above sea-level is because the rock fragments are so quickly removed that they cannot be used by the waves, and also because the constant washing by spray prevents frost action. The rock and bank pose many interesting geological problems with which we are not here concerned (Sabine, 1960). (See p. 143 for Skerryvore and Dubh Artach.)

Chapter XI

The conservation of the coast

Before the First World War, and even as late as 1939, those who visited the Highlands and Islands of Scotland were limited in numbers. Some were interested in grouse shooting, stalking or fishing; more were touring by car or bus to areas of fine scenery along the somewhat restricted road systems. There were some private cars on the lesser roads, but only at few points was the coast visited by large numbers – and numbers small compared with those of today. Climbing and mountain walking were popular, but appealed to few. For some years after the Second World War petrol rationing limited the numbers of visitors even more effectively. Once, however, car travel became unlimited, more and more people began to go to Scotland. Islands like Skye and Mull and the main tourist centres in the Highlands became rapidly more popular, and it was not long before people began to realize that parts of the coast were most attractive.

The climate, however, is to some extent a limiting factor. Parts of the west coast are very wet, and several stations record more than 80 inches (2,032 mm) in a year. The wettest area extends from the northern parts of Skye and the adjacent mainland to approximately the latitude of Glasgow. But the Outer Hebrides, Tiree and Coll (which are notably both sunny and windy) are excluded, and even the coastal fringe of the mainland has markedly less rain than the mountainous interior. The north-west and south-west are considerably less wet as measured in total annual rainfall. The sunshine records show that May and June are the most favoured months in the mid-west coast. The north and east of Scotland have a drier climate. Nevertheless not all summers are alike and the remarkably fine one of 1969 led to a great increase in tourism and encouraged many people to return in the following years. Unfortunately, from the tourist point of view, the summer climate, especially in the west, cannot be relied upon to provide long spells of dry and windless weather. Moreover, the temperatures on the whole coast are lower than in the south. The west coast is mild throughout the year and there are often spells of beautiful weather, but throughout the year winds may be persistent and often strong. The outer islands have fresh to strong winds about one day in five, and gales perhaps one day in ten, but often with much sunshine. On the north and east occasional high temperatures occur in summer, and there is far more sunny weather. But in the winter half year it is distinctly cool, and east and north-east winds are not infrequent. Nevertheless, parts of the east coast have a very pleasant climate, and that of the strip along the Moray Firth between Nairn and the Spey is excellent. Despite the variability of the weather over much of the country there is no doubt that the beauty of the coast and its relative remoteness make a great appeal to tourists and, although numbers may fluctuate somewhat from year to year, there is likely to be a continuous increase.

People are now more adventurous; caravans and even tents are much pleasanter to live in in bad weather than they were, and hotel accommodation in the High-lands and Islands has improved. Even more important is the modernization of the road system. Not only the surfaces, but also the widths and passing places on many roads have been greatly improved, and the density of motor traffic in the holiday season has increased enormously. Nevertheless some of the roads are still inade-quate for the traffic in the height of the tourist season. Since in many parts the roads follow, or are close to, the sea, it is easy to set up caravans and tents on many parts of the coast. Anyone who visits, for example, the north coast of Sutherland in summer and explores the small and beautiful bays along it will be surprised or alarmed at the numbers of caravans and tents. They do not compare with the much greater numbers in the southern parts of Britain, but the spread is indicative of an appreciation of the coast and of a desire to find more and more remote places, a desire frustrated by the better roads which engender far more and much faster traffic. Moreover, further improvements of the roads will make the matter worse!

Thus, although this may seem a reasonable development, it imposes great problems. The ideal sites for camping on the coast are in or near dunes, near good beaches, or sandy pastures or on machair land in the Hebrides. But these are also the most vulnerable areas. Achmelvich Bay on the west coast of Sutherland began to be popular immediately after the 1939–45 war; now it is well known and there has been considerable deterioration in the dunes and sandy pastures. This comes about largely as a result of driving in cars and caravans, and still more as a result of turning them. This breaks the turf cover already thinned by grazing of sheep, cattle, ponies and rabbits or by cultivation, and exposes the sand which soon begins to blow in the prevalent windy weather. There are many similar sites on the north-west coast, for example, Inverasdale and Gairloch. On the west coast at Stoer Bar and Gruinard, and near Arisaig and Morar there are beautiful beaches and sandy areas which are most popular. Many beaches have been spoiled to some degree, and unless some action is taken all may be lost both to agriculture and tourism. The same problem is present on other parts of the coast, but, if, as near St Andrews and other places on the east coast, the camps are on solid ground and the beaches are reached by a short walk, no damage is likely unless paths are made indiscriminately through dunes. On the other hand, there may be spoiling of views if camps are sited in the wrong places. The chief danger lies in the small bays and beaches reached by comparatively few cars and caravans, and where there is no effective check. Local farmers, crofters, and others often encourage campers, and in the sparsely inhabited areas it is not easy for local authorities to maintain a proper supervision.

In the islands the tourist problem is less acute, but the machair and dunes are already substantially damaged by agriculture. It is simple to cross by car ferry to the larger islands and the road systems of The Long Island, Skye, Mull and Islay are good. There are facilities to take a car to most of the larger islands of Orkney and Shetland, and some cars and tractors are present on almost all the smaller islands. Most of them are locally owned but are not infrequently used in such a way as to spoil dunes and beaches. Moreover, derelict cars are all too frequently eyesores on the islands (e.g. Barra). Sometimes, however, as at the eastern end of St Ninian's Ayre in Shetland, sand has been removed for commercial purposes. The amount of gravel removed near Kingston (Spey Bay) has considerably altered the appearance of the coast. Fortunately it has been taken well behind the beach, so no erosion has been caused. In any locality, removal of sand or gravel can be a menace if there is not a constantly renewed supply from off-shore (see p. 260, St Andrews). It can be a particularly dangerous practice on small beaches. On the machairs of The Long Island and Tiree there are already a few caravans which are harmless if not beautiful. If their numbers were appreciably increased great damage could ensue since the machair surface is so easily broken, and therefore the amenity of the landscape ruined.

There is no doubt that pressure on the coast of Scotland will increase. The recent planning application to build a hotel and chalets on Staffa illustrates the desire to exploit remote places. A hotel is to be built close to easily damaged machair in Barra. The new caravan site at Inverasdale, in Wester Ross, may give much pleasure to many but it is not an improvement to a fine sandy beach, and involves approach difficulties on a cul-de-sac road. On the other hand to build hotels and chalets on or near the coast in carefully chosen and properly sited localities is all to the good. The coast is remarkably beautiful and should be open, subject to certain local safeguards, for all to enjoy it. But the indiscriminate development that took place in much of England and Wales before the 1939–45 war must not be repeated in Scotland. Legislation certainly restricts this, but constant care is necessary, both in the vigilance of local authorities, in the control of caravans and tents, and in public enquiries to counter proposals for ill-sited developments. At Gairloch and Torridon working parties, including the Ross and Cromarty County Council, The Highlands and Islands Development Board, the National Trust for Scotland, the Nature Conservancy, the Countryside Commission for Scotland, the Crofters Commission, the Forestry Commission and the Wester Ross Tourist Association, are considering suitable tourist and recreation activities in those localities. This is an important move and one that is to be encouraged in other vulnerable places.

Along those parts of the Scottish coast near to, or within easy distance of large towns, the problem is much the same as in England and Wales. The Clyde coast is the natural outlet for Glasgow and the neighbouring towns, and there must be not only tourist and holiday development, but also industrial development. The mainland coast is almost all used for urban spread or industry much of the way from Greenock to Ayr, and more is pending. Great attention is given to this area by the Countryside Commission for Scotland, and by voluntary conservation bodies. In their second and third reports (1969 and 1970) the Commission explain the decision of the Secretary of State not to proceed with the proposal to build an oil terminal at Portencross and a refinery at Longhaugh Point. On the other hand the proposed power station at Inverkip and a deep-water terminal at Hunterston have been allowed. The important point is that these cases, which serve well as typical examples, were fully discussed and decisions were reached after public local enquiries. The Countryside Commission very properly suggested that there should be a regional plan.

We consider it essential that the Commission are fully consulted about the remaining proposals on which a decision is still to be taken, together with any future proposals for the area, so that they can all be considered in the overall context of a regional plan for the Clyde. In this respect it is helpful that a regional planning body – the West Central Scotland Plan Steering Committee – now exists for the area, and we hope that it will become sufficiently well established in time to influence the decisions still to be taken on further major proposals.

Farther south along the Clyde coast the first country park in Scotland has been established, in 1969, at Culzean. This arrangement was made between the owners (the National Trust for Scotland), the Ayr County Council, Ayr Town Council, and Kilmarnock Town Council. That agreements of this sort are possible is an excellent augury for the future.

The Countryside Commission for Scotland has been enterprising in other ways concerning the coast. The Department of Geography at the University of Aberdeen has been asked to report on many parts of the coast of the Highlands, and has produced valuable documents on Caithness, Ross and Cromarty, Inverness-shire, Sutherland, the Uists, Barra and Lewis. These writings are not merely descriptive, but are comprehensive and informative; they contain the most reliable information available about the numerous beaches, sandy or otherwise, in the northern Highlands. What is more, anyone interested in the coast from the physiographical and ecological points of view must take them into careful consideration.

The Nature Conservancy, through its Scottish Committee, has also taken a great interest in the coast. The National Nature Reserves that contain parts of the coast are the Isle of May, Tentsmuir Point, St Cyrus, the Sands of Forvie, Noss, Hermaness, Haaf Gruney, North Rona and Sula Sgeir, St Kilda, Invernaver, Inverpolly, Rhum, Loch Druidibeg, the Monach Islands and Caerlaverock. There are also numerous Sites of Special Scientific Interest all round the coast and in the islands. The Conservancy is concerned with the scientific interest of the lands under its supervision. But, in practice, scientific interest and landscape beauty are so closely related on the coasts of Scotland that it may be taken for granted that SSSIs are almost all areas that can be regarded as high-quality coast. Provided a careful watch is kept by the Conservancy and the Local Authorities concerned there is every reason to think that much of the best of the coast is under some form of protection. Unfortunately an SSSI carries no guarantee of preservation, but most of those on the Scottish coast are not under the pressure that pertains to many in England. Places such as Ailsa Craig, St Abb's Head, Siccar Point, the Mull of Galloway, Handa Island, Duncansby Head, Kerrera Island, Staffa, the Treshnish Islands, the Garvellachs, Eigg and Canna can be guarded. On the other hand Aird Torrisdale, Scourie, Morar, Benderloch, Troon Golf Links and fore-shore, the North Berwick coast and Joppa shore are examples of places which demand a greater amount of attention.

There is also one local coastal nature reserve at Aberlady Bay. Reference (see p. 270) has already been made to it. The report prepared for the East Lothian County Council is a most useful document.

The National Trust for Scotland was founded in 1931. Since that time it has acquired a considerable number of properties, but its coastal holdings, although of great value, are relatively few. There was no Enterprise Neptune in Scotland, nor was one necessary in the sense that it was in England and Wales. On the other hand a Trust property, whether in Scotland or England, is safer than any other since if it is, as most are, inalienable, any alteration in its boundaries or uses, except at the sole discretion of the Trust itself, can be made only by Act of Parliament. This is a great safeguard and is not one possessed by other bodies. Some of the more important properties under the aegis of the Trust include Fair Isle (Zetland), Rough Island (Kirkcudbrightshire), St Kilda and Shieldaig island (Ross and Cromarty).

As pressure increased the National Trust for Scotland resolved that the most practicable way of affording protection to the coastline was by means of Conservation Agreements negotiated between the owners and the Trust. The first of these, covering what have proved to be critical areas in Fleet Bay, Kirkcudbright-shire, were effected in 1943 and in respect of the West Links, North Berwick, in 1944. Subsequent agreements have afforded similar protection on nine miles (14.5 km) of the Ayrshire coast from the border of the Trust property at Culzean to the Heads of Ayr; the land in Wester Ross around Loch Shieldaig and Gairloch, and three small islands in the loch; five miles (8 km) of the north Sutherland coast, east of Durness, including Smoo Cave; and further agreements are in process of negotiation in Sutherland and on the east coast of Scotland.

Local and county trusts do not play the same individual part as they do in England because their objectives are pursued through the regional branches of the Scottish Wildlife Trust. This Trust has greatly expanded in recent years and is now of considerable influence through Scotland.

Until quite recently the pressure on the coast, except around the Clyde and parts of the Firth of Forth, has been comparatively small so that there has not been the need to form conservation or preservation societies. In view of the national activities of the Nature Conservancy, the National Trust for Scotland and the Countryside Commission for Scotland there has not been the same urgency to form local societies, but these three bodies as well as the Scottish Wildlife Trust do encourage local initiatives. The pressures are, however, increasing.

The Forestry Commission has certain extensive properties on or close to the coast. At Tents Muir in Fife, and on the Culbin sands and Burghead Bay on the Moray Firth, tree planting has obscured and stabilized some remarkable dunes. On the other hand there is still a wide strip of beach and foredunes in front of all three areas. It is desirable that these should remain.

The Scottish coast and islands are not as spoiled as is the coast in many parts of England and Wales. Nevertheless there has been much disfigurement. Natural scrub and woodland have disappeared, so too have sea-bird colonies as a result of oiling, sonic boom, and low-flying jets. The dumping of rubbish and old cars is often a serious blemish.

There is a certain amount of building near big towns and in some other places, for example, between Portknockie and Macduff, between Elie and Crail, and on the Solway between Sandyhill Bay and Rockcliffe. In the wilder and less inhabited parts the cliffs are open and unspoiled and care should be taken to make some rather more accessible, but in such a way as not to injure them. Cliff paths on the north and west coast, except locally, are unlikely to follow the indented nature of the coast. But there is the distinct possibility that private enterprise may lead to helicopters carrying passengers to many remote places. There may also be an increase in power boats so that although many people may be given the chance to see places difficult of access, there will also be the possibility of their being spoiled in one way or another. Coastal footpaths could be made with advantage between Arbroath and Stonehaven, around the coast of Caithness, between Fast Castle and St Abb's Head, and along the west coast of the Rhinns of Wigtownshire.

Because so much of Scotland is surrounded by the raised beach platforms many cliffs are fossil and do not possess the grandeur of those washed by an open ocean. Since the platform is usually narrow it may be followed by a road, often a main road. There may be occasional houses built on it in its wider parts. Where, however, there is a broad platform it presents obvious sites for development in one form or another. Moreover, the platform if covered by beach deposits is often very fertile and suitable for crofts. Parts of the western side of the Kintyre peninsula could easily be spoiled in this way and another potentially dangerous area is that near Inverbervie and Gourdon. The combination of modern planning legislation and a climate that is not as attractive to many people as that of southern Britain may suggest that it is relatively easy to conserve the coast of Scotland, but pressures will continue to increase so that every care and foresight must be used. It would be all too easy, even by slight development, to spoil places such as the many beautiful sand dunes and beaches of Aberdeenshire and those of the Moray Firth. The changes at Findhorn since the last war illustrate this. The limited but unattractive 'urbanization' of John o'Groats is regrettable.

There are some parts of the coast of Scotland which are owned by the Ministry of Defence and to which access is limited or forbidden. These include part of the fine dune area in Luce Bay, Buddon Ness, Morrich Mhór, the Hebrides range in South Uist and the bombardment range at Cape Wrath. In relation to the whole coast they are not serious restrictions, and although it may seem a pity that the Luce Bay dunes are included since they make the finest area of that nature in the south of Scotland, there is no doubt but that the dunes are largely protected under their present use.

In the Firth of Forth and the Firth of Clyde, and to a lesser extent in the Firth of Tay and a few other places, pollution of inshore waters is serious. In the Clyde the number of resorts implies that every precaution must be taken to minimize the effects of pollution. The Clyde River Purification Board is doing valuable work on this problem. In the Forth pollution is most offensive in the upper and northern reaches, but is not entirely absent in the broad outer firth. The city of Edinburgh still discharges its sewage directly into the firth. The part of the firth from Burntisland to Kinghorn is depressing, but it is short compared with similar areas in England and Wales. Recent developments related to North Sea Oil in Aberdeenshire, on the Moray Firth coast, in Cromarty Firth and in the Forth imply the possibility of pollution unless great care is taken.

Coastal conservation has reached a far more advanced stage in Great Britain than elsewhere. Unfortunately, however, the practice of conservation started late, and consequently large parts of the coast of southern Britain were spoiled before action could be taken. National bodies like the Nature Conservancy and the Countryside Commissions fortunately have to review the whole of the coast. Unless this is done, and done carefully, it is all too easy for mistakes to be made in planning and development. The acceptance of Heritage Coasts in England and Wales is significant. They are the most attractive stretches and great care is to be taken in the uses and activities which are or are not likely to be permissible in them. In Scotland this step has not been taken. But anyone who has seen the whole of the Scottish coast, and that of the islands, will be aware that many parts are spoiled and that development of one kind or another is taking place in many areas. It is surely better to consider the coast as a whole and to define those parts which are of the highest category so that steps can be taken to ensure that they are not spoiled. Enough is already known about the coast for this to be done. In addition, experience has shown that, while planning machinery operates fairly satisfactorily, there is need for intimation of development proposals and consultation between the sponsors and the conservation bodies at a very much earlier stage, and in Scotland there is an urgent case to be made for the creation of a central consultative group representative of the voluntary conservation organizations, the National

Trust for Scotland and Scottish Civic Trust in addition to the statutory authorities. What is true of the mainland is equally true of the islands. They are becoming increasingly popular and nearly all of them possess magnificent beaches which can so easily be ruined. It is pertinent to point out that trees are largely absent on the islands except in very sheltered places, and consequently buildings cannot easily be screened. On the other hand they can be very conspicuous so that great care is required in siting new structures. The significance of this point is well exemplified along the north-west coast of Lewis.

In this book I have tried to give an account of the physiography, nature and beauty of the coast of Scotland, and in this final chapter I have attempted an essay on the conservation of that coast. The constant need for the care of the coast cannot be over-emphasized. The demands on the coast are always increasing. Unless the population as a whole becomes aware of the need to appreciate the amenities of the coast, damage will continue to increase. Pollution of one kind or another is an ever present menace, and no part of the coast of the mainland and islands of Scotland can now be considered 'safe' unless under the guardianship of a public or private body capable of exercising its powers to the full in order to prevent spoliation. This may seem a gloomy view but it is not an exaggerated one.

References

NOTE. If the pagination is continuous through a volume including more than one year, the number of the volume and the first page of the reference is given.

CHAPTER I

Chapman, V. J. 1960. *The Salt Marshes and Salt Deserts of the World*. London.

George, T. N. 1965. The geological growth of Scotland. In *The Geology of Scotland*, ed. G. Y. Craig, Chap. 1. Edinburgh and London.

Gimingham, C. H. 1964. The maritime zone. In *The Vegetation of Scotland*, ed. J. H. Burnett, Chap. 4. Edinburgh and London.

Godard, A. 1965. *Recherches de Géomorphologie en Écosse du Nord-Ouest*. Strasbourg.

Hollingworth, S. E. 1938. The recognition and correlation of higher level erosion surfaces: a statistical study. *Quart. Journ. Geol. Soc.*, 94, 55.

Kennedy, W. Q. 1946. The Great Glen Fault. *Quart. Journ. Geol. Soc.*, 102, 41.

Lewis, J. R. 1957. Inter-tidal communities of the northern and western coasts of Scotland. *Trans. Roy. Soc. Edinburgh* 63, 185.

Sissons, J. B. 1967. *The Evolution of Scotland's Scenery*. Edinburgh and London.

Vevers, H. G. 1936. The land vegetation of Ailsa Craig. *Journ. Ecology* 24, 424.

CHAPTER II

Bremner, A. 1942. The origin of the Scottish river system. *Scot. Geogr. Mag.*, 58, 54.

Cloos, H. 1939. Hebung, Spaltung und Vulcanismus. *Geol. Rundschau* 30, 401 and 637.

Davis, W. M. 1895. The development of certain English rivers. *Geogr. Journ.*, 5, 127.

George, T. N. 1954–5. Drainage in the Southern Uplands: Clyde, Nith, Annan. *Trans. Geol. Soc. Glasgow* 22, 1.

1958. The geology and geomorphology of the Glasgow district. *Brit. Assoc. Adv. Science*, Glasgow meeting.

1965. The geological growth of Scotland. In *The Geology of Scotland*, ed. G. Y. Craig, Ch. 1. Edinburgh and London.

1966. Geomorphic evolution in Hebridean Scotland. *Scot. Journ. Geol.*, 2, 1.

Gregory, J. W. 1927. The fiords of the Hebrides. *Geogr. Journ.*, 69, 193.

1913. *The Nature and Origin of Fiords*. London.

Jardine, W. G. 1959. River development in Galloway. *Scot. Geogr. Mag.*, 75, 65.

Johnson, D. W. 1919. *Shore Processes and Shoreline Development*. New York and London.

Linton, D. L. 1933. The origin of the Tweed drainage system. *Scot. Geogr. Mag.*, 49, 162.

1934. On the former connection between the Clyde and the Tweed. *Scot. Geogr. Mag.*, 50, 82.

1949. Watershed breaching by ice in Scotland. *Trans. Inst. Brit. Geogrs.*, 15, 1.

1951. Problems of Scottish scenery. *Scot. Geogr. Mag.*, 67, 65.

Linton, D. L. and Moisley, J. A. 1960. The origin of Loch Lomond. *Scot. Geogr. Mag.*, **76**, 26.

Louis, H. 1934. Glazialmorphologische Studien in den Gerbirgen der Britischen Inseln. *Berliner Geogr. Arbeiten*, Part 6.

Mackinder, H. J. 1907. *Britain and the British Seas*. Oxford.

Mort, F. W. 1918. The rivers of south-west Scotland. *Scot. Geogr. Mag.*, **34**, 361.

Peach, B. N. and Horne, J. 1936. *Chapters on the Geology of Scotland*. Edinburgh.

Robinson, A. H. W. 1949. Some clefts in the Inner Sound of Raasay. *Scot. Geogr. Mag.*, **65**, 20.

Sissons, J. B. 1967. *The Evolution of Scotland's Scenery*. Edinburgh and London.

Sölch, J. 1936. Geomorphologische Probleme des Schottischen Hochlands. *Mitt. Geogr. Gesellschaft*. Vienna.

Ting, S. 1937. The coastal configuration of western Scotland. *Geogr. Annaler* **19**, 62.

Zenkovich, V. P. 1967. *Processes of Coastal Development*. Edinburgh and London.

CHAPTER III

Baden-Powell, D. F. W. and Elton, C. 1936–7. On the relationship between a raised-beach and an Iron Age midden on the island of Lewis. *Proc. Soc. Antiq. Scotland* **71**, 347.

Charlesworth, J. K. 1955. Late-glacial history of the Highlands and Islands of Scotland. *Trans. Roy. Soc. Edinburgh* **62**, 769.

Donner, J. J. 1959. The Late- and Post-glacial raised beaches in Scotland. *Anns. Acad. Scient. Fennicae* **53**, 5.

Jamieson, T. F. 1865. The history of the last geological changes in Scotland. *Quart. Journ. Geol. Soc.*, **21**, 161.

King, C. A. M. and Wheeler, P. T. 1963. The raised beaches of the north coast of Sutherland, Scotland. *Geol. Mag.*, **100**, 299.

McCann, S. B. 1961. The raised beaches of western Scotland. Thesis for Ph.D. degree, Cambridge.

Nicholls, H. 1967. Vegetational change, shoreline displacement and the human factor in the Late Quaternary history of south-west Scotland. *Trans. Roy. Soc. Edinburgh* **67**, 145.

Ritchie, W. 1966. The Post-glacial rise in sea-level and coastal changes in the Uists. *Trans. Inst. Brit. Geogrs.*, No. **39**, 79.

Sissons, J. B. 1962. A re-interpretation of the literature of the late-glacial shoreline in Scotland, with particular reference to the Forth area. *Trans. Edin. Geol. Soc.*, **19**, 83.

1966. Relative sea-level changes between 10,300 and 8,300 B.P. in part of the Carse of Stirling. *Trans. Inst. Brit. Geogrs.*, No. **39**, 19.

Sissons, J. B., Smith, D. E. and Cullingford, R. A. 1966. Late-glacial and Post-glacial shore-lines in south-east Scotland. *Trans. Inst. Brit. Geogrs.*, No. **39**, 9.

Wright, W. B. 1937. *The Quaternary Ice Age*, 2nd ed. London.

CHAPTER IV

Bailey, E. B. and Maufe, H. B. 1916. Mems. Geol. Surv. Scotland. *The geology of Ben Nevis and Glencoe.*

Boyd, A. J. 1956. The evolution of the Fionn Loch area, Sutherland. *Trans. Edin. Geol. Soc.*, **16**, 229.

Charlesworth, J. K. 1955. Late-glacial history of the Highlands and Islands of Scotland. *Trans. Roy. Soc. Edinburgh* **62**, 769.

Coles, J. M. 1964. New aspects of the Mesolithic settlement of south-west Scotland. *Trans. Dumfries and Galloway Nat. Hist. and Antiq. Soc.*, **41**, 67.

Cormack, W. F. and Coles, J. M. 1968. A Mesolithic site at Low Clone, Wigtownshire. *Trans. Dumfries and Galloway Nat. Hist. and Antiq. Soc.*, **45**, 44.

Eyles, V. A., *et al.* 1949. Mems. Geol. Surv. U.K. *The Geology of Central Ayrshire.*

Godard, A. 1965. *Recherches de Géomorphologie en Écosse du Nord-Ouest.* Strasbourg.

Heddle, M. F. 1880–1. The Hebridean Gneiss. *Min. Mag.*, **4**, 205.

Hill, J. B. 1905. Mems. Geol. Surv. Scotland. *The geology of Mid-Argyll.*

Holgate, N. 1969. Palaeozoic and Tertiary transcurrent movements on the Great Glen Fault. *Scot. Journ. Geol.*, **5**, 97.

Irvine, D. R. 1873. Mems. Geol. Surv. Scotland. *West Wigtownshire.*

Kennedy, W. Q. 1946. The Great Glen Fault. *Quart. Journ. Geol. Soc.*, **102**, 41.

1955. The tectonics of the Moine anticline and the problem of the north-west Caledonian front. *Quart. Journ. Geol. Soc.*, **110**, 357.

Kynaston, H. and Hill, J. B. 1908. Mems. Geol. Surv. Scotland. *The geology of the country around Oban and Dalmally.*

Lambert, R. St J. 1958. A metamorphic boundary in the Moine schists of the Morar and Knoydart districts of Inverness-shire. *Geol. Mag.*, **95**, 117.

1959. The mineralogical metamorphism of the Moine schists of the Morar and Knoydart districts of Inverness-shire. *Trans. Roy. Soc. Edinburgh* **63**, 553.

Lamont, A. 1945. The migration of beach material in the Kyles of Bute and Loch Striven area, and in North Wales. *Trans. Buteshire Nat. Hist. Soc.*, **12**, 84.

Lee, G. M. 1920. Mems. Geol. Surv. Scotland. *The Mesozoic rocks of Applecross, Raasay and North-east Skye.*

McCallien, W. J. 1919–26. The structure of South Knapdale. *Trans. Geol. Soc. Glasgow* **17**, 377.

1926–8. The geology of Gigha. *Trans. Roy. Soc. Edinburgh* **55**, 395.

1928–31. Preliminary account of the Post-Dalradian geology of Kintyre. *Trans. Geol. Soc. Glasgow* **18**, 40.

1928. The surface features of Kintyre. *Scot. Geog. Mag.*, **45**, 219.

1939. The rocks of Bute. *Trans. Buteshire Nat. Hist. Soc.*, **12**, 84.

McCallien, W. J. and Lacaille, A. D. 1940–1. The Campbelltown raised beach and its contained stone industry. *Proc. Soc. Antiq. Scotland* **75**, 55.

Macculloch, J. 1819. *A description of the Western Isles of Scotland.* 3 vols. London.

McCann, S. B. 1961. Some supposed 'raised beach' deposits at Curran, Loch Linnhe, and Loch Etive. *Geol. Mag.*, **98**, 131.

1963. The Interglacial raised beaches and readvance moraines of the Loch Carron area, Ross-shire. *Scot. Geogr. Mag.*, **79**, 164.

1966. The limits of the Late-glacial Highland, or Loch Lomond, re-advance along the West Highland seaboard from Oban to Mallaig. *Scot. Journ. Geol.*, **2**, 84.

Marshall, D. N. 1939. A survey of the caves of Bute and the Cumbraes. *Trans. Buteshire Nat. Hist. Soc.*, **12**, 113.

Marshall, J. R. 1962*a* The physiographic development of Caerlaverock merse. *Trans Dumfries and Galloway Nat. Hist. and Antiq. Soc.*, **39**, 102.

1962*b* The morphology of the Upper Solway salt marshes. *Scot. Geog. Mag.*, **78**, 81.

Morss, W. L. 1925–6. The plant colonization of the merse lands in the estuary of the river Nith. *Trans. Dumfries and Galloway Nat. Hist. and Antiq. Soc.*, **13**, 162.

Peach, B. N. and Horne, J. 1899. Mems. Geol. Surv. U.K. *The Silurian Rocks of Britain*, Vol. 1, *Scotland.*

1907. Mems. Geol. Surv., Scotland. *The geological structure of the North-west Highlands of Scotland.*

Peach, B. N. and Kynaston, H. 1909. Mems. Geol. Surv. Scotland. *The Geology of the seaboard of Mid Argyll.*

Peach, B. N., Horne, J., *et al.* 1910. Mems. Geol. Surv. Scotland. *The geology of Glenelg, Lochalsh and the south-east part of Skye.*

Peach, B. N. *et al.* 1911. Mems. Geol. Surv. Scotland. *The geology of Knapdale, Jura and North Kintyre.*

Perkins, E. J. 1968. The marine fauna and flora of the Solway Firth area. *Trans. Dumfries and Galloway Nat. Hist. and Antiq. Soc.*, **45**, 15.

Reid, R. W. K. *et al.* 1966–7. Prehistoric settlements in Durness. *Proc. Soc. Antiq. Scotland* **99**, 25.

Richey, J. E. 1961. *British Regional Geology Scotland: The Tertiary Volcanic Districts* (revised by A. G. Macgregor and F. W. Anderson). HMSO.

Richey, J. E., Thomas, H. H., *et al.* 1930. Mems. Geol. Surv., Scotland. *The geology of Ardnamurchan, North-west Mull and Coll.*

Richey, J. E. *et al.* 1930. Mems. Geol. Surv. U.K. *The geology of north Ayrshire.*

Ritchie, J. N. G. 1966–7. Keil Cave, Southend, Argyll. *Proc. Soc. Antiq. Scotland* **99**, 104.

Ritchie, W. and Mather, A. 1969. *The Beaches of Sutherland.* Dept of Geography, Aberdeen, for The Countryside Commission of Scotland.

Scott, J. F. 1928–31. The general geology and physiography of Morvern. *Trans. Geol. Soc. Glasgow* **18**, 149.

Smith, J. 1889–93. The sandhills of Torrs Warren, Wigtownshire. *Trans. Geol. Soc. Glasgow* **9**, 293.

1896. The geological position of the Irvine Whalebed. *Trans. Geol. Soc. Glasgow* **10**, 29.

Sutherland, A. 1925. The shore vegetation of Wigtownshire. *Scot. Geogr. Mag.*, **41**, 1.

Sutherland, D. 1926. The Vegetation of the Cumbrae Islands and South Bute. *Scot. Geogr. Mag.*, **42**, 272 and 321.

Tomkieff, S. I. 1953. 'Hutton's Unconformity', Isle of Arran. *Geol. Mag.*, **90**, 404.

Tyrrell, G. W. 1926. The igneous rocks of the Cumbrae Islands, Firth of Clyde. *Trans. Geol. Soc. Glasgow* **16**, 244.

1928. Mems. Geol. Surv. Scotland. *The geology of Arran.*

Vevers, H. G. 1936. The land vegetation of Ailsa Craig. *Journ. Ecology* **24**, 424.

Walker, F. 1924–31. A dry valley at Onich, Inverness-shire. *Trans. Edin. Geol. Soc.*, **12**, 114.

Weir, J. A. 1968. Structural history of the Silurian rocks on the coast west of Gatehouse, Kirkcudbrightshire. *Scot. Journ. Geol.*, **4**, 31.

Wright, W. B. 1937. *The Quaternary Ice Age*, 2nd ed. London.

Zenkovich, V. P. 1967. *Processes of Coastal Development.* Edinburgh and London, pp. 480 ff. (Since this typescript was handed to the Press, the Department of Geography, Aberdeen, has produced two more volumes for The Countryside Commission of Scotland: A. Mather and R. Crofts, *The Beaches of West Inverness-shire and North Argyll* (1972); and *idem.*, *The Beaches of Western Ross* (1972).)

CHAPTER V

Anderson, F. W. and Dunham, K. C. 1966. Mems. Geol. Surv. Scotland. *Northern Skye.*

Bailey, E. B. 1914. The Sgurr of Eigg. *Geol. Mag.*, **1**, 296.

Bailey, E. B. and Anderson, E. M. 1925. Mems. Geol. Surv. Scotland. *The Geology of Staffa, Iona and West Mull.*

Bailey, E. B. *et al.* 1924. Mems. Geol. Surv. Scotland. *The Tertiary and Post-Tertiary Geology of Mull, Loch Aline and Oban.*

Barkley, S. Y. 1925. The vegetation of the Island of Soay, Inner Hebrides. *Trans. Proc. Bot. Soc. Edinburgh* **36**, 119.

Barrow, G. 1908. In Mem. Geol. Surv. *The Geology of the Small Isles of Inverness-shire* (pp. 20–6).

Clough, C. T. and Harker, A. Mems. Geol. Surv. Scotland. *The Geology of West central Skye with Soay.*

Cunningham Craig, E. H. *et al.* 1911. Mems. Geol. Surv. Scotland. *The Geology of Colonsay and Oronsay.*

Geikie, A. 1897. *The Ancient Volcanoes of Great Britain.* 2 vols., London.

George, T. N. 1966. Geomorphic evolution in Hebridean Scotland. *Scot. Journ. Geol.,* **2,** 5.

Gillham, M. E. 1957. Coastal vegetation of Mull and Iona in relation to salinity and soil reaction. *Journ. Ecology* **45,** 757.

Haldane, D. 1931–8. Notes on the Nullipore or Coralline sands of Dunvegan, Skye. *Trans. Edin. Geol. Soc.,* **13,** 442.

Harker, A. 1904. Mems. Geol. Surv. U.K. *The Tertiary igneous rocks of Skye.*

1908. Mems. Geol. Surv. Scotland. *The geology of the Small Isles of Inverness-shire.*

1914. The Sgurr of Eigg: some comments on Mr. Bailey's paper (1914). *Geol. Mag.,* **1,** 306.

Hossack, W. 1930. A sketch of the geology of Trotternish, Skye. *Scot. Geog. Mag.,* **46,** 337.

Hunter, W. R. and Muir, D. A. 1952–6. On the situation and geological structure of the Garvellach Isles. *Glasgow Naturalist* **17,** 129.

McCann, S. B. 1961. The raised beaches of western Scotland. Thesis for Ph.D. degree, Cambridge.

1964. The raised beaches of north-east Islay and western Jura, Argyll. *Trans. Inst. Brit. Geogrs.,* No. 35, 1.

1966. The main Post-glacial shoreline of western Scotland from the Firth of Lorne to Loch Broom. *Trans. Inst. Brit. Geogrs.,* No. 39, 87.

1968. Raised shore platforms in the Western Isles of Scotland. In *Geography at Aberystwyth: Departmental Jubliee.* Cardiff.

McCann, S. B. and Richards, A. The coastal features of the Island of Rhum in the Inner Hebrides. *Scot. Journ. Geol.,* **5** (1969), 15.

Peach, B. N. and Kynaston, H. 1909. Mems. Geol. Surv. Scotland. *The geology of the seaboard of Mid Argyll.*

Peach, B. N., Horne, J. *et al.* 1910. Mems. Geol. Surv., Scotland. *The geology of Glenelg, Lochalsh and the south-east part of Skye.*

Richards, A. 1969. Some aspects of the evolution of the coastline of north-east Skye. *Scot. Geog. Mag.,* **85,** 122.

Richey, J. E. 1961. *British Regional Geology Scotland: The Tertiary Volcanic Districts* (revised by A. G. Macgregor and F. W. Anderson). HMSO.

Richey, J. E. and Thomas, H. H. 1930. Mems. Geol. Surv. Scotland. *The geology of Ardnamurchan, north-west Mull and Coll.*

Robinson, A. H. W. 1949. Some clefts in the Inner Sound of Raasay. *Scot. Geog. Mag.,* **65,** 20.

Ryder, R. H. 1968. Geomorphological Mapping of the Isle of Rhum, Inverness-shire. Thesis for degree of M.Sc., Glasgow.

Ting, S. 1936. Beach ridges and other shore deposits in south-west Jura. *Scot. Geog. Mag.,* **52,** 182.

Vose, P. B., Powell, H. G. and Spence, J. B. 1959. The machair grazings of Tiree, Inner Hebrides. *Trans. Proc. Soc. Bot. Edinburgh* **37,** 89.

Walker, F. 1930*a.* The geology of the Shiant Islands. *Quart. Journ. Geol. Soc.,* **85,** 355.

1930*b.* The trap isles of the North Minch. *Scot. Geogr. Mag.,* **52,** 182.

1924–31. An olivine sand from Duntulm, Skye. *Trans. Edin. Geol. Soc.,* **12,** 321.

1929–31. The dolerite isles of the North Minch. *Trans. Roy. Soc. Edinburgh* **56,** 753.

Wilkinson, S. B. *et al.* 1907. Mems. Geol. Surv. Scotland. *The geology of Islay.*

CHAPTER VI

Baden-Powell, D. F. W. 1938. On the Glacial and Interglacial marine beds of northern Lewis. *Geol. Mag.*, **75**, 395.

Baden-Powell, D. F. W. and Elton, C. 1937. On the relation between a raised beach and an Iron Age midden on the island of Lewis. *Proc. Soc. Antiq. Scotland* **71**, 347.

Campbell, M. S. 1945. *The Flora of Uig.* Arbroath.

Elton, C. 1938. Notes on the natural history of Pabbay and other islands in the Sound of Harris. *Journ. Ecology* **26**, 275.

Geikie, J. 1873. On the Glacial phenomena of The Long Island, Outer Hebrides. *Quart. Journ. Geol. Soc.*, **29**, 532 and *ibid.* (1878), **34**, 819.

Gimingham, C. H., Gemmell, A. R. and Greig-Smith, P. 1949. The Vegetation of a sand dune system in the Outer Hebrides. *Trans. Proc. Bot. Soc. Edinburgh* **35**, 82.

Godard, A. 1965. *Recherches de Géomorphologie en Écosse du Nord-Ouest.* Strasbourg.

Gregory, J. W. 1927. The Fiords of the Hebrides. *Geogr. Journ.*, **69**, 193.

Jehu, T. J. and Craig, R. M. The geology of the Outer Hebrides. *Trans. Roy. Soc. Edinburgh:* Pt I (Barra and Southern Islands), vol. **53** (1921–5), 419; pt II (South Uist and Eriskay), vol. **53** (1921–5), 615; pt III (North Uist and Benbecula), vol. **54** (1923–6), 467; pt IV (South Harris), vol. **55** (1926–8), 457; pt V (North Harris and Lewis), vol. **57** (1931–3), 839.

MacLeod, A. M. 1951. Some aspects of the plant ecology of Barra. *Trans. Proc. Bot. Soc. Edinburgh* **35**, 67.

Ritchie, W. 1966. The physiography of the machair of South Uist. Thesis for degree of Ph.D., Glasgow.

1967. The Machair of South Uist. *Scot. Geog. Mag.*, **83**, 161.

1968. *The coastal geomorphology of North Uist.* O'Dell Memorial Monographs No. 1, Univ. of Aberdeen.

Ritchie, W. and Mather, A. 1970. *The beaches of Lewis and Harris.* Dept. of Geography, Aberdeen.

CHAPTER VII

Allen, J. 1947. Report on Firth of Tay Model for Dundee Harbour Board.

Bagnold, R. A. 1937. In discussion on The Culbin Sands and Burghead Bay. *Geogr. Journ.*, **90**, 523.

Balsillie, D. 1915–24. Descriptions of some volcanic vents near St Andrews. *Trans. Edin. Geol. Soc.*, **11**, 69.

Campbell, R. 1970. The Geology of Inchcolm. *Trans. Edin. Geol. Soc.*, **9**, 121.

Crampton, C. B. and Carruthers, R. G. 1914. Mems. Geol. Surv. Scotland. *The geology of Caithness.*

Crawford, R. M. M. and Wishart, D. 1966. A multivariate analysis of the development of dune slack vegetation in relation to coastal accretion at Tents Muir, Fife. *Journ. Ecology* **54**, 729.

Cumming, G. A. 1931–8. The structural and volcanic geology of the Elie–St Monance District, Fife. *Trans. Edin. Geol. Soc.*, **13**.

Davidson, C. F. 1932. The Arctic clay of Errol, Perthshire. *Trans. Perth Soc., Nat. Sci.*, **9**, 55.

Davies, L. M. 1936. The geology of Inchkeith. *Trans. Roy. Soc. Edinburgh* **58**, 753.

East Lothian County Council, County Planning Department. 1970. *Dune Conservation.*

Flett, J. S. 1915–24. The submarine contours around the Orkneys. *Trans. Edin. Geol. Soc.*, **11**, 42.

Geikie, A. 1863. Mems. Geol. Surv. Scotland. *The geology of East Berwickshire.*

    1902. Mems. Geol. Surv. Scotland. *The geology of Eastern Fife.*

Gimingham, C. H. 1951. Contributions to the maritime ecology of St Cyrus, Kincardineshire. *Trans. Proc. Bot. Soc. Edinburgh* **35**, 387.

    1953. The salt marsh. *Ibid.* **36**, 137.

Graham, A. 1961–2. Morison's Haven. *Proc. Soc. Antiq. Scotland* **95**, 300.

    1963. Cove Harbour. *Proc. Soc. Antiq. Scotland* **97**, 226.

    1966–7. The old harbour of Dunbar. *Proc. Soc. Antiq. Scotland* **99**, 173.

Grove, A. T. 1953. Tents Muir, Fife; soil blowing and coastal changes. Unpublished paper lodged with The Nature Conservancy in Edinburgh.

    1955. The mouth of the Spey. *Scot. Geogr. Mag.,* **71**, 104.

Hickling, G. 1908. The Old Red Sandstone of Forfarshire. *Geol. Mag.,* **5**, 396.

    1912. On the geology and palaeontology of Forfarshire. *Proc. Geol. Assoc.,* **23**, 302.

Hinxman, L. W. 1907. The Rivers of Scotland: the Beauly and Conon. *Scot. Geog. Mag.,* **23**, 192.

Holgate, N. 1969. Palaeozoic and Tertiary transcurrent movements on the Great Glen Fault. *Scot. Journ. Geol.,* **5**, 97.

Horne, J. and Hinxman, L. W. 1914. Mems. Geol. Surv. Scotland. *The geology of the country round Beauly and Inverness.*

Howden, J. C. 1866–9. On the superficial deposits at the estuary of the South Esk. *Trans. Edin. Geol. Soc.,* **1**, 138.

Kirk, W. 1953. Prehistoric site at the Sands of Forvie, Aberdeenshire. *Aberdeen Univ. Rev.,* **35**, 150.

    1958. The Lower Ythan in prehistoric times. In *A History of the Burgh and Parish of Ellon,* ed. J. Godsman, Chap. 4. Aberdeen.

Landsberg, S. Y. 1955. The morphology and vegetation of the Sands of Forvie. Thesis for degree of Ph.D., Aberdeen. [See also map and summary, Chap. 4 by C. H. Gimingham, in *The Vegetation of Scotland,* ed. J. H. Burnett, 1964, Edinburgh and London.]

Lee, G. W. 1925. Mems. Geol. Surv., Scotland. *The geology of the country around Golspie, Sutherlandshire.*

Lyell, Sir C. 1867. *Principles of Geology,* vol. 2, 10th ed. London, p. 508.

MacKenzie, D. H. 1954–6. A structural profile south of Eyemouth, Berwickshire. *Trans. Edin. Geol. Soc.,* **16**, 248.

Mackie, W. 1893–8. The sands and sandstones of Eastern Moray. *Trans. Edin. Geol. Soc.,* **7**, 148.

McLennan, J. S. 1892–6. The coastline of Berwickshire from Eyemouth to Marshall Meadows Bay. *Trans. Geol. Soc. Glasgow* **10**, 337.

Ogilvie, A. G. 1914. The physical geography of the entrance to Inverness Firth. *Scot. Geogr. Mag.,* **30**, 21.

    1923. The physiography of the Moray Firth coast. *Trans. Roy. Soc. Edinburgh* **53**, 377.

Read, H. H. 1923. Mems. Geol. Surv. Scotland. *The geology of the coast round Banff, Huntly and Turriff.*

Rice, R. J. 1960. The glacial deposits of the Lunan and Brothock valleys in south-eastern Angus. *Trans. Edin. Geol. Soc.,* **17**, 241.

    1962. The morphology of the Angus coastal lowlands. *Scot. Geogr. Mag.,* **78**, 1.

Ritchie, W. and Mather, A. 1970. *The Beaches of Caithness.* Dept. of Geography, Aberdeen, for the Countryside Commission of Scotland.

Robinson, E. T. 1951. Contributions to the maritime ecology of St Cyrus, Kincardineshire: the cliffs. *Trans. Proc. Bot. Soc. Edinburgh* **35**, 370.

Robinson, E. T. and Gimingham, C. H. 1951. *Ibid. Trans. Proc. Bot. Soc. Edinburgh* **35**, 387

Shiells, K. A. G. and Dearman, W. R. 1963–4. Tectonics of the Coldingham Bay area of Berwickshire. *Proc. Yorks. Geol. Soc.*, **34**, 209.

Simpson, W. D. 1940–1. The red castle of Lunan Bay. *Proc. Soc. Antiq. Scotland* **75**, 115.

Simpson, S. and Townshend, G. K. 1945–51. The tunnelling stream and the melt-water channel at Muchalls, Kincardineshire. *Trans. Geol. Soc. Edinburgh* **14**, 396.

Sissons, J. B. 1966. Relative sea-level changes between 10,300 and 8,300 B.P. in part of the Carse of Stirling. *Trans. Inst. Brit. Geogrs.*, **39**, 19.

Sissons, J. B. and Smith, D. E. 1965. Peat bogs in a Post-glacial sea and a buried raised beach in the western part of the Carse of Stirling. *Scot. Journ. Geol.* **1**, 247.

Sissons, J. B. *et al.* 1965. Some pre-carse valleys in the Forth and Tay basins. *Scot. Geogr. Mag.*, **81**, 115.

Smith, D. B. and Parsons, T. V. 1968. The investigation of spoil movements in the Firth of Forth, using radioactive tracers. *Symposium, Proc. Inst. Civ. Engs.*, 18 Oct. 1967, 47.

Smith, J. W. 1950–2. A Buchan burn: the Cruden water. *Trans. Buchan. Field Club* **17**, 49.

Smith, S. G. M. 1963. Some notes on post-glacial gullying. *Scot. Geogr. Mag.*, **79**, 176.

Steers, J. A. 1937. The Culbin Sands and Burghead Bay. *Geogr. Journ.*, **90**, 498.

    1938. Minor changes in shingle spits and the value of regular mapping. *C.R. Congrès Internat. de Géog. Amsterdam*, vol. 2, section 1, 162.

Stevenson, R. 1811–16. Observations upon the Alveus or general bed of the German Ocean and British Channel. *Mems. Wernerian Nat. Hist. Soc.* **2**, 464.

Usher, M. B. 1967. *Aberlady Bay local nature reserve, description and management plan.* E. Lothian C.C. Planning Department.

Wallace, T. 1896. Recent geological changes and the Culbin Sands. *Trans. Inverness Sci. Soc. and Field Club* **5**, 105.

Walker, F. 1936. The geology of the Island of May. *Trans. Edin. Geol. Soc.*, **13**, 275–85.

Walton, K. 1956. Rattray; a study in coastal evolution. *Scot. Geogr. Mag.*, **72**, 85.

    1959. Ancient elements in the coastline of north-eastern Scotland. In *Geographical Essays in memory of A. G. Ogilvie*, Chap. 4. London.

    1963. The site of Aberdeen. *Scot. Geogr. Mag.*, **79**, 69.

Zenkovich, V. P. 1967. *Processes of Coastal Development.* Edinburgh and London.

CHAPTER VIII

Childe, V. G. 1951. *Skara Brae.* London.

Flett, J. S. 1915–24. The submarine contours around the Orkneys. *Trans. Edin. Geol. Soc.*, **11**, 42.

Stevenson, R. 1811–16. Observations upon the Alveus or general bed of the German Ocean and British Channel. *Mems. Wernerian Nat. Hist. Soc.*, **2**, 464.

Waterston, G. 1946. Fair Isle. *Scot. Geogr. Mag.*, **62**, 111.

Wilson, G. *et al.* 1935. Mems. Geol. Surv. Scotland. *The geology of the Orkneys.*

CHAPTER IX

Flinn, D. 1961. Continuation of the Great Glen Fault beyond the Moray Firth. *Nature* **191**, 589.

    1964. Coastal and submarine features around the Shetland Isles. *Proc. Geol. Assoc.*, **75**, 321.

Hamilton, J. R. C. 1956. *Excavations at Jarlshof.* HMSO.

Hoppé, G. *et al.* 1965. Submarine peat in the Shetland Islands. *Geogr. Annaler*, **47a**, 195.

Jelgersma, S. 1966. Sea-level changes during the last 10,000 years. *Roy. Met. Soc. Proc. Internat. Symposium of World Climates from 8,000 BC to 0 BC.*

Pirkis, D. H. B. 1963. The coastline of Foula. *Brathay Exploration Group: Annual Report and Account of Expeditions in 1963*, 44.

Zenkovich, V. P. 1967. *Processes of coastal development*. Chap. 7. Edinburgh and London.

CHAPTER X

Cockburn, A. M. 1933–6. The geology of St. Kilda. *Trans. Roy. Soc. Edinburgh* **58**, 511.

Mathieson, J. *et al.* 1928. St. Kilda. *Scot. Geogr. Mag.*, 44.

Sabine, P. A. 1960. The geology of Rockall, North Atlantic. *Bull. Geol. Surv. Gt. Brit.* No. 16, 156.

Stewart, M. 1932. Notes on the geology of North Rona. *Geol. Mag.*, **69**, 179.

1933. Notes on Sula Sgeir and the Flannan Islands. *Geol. Mag.*, **70**, 110.

1938. Notes on the geology of Sule Stack, Orkney. *Geol. Mag.*, **75**, 135.

Walker, F. 1931. The geology of Skerryvore, Dubh Artach and Sule Skerry. *Geol. Mag.*, **68**, 320.

Index